An Era of Expansion

Changing conditions in Higher Education and national funding regimes preceded a proliferation of construction projects in universities between 1996 and 2006. This book traces and comments on the expansion programme at Cambridge including nine detailed case studies from the author's time in charge of capital projects at the University of Cambridge which illustrate how these projects were conceived, argued for, designed, procured, managed, constructed and passed on to building users. Readers with an interest in project management, estate management, University management or the history of the University of Cambridge will find this fascinating and wide-ranging book to be uniquely valuable.

David Adamson is a consultant on sustainable construction, and lectures at the University of Cambridge. He was Director of Construction Policy for UK HM Treasury between 2006 and 2007, and Director of University Estates at the University of Cambridge from 1998 to the end of 2005. David was educated in Edinburgh and at King's College, Cambridge, and then served around the world in the Royal Engineers, commanding a field regiment and serving as a staff-officer in NATO HQ Europe. He has previously co-authored *Change in the Construction Industry 1993–2003*, Routledge, 2006 and two books on the underground use of foamed concrete. He is a visiting professor at UCL and UWE.

An Era of Expansion

Construction at the University of Cambridge
1996–2006

David M. Adamson

Routledge
Taylor & Francis Group

LONDON AND NEW YORK

First published 2015 by Routledge

2 Park Square, Milton Park, Abingdon, Oxon, OX14 4RN

605 Third Avenue, New York, NY 10017

Routledge is an imprint of the Taylor & Francis Group, an informa business

First issued in paperback 2020

Copyright © 2015 David M. Adamson

The right of David M. Adamson to be identified as author of this work
has been asserted by him in accordance with sections 77 and 78 of the
Copyright, Designs and Patents Act 1988.

All rights reserved. No part of this book may be reprinted or reproduced
or utilised in any form or by any electronic, mechanical, or other means,
now known or hereafter invented, including photocopying and recording,
or in any information storage or retrieval system, without permission in
writing from the publishers.

Notice:
Product or corporate names may be trademarks or registered trademarks,
and are used only for identification and explanation without intent to infringe.

British Library Cataloguing-in-Publication Data
A catalogue record for this book is available from the British Library

Library of Congress Cataloging in Publication Data
Adamson, David M., 1943-
An era of expansion : construction at the University of Cambridge
1996-2006 / David M. Adamson.
pages cm
Includes bibliographical references and index.
ISBN 978-1-138-85063-7 (hardback : alk. paper) -- ISBN 978-1-315-
72459-1 (ebook) 1. University of Cambridge--Buildings. 2. College
buildings--England--Cambridge--Design and construction 3. Campus
planning--England--Cambridge. I. Title.
LF110.A23 2015
378.1'90942659--dc23
2014049444

ISBN: 978-1-138-85063-7 (hbk)
ISBN: 978-0-367-73791-7 (pbk)

Typeset in Bembo
by Fakenham Prepress Solutions, Fakenham, Norfolk NR21 8NN

This book is dedicated to Oscar, Kaspar and Ryan:
may you do well whatever ups and downs you go through.

Contents

Figures

Tables

Foreword

The medieval Cambridge Colleges cluster around the centre of Cambridge as if they were placed by some master planner, but of course there was no master planner. It was merely that there was little else in medieval Cambridge so the colleges were built comfortably side by side and this continued for hundreds of years. In the nineteenth century an increasing number of University buildings appeared where scholars of a given discipline from all of the colleges gathered together to pursue their interests. This began with the scientists because their laboratories were too expensive for a single college. Subsequently all disciplines gained their own buildings although this took until the end of the twentieth century when new buildings for law, divinity and English were built. This was a time when a large number of modern research facilities were also being built for medicine, engineering and the sciences, resulting in the most rapid expansion the University had ever seen. The administration of the University was stretched to the limit of its capacity in realising this expansion and was at last persuaded that it was necessary to develop a long-range plan.

David Adamson tells the story of the buildings of Cambridge University from the very beginning when the scholars came to Cambridge in 1209 to escape the town riots in Oxford. The early history is fascinating but is not the main purpose of this book. The major purpose is to record how the University successfully provided the facilities and buildings it needed to maintain its position as a world-leading university towards the end of the twentieth century. This was a period of intense international competition as US universities with their greater resources raced ahead of most of the world's universities, but Cambridge maintained, even enhanced, its position, which was already close to the top of the international ranking tables. This could not have been done without the unprecedented expansion of the University estate. An expansion which was made more difficult because many of the new buildings had to house highly sophisticated scientific equipment and all had to meet increasingly severe environmental and climate change standards. To succeed it was necessary to find new ways for architects, engineers and contractors to work together. It was also necessary to reform the University's decision-making processes and to raise immense additional funds. Lastly, and of equal importance, the University's Estate Management Department had to develop the skills to find the way through the UK's tiresome planning regulations, a task made more difficult by the continual revision of these regulations.

It had become clear much earlier in the twentieth century that more space was needed and colleges and university departments had already built to the north, west and south of the expanding city centre. This unstructured expansion continued, especially for the sciences, medicine and engineering, but without a sufficiently comprehensive long-term plan. There was in the early 1960s an unsuccessful plan to expand in the centre of Cambridge. Three large tower blocks were to be built near the old Cavendish laboratory but this project failed to gain planning permission, fortunately, because these towers would have dwarfed and obscured the precious and well-proportioned medieval buildings, much as the concrete buildings built in the same era obscured St Paul's Cathedral.

The first attempts to look further ahead were in the 1970s. For example, Peter Swinnerton Dyer who was Vice-Chancellor from 1979 to 1981 identified an ellipse that enclosed a large area of land to the west of the city and proposed that the University should expand to the west. In 1971 the Cavendish Laboratory began to move to the closest end of what was to be called the West Cambridge site, a 66-hectare rectangle that lies along the southern side of the Madingley Road just north of the ellipse. The University had progressively acquired this land starting in 1923 and departments had moved there in the 1970s but it was not until 1996 that a public inquiry finally gave approval for a major expansion of the University on this site and a master plan was commissioned. The plan described in broad terms how the site would be developed. It was to be mainly for science and engineering but some housing and limited catering facilities were included. A large proportion of this expansion has now been completed.

Soon after this, in 2002, the University made the case to the county and city planners that it should be able to expand into the triangle of land bounded by Huntington Road, Madingley Road and the M11, known as the North West Cambridge site. This land was within the green belt but the University's case prevailed and permission was given to go ahead on the basis that the University said that it would include housing, retail shopping and a school, in addition to the development of the University and its colleges. Gaining this planning permission was a major achievement for the University as it ensures space to fulfil its ambitions for many decades into the future. There are few universities in the world that have this much space available adjacent to their existing facilities.

Adamson describes in detail how these strategic advances were made, and how at the same time £750M of new buildings were completed. He came to Cambridge from Bristol University with extensive experience in university estate management and this enabled him to stay ahead of what became, as we entered the twenty-first century, an almost overwhelming onslaught of activities. He reformed the way buildings were planned and constructed, turning away from the concept of having an iconic architect dominate the whole process, to one where the users of the building, the engineers and the contractors were involved from relatively early in the process. He introduced the concept of the representative user, where a single point of contact for a department or faculty was identified to present a coherent interface to the designers and builders. He also insisted on optimising cost in discounted cash terms over a ten-year period to avoid what had happened in the past where maintenance costs had been hopelessly under-estimated. Finally he ensured that post-occupancy evaluations were carried out to see how well a building met the needs of its users. The net result of these changes was that buildings began to be built on time and to budget, unlike those built in the 1960–1970s where there was general overspend in the public and private sector and projects almost always ran behind schedule. It also meant that the buildings more closely met the needs of the users and were better suited to the activities they were to house. Not everything worked out as hoped but the new procedures produced feedback on what had gone wrong and allowed everyone to do better the next time around.

The ability to build high quality infrastructure economically is setting nations apart. We often hear that China completes projects in a fraction of the time we take in Europe. This book describes how we can close the gap thereby enabling us to take full advantage of our world-leading scholarship and research. Anyone who knows Cambridge will enjoy the book because it reveals what was going on behind the scenes during the rapid expansion of the city and the University. There is much to look back on and much to learn about what is going to happen in the future. For those in the University it throws light on why things happened the way they did and the reasoning behind the future planning. It also describes the governance changes within the University that improved its ability to plan its future. Anyone interested in construction, and particularly in building programmes for universities, will find this book a fascinating read and a valuable reference book full of ideas about how to improve performance, and especially how to provide buildings and facilities that meet their users' expectations. No university can compete in today's research environment without state-of-the-art facilities and this book provides a plethora of examples about how, and how not, to go about building them.

Professor Lord Broers FREng FRS

Preface

When Alec Broers stepped down as Vice-Chancellor of the University of Cambridge in 2003 he suggested that I write up the hitherto unprecedented expansion of the University. The building and property programmes had been moving too fast to record the back-stories: why and how new buildings and property deals came about. 1996–2006 was also a decade of substantial change as the University developed new policies for management and capital procurement while funding for projects came flooding in. It was interesting to observe how people in university management and people in estate management actually operated, and what was the outcome, and I hope that it will be of interest to a wider audience, now and in the future. A shrewd architect walking in the middle of Cambridge suddenly stopped and said 'if only someone had written such an account of University building development 20 or 50 years ago'. Colleges and other universities have written records of their estate development as useful references, and for this decade of change and expansion it seemed a good time for the University to do that. I waited six years before starting this assessment so as to allow enough time for a mature view to emerge after the end of the decade, 2006. I couldn't, though, wait too long as many of those involved were moving away or losing clarity of memory about that decade, and records were getting lost or becoming inaccessible during high turnover of staff, and changes in office systems. It also seemed timely to document how the path was prepared for the next surge of expansion, at North West Cambridge.

More generally, this is a good time to assess how national reforms of the construction industry in the mid/late 1990s have been working through in the development programme of a client which is in the public sector but has many of the characteristics of a private sector client. Individual projects often get written up, usually soon after the new building is handed over, but very rarely is there a review of a whole programme of work some years later.

There are thousands of facts and figures in this account, mostly taken from internal management control documents retained from the period. Where records are not available or there are gaps I have used the best information available. Definitions of costs and dates in assessments of capital programmes are not straightforward in the construction industry; costings quoted generally exclude associated maintenance and minor works jobs except when subsumed into the main project; start dates are when preliminary works began if those were substantial, and completion dates as handover even if some fitting-out was still being done. However, I hope that the general picture is clear enough to be of some interest and value.

I have tried to colour and illustrate the factual narrative with personal observations from my former position as Director of Estates responsible for planning and managing the University estate at Cambridge. They are just personal views. I hope they are fair, and I apologise for any that are not.

DMA Nov 2014

Note

References to 'Reporter' are to Cambridge University Reporter.

Acknowledgements

I would like firstly to thank Lord Alec Broers of Cambridge for suggesting to me that I should write an account of the physical expansion of the University over the decade of its biggest expansion to date, and for encouraging me during the drafting; I would also like to thank him for his inspiring leadership, and his confidence, sometimes justified, in the University's capital procurement programme.

It has been good working with Benedicte Foo who has developed case-studies of nine buildings; her shrewd architect's eye and her assessments have enhanced this account, and I thank her very much.

I would also like to thank Geoffrey Skelsey, Chef de Cabinet to successive VCs from 1977 to 2003, for his knowledge and advice from the start, Alison Wilson for her wise counsel and her experienced editing, and Alan Franklin for his comment and technical help.

I am especially grateful to well over 50 academics and administrators in the University of Cambridge who through interviews and other discussions have given me their views and assessments, and have provided information directly and by access to documentation, and to the leaders of the local community and of the construction industry who also have given me their views and have allowed me to quote them. I am also grateful to Wayne Boucher and BDP who have kindly let me include several of their photos.

I want to thank especially the staff of the Estate Management and Building Services during the period 1998–2005: for their professionalism, their hard work and their cheerful willingness to commit to team-working. In the football phrase, they kept at it and they done good.

I am grateful to King's and to Murray Edwards Colleges for lots of things, and to the Universities of Bristol and Cambridge for the wonderful professional opportunities to look after the various aspects of their estates.

DMA Nov 2014.

1 A history of the estate of the University of Cambridge

Hinc lucem et pocula sacra.
From this place, we gain enlightenment and precious knowledge.

University of Cambridge motto

History always matters; to understand how the University of Cambridge meets its current challenges, understanding its history is crucial. How the University managed its estate in response to the sudden burst of building opportunities and demands at the end of the twentieth century is rooted in the way the estate was then configured and how people thought about it. The locations and nature of University buildings were shaped by centuries of decisions by small groups of highly motivated and highly intelligent academics dedicated to developing their own subject interests; only rarely were these decisions made within any sort of overall University-level planning. Further, a higher priority was given to implementing current academic leaders' needs as current academic leaders saw them, than to anticipating the needs of those who would succeed them. So, while at the start of the decade documented in this book, some of the University estate was effective by current criteria, large chunks of it were far short of the top international standards of its researchers, teachers and students, and overall it was managed in a fragmented way serving laudable, and justifiable, but rather immediate aims. The estate challenge suddenly facing the University as it approached the Millennium was how to stitch together the best aspects of that fragmented past with the huge abilities of its current academics, and to meet the need to modernise and greatly expand its stock of buildings and infrastructure; and to fund and manage the tsunami of construction projects that swept through the University.

The early days

In the beginning there was Oxford and over to the east near the edge of the coastal swamps a market settlement around the bridge over the River Cam. As in other settlements in the Fens, diseases there were rife and strangers unwelcome, but when a small group of young men appeared saying that they'd come from Oxford fearing for their lives after a tavern brawl in which a young woman, maybe a bar-girl, was killed and a couple of the students involved were hanged by the townsfolk, they were allowed to settle. And settle they did, supported by the many religious foundations. And thus started the chain of events that led to the University of Cambridge.

That was in or around 1209. The Cambridge area had long been settled. Recent excavations in the grounds of Fitzwilliam College have unearthed flint tools and pottery from a farm that was flourishing around 3,500BC. The first known bridge over the Cam (or Granta as it was then called) was built by 875AD as a key link for the market town and, as the Domesday Book of 1086 noted, there were already prosperous residences and businesses, and many religious institutions with a lot of power and independence since the church had stood up to the King after the murder of Becket four decades earlier; it was into these institutions that the six scholars from Oxford settled.[1] One can only imagine what it must have been like for these young lads: tired and a bit frightened, rather confused, and probably rather unwelcome. But they set about continuing their studies, and increasingly teaching others: the logic of closely pairing research and teaching has long been one of the great strengths of the University, and has been stoutly defended through the ages.

The subjects that young students of those early days were studying were streamed: first a grounding course in grammar, logic and 'disputation', music, arithmetic, geometry and astronomy. Some students, on completing those studies, went on to study law, divinity and medicine at what we would now see as 'University level'. The 'scholars' (students) were 'clerks', training to be clergymen. The considerable breadth of studies reflected well the breadth of what was seen as the world's knowledge. (It is interesting to compare that breadth of education with the recent trend back towards inter-disciplinarity.)

There grew, certainly by 1226, an organised network of classes of scholars, the senior of whom acted as supervisors, or 'Masters'; their leader, from 1412, was called 'the Chancellor'. The scholar market soon became more regulated: King Henry III set out rules to protect scholars from the landlords who were ripping them off for rent, and to legislate that only scholars enrolled by recognised Masters could stay in town. Soon there was a body of Masters who, with the Chancellor, and later his deputy the Vice-Chancellor, were regulating examinations as well as classes, with different levels of scholars, then as now, differentiated by the length of their gowns and colour of their caps and hoods.

Extensive records of property, either privately owned or owned by the religious foundations, still exist. Over the centuries, property of both types were transferred to the nascent Colleges, from the thirteenth century onwards, and especially after the Dissolution of the Monasteries from 1536. An early significant property transfer, in 1284, by the Bishop of Ely (Hugh Balsham) was St Peter's House, set up as Peterhouse, the first College to survive to the present day. From then on, the Colleges continued to grow in number and size. At times, the growth was stimulated and achieved by royal patronage (especially of Lady Margaret Beaufort and Henry VIII), at times stimulated when monarchs and governments had particular suspicion of Oxford. It is suggested, for example, that Henry VI's worries about 'Oxford' influence in his court led him to the surprising decision to have King's College set up in Cambridge; and later, for a while during the Reformation, Oxford was seen as a 'hotbed of Lollardry', though to be fair, a lot of key reformers were also in Cambridge.

'The University' as an organisation is not be conflated with the Colleges *per se* which were and are legally separate entities, albeit closely and generally amicably entwined with the University; they hold their own estates quite separate from the University estate with which this book is concerned: matters of College estates are only mentioned in this record when they affect the University estate directly.

The first building to be erected specifically for teaching in the University was the (Old) Divinity School. The site, a slight mound of gravel, later used by King's College and the Old Schools, was bought in 1278 (so, the start of the University estate just before the foundation of the first College to survive), but it was not until about 1350 that the first building was started, to be completed around 1400 (the construction time hence being about five times as long as that of the 100-odd projects that were to expand the University by 33 per cent from 1996 to 2006). That first building, with its windows of irregular shape, was built for the purpose of teaching Divinity. There were further buildings on the site: in 1430–1460 for the teaching of canon law, and in 1457–1470 for teaching civil law and philosophy, with a library. The West Court of the Old Schools (including the Syndicate Room and offices for the Registrary and others) includes the Council Room (finished in 1466). Its ante-room, known as the Dome (the VC's office, since 1975), was formerly part of King's College: it was above the porters' lodge of the original court of King's College.

Figure 1.1   The Old Schools
By courtesy of info@Cambridge 2000

It was during the mid fifteenth century that the University started to develop its estate. Land around the current Senate House was bought to put up buildings for teaching and 'disputations', a chapel and a library, and a 'treasury' for chests to hold the money paid by scholars. The University's financial assets are still known as 'The Chest', and one such 'chest' lies in the office of the Registrary (Registrar in other universities). As long as the University estate was just the buildings around the Old Schools, it was relatively easy for its management to be sensibly controlled by a small group of the Masters under their Chancellor. Later however, when teaching and research requirements increased, and the actual and perceived autonomy of emerging Faculties (departments) grew (in a manner that in some ways reflected the well-established autonomy of the Colleges), it became more and more difficult to develop and maintain an overall plan of what the University owned and have some idea of how that should best be developed and managed. Indeed, it was not until nearly 800 years after the start of the University that it developed and agreed a coherent schedule of its estate, by then worth £2.4bn, and began to analyse how it should be developed.

The University estate grew non-contiguously, and in a manner which could be called haphazard, but which equally and more constructively can be seen as a series of generally sensible decisions by clever men who usually discussed matters sensibly: sometimes at huge length and repeatedly, sometimes briefly, often in caucuses deliberately limited and shaped to suit the dominant players. The decision-makers were, as now, leaders in their departments and in their Colleges: all academics for much of the University's history had to have College Fellowships. (In recent decades with rapid growth in academic numbers but little increase in the number of Colleges, the proportion of academics with College Fellowships has fallen considerably.) It was from the mid-sixteenth century that the Colleges as such started to play a central role in how the University itself developed, with their Heads of House taking key University roles.

Although there was little expansion beyond the area of what now is the Old Schools, maintenance of those buildings was documented. An account of a contract dated 25 June 1466 notes:[2]

> A contract for Indenture of covenant for carpenter's work on the old Schools [to] ... supply, carriage and workmanship of timber for the floors and roof of the new Schools before Lammas Day for payment of £23 6s 8d in addition to the £10 paid at the making of the agreement.

A church existed on the site of Great St Mary's (properly, the church of St Mary the Great) by 1205; however there was a major fire in 1290, with re-construction, and then re-consecration in 1351. An attack during the Peasants' Revolt in 1381 led to the Statutes and Charters being taken out and burnt in the Market Place (later Square). The church was also the administrative centre of the University until 1730 when Senate House became the centre for meetings and discussions. Further work was done on the church from 1478, a tower being built from 1491 and the nave roof finished in 1508. Archbishop Parker, Master of Corpus Christi College, had much of these works done and left enough money for their completion. Nearly 500 years later, a scheme to erect a spire, which had probably been the original intention, did not come to fruition.

By the end of the sixteenth century there was an organisation of Proctors and two Esquire Bedells whose purpose was leading at official University functions and certifying University accounts (the Proctors held one of the two keys needed to gain access to the University Chest, where its money was kept, the Vice-Chancellor had the other), and also for keeping good order generally around the streets and Colleges. The University Marshall had a lesser ceremonial and administrative role. An annual report from 1575 includes a note of the work of Junior Proctors on 'repairs – upkeep of building, street cleaning and night wonderers [sic]'. In 1630, the managers of the University estate decreed that bear-baiting should not take place on its premises.

From 1700

By the late seventeenth century, the well-being of local bears thus safeguarded, construction of the current Senate House was commissioned; this was required to release space for expansion of shelving in the University Library in the Old Schools, for the rapidly increasing number of printed books. (The University Library had begun as a collection of books stored in the tower of Great Saint Mary's University Church in the mid-fourteenth century.) The procurement of the Senate House building was long and tortuous with endless argument about the specification of the building. A design idea was developed in 1713–1714 by Nicholas Hawksmoor, who was an assistant to Wren. That scheme got dumped, but in 1722 a design by James Gibbs (also the architect for King's College Fellows' Building) was accepted, and work started that year; but then part way through construction, money ran out and work stopped. Most of the Senate House was finally completed in 1730, the west end being finished 38 years later. It is an appropriate building, dignified, well placed in its setting and still big enough for today's purposes. As might be expected, the heating system was primitive (it had a charcoal-burning stove and needed a lot of expensive upgrading over the years). Also, the acoustics were not good for the purposes for which the building would later be used. When electricity came along, it was provided by the Bailey Grundy Electric Company which generated electricity in a building behind Kenmare, a house on Trumpington Street (currently the home of Estate Management and Building Services), and transmitted at low voltage along underground cables to Senate House. There were various nugatory schemes to extend the area of the Old Schools: one was aborted by a counter-proposal from the Chancellor, which was then itself voted down.

An Observatory was built in 1822–1823 on land sold to the University by St John's College, on its present site off Madingley Rd; an earlier construction in

Trinity College was never completed. From 1824, a site was assembled by the University for the Pitt building, largely by buying (for £12k) and demolishing two houses as close to Pembroke College as possible. The building's cost, some £11k, was partly funded by public subscription in memory of William Pitt the Younger, MP for Cambridge, and an alumnus of Pembroke College. It was to be used for printing books, at first largely of a religious nature, by Cambridge University Press. (Printing started in the University in 1583 following 'a royal charter' by way of 'letters patent' to the University by Henry VIII in 1534.)[3] The foundation stone was laid with great ceremony in 1831 and the Press with its high central tower and tall windows was opened in 1833: there is a wonderful description in Council papers of how 'the University' formed up by the Senate House and, led by the VC, 'processed' down for the opening.[4]

From the mid nineteenth century

Estate development proceeded piecemeal, albeit purposefully. In 1853 a four-acre site (land once owned by the Austin Friars) around the area of the current New Museums Site was bought for £3,448 from the Botanic Gardens Trust; it had been given in trust to the University by Dr Richard Walker, Vice-Master of Trinity College, in 1762 to build a botanic garden. In 1831 it was decided that the area was too small for the botanic garden so the current 38-acre site south of the city to the east of Trumpington Street was bought from Trinity Hall: the move there started in 1844. The site was purchased principally to provide accommodation for science teaching following the introduction of the Natural Sciences Tripos in 1848. (The Mechanical Sciences (Engineering) Tripos was set up in 1894.)

The first buildings along the Free School Lane Site were put up (internally in the white brick which was then common around Cambridge) in 1863–1865 and 1876–1890.

On 14 June 1853 'The Cambridge University and Town Waterworks Company' received Royal Assent following years of petitioning on behalf of both Town and University. A well 48 feet (14.4m) deep was dug out at Cherry Hinton and, using a beam engine, water was pumped up into a reservoir at the top of Lime Kiln Hill. This company metamorphosed into the current Cambridge Water Company, which now supplies 72 million litres daily, to 320,000 people. In 1894 the main sewerage pumping station for Cambridge was built off Newmarket Road by the River Cam.

Despite weaknesses in the management of procurement having been apparent, a plethora of uncoordinated committees and syndicates came into being,[5] some of these were for purposes of looking after and developing University buildings: for example, separate syndicates were set up for heating the Old Schools in 1790 and for heating the Senate House in 1850. Syndicates, though usually well intentioned, had little if any professional input,[6] and by their independence they lost the benefits of the continuity of management achieved when concept development, design, construction and maintenance of a building are done under one continuous management system. Also, syndicates were often set up as a mechanism just to reduce input from and responsibility to 'the rest of the University', and this often led to adverse results; syndicates were not part of the annual rounds of committee reviews of priorities and resource allocation. The degree of answerability to 'the University' versus sensible independence is a fine balance to this day. Also, lacking central co-ordination, syndicates sometimes overlapped and clashed: the Reporter of 21 January 1908, page 543, notes that in the Council of the Senate report on the establishment of a Buildings Committee, Mr R T Wright is reported as saying,

> He did not think it was the business of the Sites Syndicate to interfere with what Buildings Committee and Buildings Syndicates [sic] had to do; it was their business to assign sites, and they had already found that quite enough to do. He strongly objected to their putting their fingers into bricks and mortar ... He thought that the proposal would have the effect of giving members of the Buildings Committee a very undue amount of preponderance in the policy of the University as concerned buildings.

A Royal Commission initiated by the Government in the 1850s led eventually to the setting up of the Financial Board and the General Board to support Council and to bring about more coherent planning and administration in the University. There was a University Property Syndicate made up of interested academics from 1857 until 1881 when the Financial Board subsumed it, stating that it 'had too much to do'. The Financial Board, however, also had no members who nowadays would be considered as professional in property or procurement matters.

The year 1837 saw the start of construction of a purpose-built extension (designed by C. R. Cockerell) to the University Library beside the Old Schools, subsequently occupied by the Squire (Law) and Seeley History Libraries. The Fitzwilliam Museum, after various amazingly grand designs had been rejected, was completed in November 1837, and nine years later Fenners Cricket ground opened. In 1872–1874 William Cavendish, the 7th Duke of Devonshire, Chancellor 1861–1891, made a gift of £6,300 to fund a physics lab, 'the Cavendish', which was opened in 1874 under the direction of James Clerk Maxwell, the University's first Cavendish Professor of Experimental Physics. On the site, once a medieval hospital and graveyard, opposite St John's College, an elegant brick and limestone 'Divinity School' was completed in early Tudor style in 1879 at a cost of £11,074; the site was purchased from St John's for £3,790, (partly funded by Professor William Selwyn, brother of Bishop George Selwyn after whom Selwyn College is named). It was used until a new Divinity School was built on the Sidgwick Site in 2000. During recent refurbishment a cemetery housing 1,300 individuals was discovered under the foundations. The Union Society, later to be shaken by one of a stick of bombs dropped by the Luftwaffe, was constructed nearby in 1886.

Infrastructure to the west of the River Cam was slowly developing. In the 1850s it was possible to build Queen's Road, named in honour of Queen

Victoria, as the river was by then better constrained within its present banks. The river had for centuries been the main channel for commerce (as far as Queens' College and Newnham Mills) and had been seen as the natural boundary of the University. Routes along what now are West Road and Sidgwick Avenue (and Pightle Walk, which was integrated into Sidgwick Avenue) became increasingly used as Selwyn and Newnham Colleges and Ridley Hall opened up nearby in 1881–1882. From 1861, Colleges started to relax the ban (imposed originally by Queen Elizabeth I) on marriage of Fellows who had no University appointment, and as the dons were able to live with their now-official wives in the University area, large detached family houses began to appear along Queen's Road and West Road. From 1856, appointments began to be made to University posts directly, rather than (as is still the case at the University of Oxford), appointment to a College and simultaneously to a University post, and that also made for less dependence on Colleges and helped create a desire for local house-ownership.

As 1914 approached, a lane opened up from West Road to the site that was to accommodate a large military hospital during and after the First World War, and later the University Library.

The ADC (Amateur Dramatic Club) theatre is the longest-running student theatre in the UK. The first student play by the Cambridge University Amateur Dramatic Society, in 1855, was in a pub on the site after it leased a couple of rooms. The property, in Park Street, suffered a bad fire in 1933 and was rebuilt by 1935; it was leased by the University in 1973/4 from the ADC, its freehold owners, in large part because the new VAT regulations required much accounting (which, it was felt, would be done with greater dedication by the University administration than by the ADC students).

On 4 June 1897, the Syndics of the Press reported to Senate that they had

> purchased the freehold property in Mill Lane known as the 'Wagon and Horses', which immediately adjoins the Press premises; the total cost of this purchase has been defrayed from the funds of the University in the hands of the Syndics ... the Syndics recommend that the purchase be approved.[7]

The report goes on to note that the Syndics had entered into a contract to purchase 'an important block of Land situate between Mill Lane and Silver Street'. Purchases by Syndicates with subsequent requests to the Senate for their retrospective approval seem, surprisingly, to have been common.[8] In 1891, Arthur Balfour, the University Chancellor, criticised the 'abominable system of managing everything through syndicates and committees – ingenious contrivances for making the work of 10 wise men as if it were inferior to the work of one fool.' (And the aphorism, 'And Committee shall speak unto Committee' was applicable to the University over the centuries, and illustrated aptly in C. P. Snow's later novels.) As time went on, the number of syndicates, and hence the problems they caused by their fragmentation, was reduced, and by 2005 nearly eliminated, other than the 'permanent syndicates' which have a specific and continued and relatively independent operating function, such as the Press, the University Library, Fitzwilliam Museum and the University Centre.

A report to the Senate on 21 January 1907 recommended setting up a permanent Sites Syndicate of the VC and 12 members: that functioned only until 9 May 1912 (nothing on its file thereafter): a Buildings Syndicate was set up in December 1912,[9] it resolved at its first meeting on 24 January 1913 to meet every second Tuesday at 4pm. Its first Minute noted 'the possibility of fire in Professor Nuttall's Laboratory'; Minute 5 noted a request for an 'animal house in the neighbourhood of the new Physiological Laboratory'. At the meeting on 3 March 1914 (attended by the VC, the Masters of St John's and Gonville and Caius Colleges, one Dr and 4 Mr's, and resident MA's), Item 2 was 'The Use of University Rooms by Ladies'. The VC reported that 'the actions of the University Buildings Syndicate in granting the use of a room for a course of lectures to be delivered by a Lady had been called in question by a Member of the Council of the Senate'. The VC asked the Syndicate to assent to the following Minute:

> The University Buildings Syndicate requested the Vice-Chancellor to inform the Council that they had acted *ultra vires* in granting the application, they intended in future to refer such applications to the Senate. Proposed by the Master of St John's, seconded by Mr Lock, Ayes 7, recom carr.

Its third report, in 1916, concentrated mostly on painting of walls, fire matters, YMCA use of the Engineering Drawing Room, billeting troops and Red Cross work, and Belgian and Refugee Committee work.

Heading into the twentieth century and up to 1939

New Museums Site

In 1864–1865 buildings had been erected for the museums of zoology and comparative anatomy, extended in 1876, and for the Cavendish Laboratory in 1873. A library and an examination hall were added in extensions 1880–1884. On 13 May 1896, the University, advised by Mr Bidwell, bought from the Perse School (which had given the street its name, Free School Lane) land running up to the Corn Exchange, including the stables for Nos 4 and 5 Free School Lane. Then another Syndicate was set up to prepare a scheme 'for the future appropriation of the site to be purchased from Messrs Mortlock & Co [the bank merged into Barclays, which stayed on the reduced site for well over a century] and from Downing College and of other sites, if any, at the disposal of the University'. These, along with previous purchases (notably from 1762 for the Botanic Garden) developed into the current 'New Museums' Site. The site then saw a series of rather uncoordinated buildings go up for departments, the majority of which have since moved elsewhere. The Perse School premises were converted for the teaching of engineering 1891–1894. An engineering workshop had been opened along the Lane in 1878 with extensions in 1882. Perhaps the most exciting development was over the long summer vacation of 1884 when a roof, 50 tons in weight and 110 feet long, over the herbarium and the then Department of Mineralogy building, was jacked up, and brick walls built under the new position: it was recorded that there was no damage at all to the roof. Good engineering starts at home.

The development of the New Museums Site as fragmentally planned in the nineteenth century was virtually complete by the First World War. Despite the world-famous pre-war scientific developments such as the discovery of the electron in 1897, and the splitting of the atom in 1917, not much was added physically on the site between the wars except an extension to the Zoological Department in 1932 (architect Murray Easton), the Mond building 1932–1933 (architect H. C. Hughes) and the Austin Wing for Physics in 1940 (architect Charles Holden). Later within that site the 'the secret of life,' the DNA double helix was discovered.

From 1958 there was a review of University development, published in the 'Report of the Council of the Senate on the provision and development of sites for University use. 24.11.58. Approved 13.12.58'.[10] This was about a 'policy of concentration' on central sites with 'one or two buildings of substantially more than seven storeys'. Surrounding buildings would be much lower with more open space. The report mentioned all the outlying sites owned by the University and the possibility of acquiring Addenbrooke's in the future.

Denys Lasdun was appointed in 1959 to draw up a comprehensive plan for re-housing the Cavendish Laboratory, established on the western part of the New Museums Site in the 1870s. The various schemes published in 1961 envisaged three high tower blocks which would have been visible for miles, two of which, for Physics, would have risen fifteen storeys to 61.5m (205 feet), with a lower ten-storey tower for mathematics. These were rejected by the planning authority. Revised, at much further cost, in November 1962, the plan then proposed one thirteen-storey tower for Mathematics (it having been decided that Physics would move to the Old Addenbrooke's Hospital Site), but this plan too was rejected. The cost of the nugatory work while the University thrashed around without any co-ordinated estate plan was huge, and achieved little. In due course a much-revised and scaled-down plan for the eastern edge of the site was adopted, with a more modest 30m (100 foot) seven-storey tower for the Metallurgy and Computer Science departments, designed by Philip Dowson of Arup Associates in an uncompromising and brutal style which catered inadequately for their academic users having limited floor area on each floor: like most people, academics communicate quite well horizontally but not at all well vertically. (There was a window of opportunity in 2004–5 when the Arup Tower probably could have been demolished in favour of a much improved scheme at low cost, as noted in Chapter 8.)

It is worth noting for this site, and for the Downing and Sidgwick Sites, that had the 'plan-led' planning system as described in Chapter 3 become Town and Country Planning law 80 years earlier, the fragmented and disorganised development of those sites would never have occurred because the University would have been required to develop site master plans which would have been the basis in the Local Plan for Cambridge for future development. In the absence of that requirement for forethought and planning, and consultation with the Council and the citizens of the city, those sites continued to develop higgledy-piggledy, driven by the vagaries of brilliant and powerful academic leaders (sometimes referred to unfairly as academic 'barons'), disaggregated syndicates and the spasmodic funding these could muster.

In 1959 the first University building for the study of chemical engineering was built. The architectural aspects of the design were done by the practice Easton and Robertson; they had a hard task because the two leading professors, Fox and Armstrong, were formidable clients. (The University *per se* had little to do with the design or the procurement.) A central concept was a wide and therefore flexible space. The architect's drawings when proudly presented at one meeting showed a line of stark and obstructive columns. When the professors demanded their removal, the architect assured them that structural calculations proved them necessary. At that point, Professor Fox produced from his pocket structural calculations which proved the contrary.[11] The professors preferred the doors recently installed in the new chemistry building and demanded, and got, similar.

As Engineering, Chemistry and Physics departments moved out, their buildings were taken up largely by Arts Faculties, and the site filled up with a motley collection of hugger-mugger buildings, some efficient and some not, all responding to the massive academic research and teaching progress being so brilliantly achieved. The buildings on the site by 1996 were:

- Babbage Lecture Theatre
- Computer Laboratory – relocated to West Cambridge in 2001
- Central Science Library
- Cockcroft Lecture Theatre
- Department of Chemical Engineering
- Department of History and Philosophy of Science and Whipple Museum
- Department of Materials and Metallurgy – relocated to West Cambridge in 2013
- Department of Social Anthropology
- Department of Social and Developmental Psychology
- Department of Sociology
- Department of Zoology
- Old Examinations Hall (formerly Arts School)
- Old Cavendish Laboratory – former physics laboratory
- Phoenix – former University mainframe
- University Computing Service

- Whipple Museum of the History of Science
- Zoology Museum.

A full and interesting history of the site is set out in a report by Beacon Planning.[12]

Downing Site

After the previous syndicate for the New Museums Site lapsed in 1898, a new syndicate was set up for the purchase of what was to develop into the 'Downing Site': between 1895 and 1901, the University bought from Downing College about nine acres for £45,000. From 1900 onwards, a further 8.65 acres were bought that were to allow for the final development of the 'Downing Site' area. This acreage was purchased in three phases for a total of £42,000. The Law School was soon set down on the Downing Site with a new building for Law, Geology and Archaeology facing Downing Street (Sir T. G. Jackson, architect 1904–1911). The Sedgwick Museum of Geology and the Solar Physics Observatory were completed in 1911. Buildings for the Departments of Pathology, Biochemistry and Geography were achieved by 1927.

The site was fully developed over the following four decades, especially after negotiations with and funding by John D. Rockefeller and the International Education Board which led to a new plan for buildings, including for the study of botany, physiology and zoology at a budget of £1.179M with funding also by the Ministry of Agriculture and Fisheries (£50k); The Empire Marketing Board (£50k); and The Royal Agricultural Society (£1k). Other major buildings on the site echoed contemporary scientific discoveries, including laboratories for Biochemistry (1924), Pathology (1927) and Anatomy (1938); the Veterinary Anatomy building was completed in 1958. In the late 1980s, after some controversy, a new building to re-house the Biochemistry Department, previously crowded into four buildings, was built. One of the controversial but better buildings replacing a series of ungainly huts was the MacDonald Institute in 1990 following a benefaction by Dr D. M. McDonald. It poked into the large space in the middle of the Downing Site, designed by Casson Conder, the practice that did the excellent original master plan of the Sidgwick Site.

So, like the New Museums Site, the Downing Site filled up chunk by chunk as needs and capital finance came together. By the period of this book, buildings on the Downing Site were:

- Archaeology and Anthropology
- Biffen Lecture Theatre (re Genetics)
- Biochemistry
- Craik-Marshall Building (Experimental Psychology and Anatomy)
- Earth Sciences
- Experimental Psychology

- Genetics
- Geography
- McDonald Institute for Archaeological Research
- Museum of Archaeology and Anthropology
- Physiology, Development and Neuroscience: Anatomy Building
- Physiology, Development and Neuroscience: Physiology Building
- Pathology
- Plant Sciences
- Sedgwick Museum of Earth Sciences.
- Sir William Hardy Building (Experimental Psychology, Geography and Landscape Modelling)
- Veterinary Anatomy.

Mill Lane

At the junction of Mill Lane and Trumpington Street lies 'Kenmare', a Grade II listed building built from 1768, bought by the University and subsequently occupied by the new Department of Estate Management. In 1858 the Local Examinations Syndicate was established, and in 1886 it set up in a major building: the red-brick house, with Dutch gables, in Mill Lane, which was extended in 1893. In 1925 an elegant building with wrought iron gates, nearby Stuart House, was opened for the use of the 'Extra-Mural Studies Department', which had bought the land from Mr W. Sindall. Later it metamorphosed into the Institute of Continuing Education and re-located to Madingley. It greatly improved the built environment of Mill Lane with its impressive classical facade and garden frontage. From 1930 to 1933 a block of rather 'bricky' lecture theatres was added next door to the University's Estates Department. The two Maths departments moved into the former printing shop and warehouse of the Cambridge University Press in Mill Lane in 1967/1968. Meanwhile, down by the river-front in a really wonderful site, the new University Centre was initially conceived in 1964 for graduates without college affiliations (a worthy purpose) funded kindly by the Wolfson Foundation; a building much appreciated by its users although there is a wide spectrum of views about the attractiveness of its 'brutalist' concrete facade staring over the River Cam.

Department of Engineering Site

The Engineering Department started to move to a site bordering Coe Fen and Trumpington Street after the 1919–1920 purchase of three acres at Scroope House (later in 1961 to be demolished) from Gonville and Caius College, for £14,900 following a benefaction of £25k from Sir Dorbji Tata, an alumnus of Caius College and founder of the Tata Group conglomerate. The first new building was the Inglis Building; the site was levelled and the foundations dug in 1919–1920 at a cost of £2,000, and then the first main building went up in 1920–1922, with extensions added until 1948.

The Inglis Building was the University's biggest mid-wars building. After the Second World War the Engineering Department Site became the subject of the first of many major planning controversies involving the University, a consequence both of tighter development controls and also the growth of strong conservation lobbies worried about development of the green areas around Cambridge. The E-shaped Baker building was, however, added, the structural steelwork of which was designed by Prof Baker with 20 per cent less steel than in a standard design, and was opened by HRH the Duke of Edinburgh and the Chancellor, Lord Tedder, on 13 November 1952. Construction was by Rattee and Kett, and W. Sindall Ltd; the cost was £225k.[13]

Lensfield Road development

Following the death of its owner, a doctor at Addenbrooke's, the land of a house named 'Lensfield' (the name came from Dr John Lens, Sergeant-at-Law who originally owned the plot) was bought in 1927 to create the Lensfield Road Site. The funding came largely from a payment of £210k (equivalent in purchasing power to about £8.4M in 2015) from British oil companies in 1919. The first building on the site was the Scott Polar Research Institute, funded by subscriptions in memory of Captain Scott, designed by Sir Herbert Baker and constructed 1933–1934.

The teaching of chemistry was changing in a way that had considerable estate implications. The need for expansion of space was clear: the premises in Free School Lane and Pembroke Street were far too small and out of date, and in 1944 expansion was promised. In the mid-1940s chemistry was being taught in two separate Departments, Organic[14] and Physical. Any thoughts of bringing these two departments together long foundered on the personal relationships (described as being 'cool') between Professors Norrish and Todd. In 1956, however, the two departments were brought into one building with one entrance (although they still kept quite separate organisationally, with only the library as 'common' space): this 'coming together' was effected by a move to new premises on Lensfield Road. The first phase of the present building began in 1956, officially 'opened' in 1958. It was a large building, larger than most expected. Under the pseudonym of pH6 (i.e., slightly acidic) a wag, having a go at a perceived egotism in Professor Todd and his relationship with the physicist Professor Mott, wrote:

> Lord, give me leave to build a lab,
> So large that when I've trod
> Its lofty aisles and nave I'll think
> I'm in thy house, O Todd.

Another wag wrote

> I'm God said Todd,
> You're not, said Mott.[15]

It is also worth noting that a Health Centre was built near the cricket ground at Fenners in 1951; this was in part for the screening of staff and students for TB, mutating later into the Department of Human Ecology, then the Department of Occupational Health.

Addenbrooke's site

A second big area of expansion of the estate was the Medical School in Trumpington Street. That story had started a couple of centuries earlier when in 1756, Dr Addenbrooke, a Fellow of St Catharine's College, gave £4,727 (about £620k equivalent purchasing power today) for the provision of 20 beds and treatment of 11 patients. The hospital thus funded was extended in 1824, 1834 and, extensively, in 1856. In 1915 a top floor was added, followed by an outpatients' department and a private wing. In 1948 planning started in the new NHS (whose genesis was in the war-time Emergency Medical Service) for a major replacement hospital on a site to the south of Cambridge, and in 1962 Her Majesty the Queen opened the first part of 'New Addenbrooke's'. After the hospital had fully de-camped to its new 'edge of city' site, controversy increased about use of the original site. In the important policy document 'Planning for the Late 1980s', written by the University in response to the University Grants Committee Circular Letter 12/85 (which set out further cuts) Annex A noted that,

> Because it will be many years before the University can expect to have the necessary capital to develop the whole [Addenbrooke's] site, it is antici- pated ... that the northern half of the site containing most of the existing hospital buildings will be let commercially. The southern part will ... be developed primarily for biological departments.[16]

What happened in the end was that in the early 1990s the small added buildings behind and at each side were demolished or re-ordered for academic/administrative use, and the clinical building on the southern side was sold to Browns restaurant chain (the original 'Browns' being in Oxford; the next was in Bristol, and then more in other University cities). The main, original building was the challenge, and a battleground between those who wanted to see the building kept much as it was, those who wanted to keep the facade as it was but gut the building behind ('facadism') and those who would cheerfully see the bulldozers move in. This 'town planning' controversy was said by some academics to have led to the resignation of the then Director of Estates, and that may have been one factor: planning and conservation matters can always get frenzied, and after the desecration of Petty Cury a couple of decades earlier, tempers could flare in the University, among people in the city and in its Council. Petty Cury had been a wonderful old street leading into the market; it was rich with old former coaching houses in Lion Yard behind it, and buildings stepped out over the street below. In the late 1960s and early 1970s it was ruined in the development of the Lion Yard shopping centre.

There was a strong view among many of the Councillors that as a public-realm building the 'Old Addenbrooke's' site should remain 'public sector'; the University was not seen as 'public sector', as it of course is. Those Councillors wanted the building (somehow) to be used as 'affordable housing' given the great extent to which housing provision in Cambridge had fallen far behind the rocketing employment increases as a consequence of the green belt having been drawn far too close around the city. (See Chapter 3.) Finally it was decided to leave the facade, add a floor and re-order the interior without destroying its structural design integrity. The re-design was spectacular with a multi-coloured tapestry of coloured bricks and coloured concrete in the elevations, and wide bridges over a large atrium in the equally multi-coloured interior. The way the work was organised was singularly bad. The benefactor put pressure on the University on selection of the architect. Once appointed, the architect, John Outram, proved his capacity for remarkable design, but his practice had also been appointed to manage the project as well as to design it, and that promptly ran into terrible problems: working drawings came late and allegedly substantially incomplete. It was said that on one occasion when the tough contractors, Laing, the UK's oldest family construction company, asked for working drawings they were given water-colour paintings: probably an apocryphal anecdote. Work soon was badly behind schedule, the contractor was paid a very large sum to continue on site. The workmanlike architects, Fitzroy, were appointed to take over management and working drawing production. (This was somewhat redolent of the system common in the USA of appointing firstly an architect to produce the concept and the vision of the new building, followed by the 'architect of record' who does the detailed architectural work. This concept has many supporters in the UK, but the RIBA has consistently opposed it on the grounds that any good practice should be capable of doing both and, further, it is better professionally for architects to see jobs through. Design and management are very different skills and it is rare to find in one person both the high-level architectural vision and also the application to endless working drawings. Some architects are good at directing work on site with firmness; some are too nice, so there is much to be said for split of responsibilities.) In any event, the procurement lessons were soon learnt and in this project a relatively good building, the Judge School of Management, was finally produced. Legal matters were settled several years later in 1999.

Sidgwick Site

The Sidgwick Site development has its academic origins in the early 1920s. When earlier the University was deciding where best to locate a new library, one of the two short-listed sites was the cricket ground and garden of Corpus Christi College off Sidgwick Avenue. The site which in the end was selected for the University Library was much cheaper than the £25,000 price set by

Corpus, and so the Corpus site remained a sports ground till its later purchase by the University (for the same price, much less in real terms) in 1948 for facilities for the Arts, Social Sciences and Humanities: the main Lecture Room block there was completed in 1958 and the Raised Faculty Building in 1960. (Chapter 6 sets out subsequent site development history.) The large concert hall extension to the School of Music villa on West Road was built in 1977, not pretty but effectively functional for teaching and good as a concert hall (and much enhanced by internal refurbishment 25 years later).

University Library

Meanwhile, political discussions and fund-raising were leading to consideration of provision of a large-scale library for the University generally, and the site for a new University Library followed the purchase in 1922–1925, after much and expensive legal activity, of a 7.8 acre cricket grounds site from King's and Clare Colleges for £12,483. Interestingly, the site had been requisitioned by the government during the First World War to build a huge hospital of 1,200 beds in 24 huts, and after the war the site was used for emergency housing for local people; the huts were made habitable with blockwork walls, the average occupancy being for over two years. The site was progressively cleared and the last of the huts sold off to the public in 1929. In 1928 it was decided that the University should itself pay half of the cost of the new library. The appeal for the balance of funding of the building was started in 1929 by a group of University leaders and friends, particularly the Rockefeller Foundation, known as 'The London Committee'; this delegated the task to a sub-committee chaired by Lord Melchett (whose fortune came from the chemical industry). That project followed controversy within the University as to who, if anyone, would take the trouble to go over to the west side of the river,[17] and that decision to site the building there was crucial in the later development of the University estate: had the wrong decision been made, to develop it close to the existing estate (as had, perhaps regrettably, been done earlier at Oxford), then the estate would have continued to be cramped rather than being in its current position, nearly unique among the top world universities, of having contiguous land sufficient for many decades of development. The University Library is a copyright, or legal-deposit, library (one of 5 in the UK) having an option to obtain a copy of all books published in the UK, and therefore needing ever more space. The architectural design of a library with a budget of £500,000 was by Giles Gilbert Scott, the designer also of Battersea Power Station, Liverpool Cathedral and the British telephone box, a goodly mix. Funding started to come in: £25,000 from the Local Examinations Syndicate, £65,000 from a bequest from Rev. Ellis (late of Trinity College) and the remainder of the planned budget was to come from a 50-year loan. Another £204k came via the Cambridge University Association (the predecessor of the current Cambridge University Development and Alumni Office).[18] Work started on the University Library in 1931, and it was formally opened in October

1934 by King George V; it was said to be the largest open-access library in Europe, and the first building with aluminium windows. The University Library may not be a gracious building: the tower, a concept brought in as an afterthought,[19] dominates the Cambridge skyline in a rather industrial way (as might be expected from the architect whose design of Battersea Power Station brought him such regard); but the design has proved to be good for its purpose, and indeed to be flexible for the expansions which started 35 years later. The first big extension to the Library was in 1966, built by Rattee and Kett, (later to be taken over by Mowlem, which in turn was later taken over by Vinci).

Farmland purchases

Madingley Hall was built in 1543, and its estates were sold off in 1871 and subsequently bought by the University in 1948.[20] The manorial estate at Madingley was some 3,000 acres and over the years about a quarter of it was bought by the University; for example, 402 acres for the then School of Agriculture in 1923; Catch Hall Farm was bought from Messrs Chivers and Sons in 1998. On the south-west side of Cambridge, Laundry Farm was bought in 1947 from St Catharine's College for development, but planning permission was denied. (Now it is the home of the University's maintenance department.) In 1943 the University gave a chunk of land off the Madingley Road 'to the people of the United States of America' to site first a temporary then a permanent US services cemetery which opened in 1956 with 3,812 graves, largely of airmen and sailors, there being a US military cemetery in Brookwood, south of London.

The history of the West Cambridge Site

Although not developed until the end of the twentieth century, the history of this University site goes back a long way. Before and during the Second World War, the fields to the west of Cambridge had been systematically acquired. This area of Cambridge had never been developed, largely because of the timing of the series of Enclosure Acts: in general, Enclosure Acts for Cambridge came later than they did in many other areas of the UK. Land plots to the north, east and south of Cambridge were consolidated during the seventeenth, eighteenth and nineteenth centuries, facilitating land-assembly and hence development there. Enclosure Acts leading up to those of 1802 and 1807 allowed just limited enclosure around Castle Hill and Newnham. So the massive expansions in Cambridge, especially new sprawling housing estates when the 'Eastern Counties Railway Company'[21] railway came in 1845–1846, were mainly to the east and south because those areas were ready for development, unlike the land to the west or north.[22] The pattern of subsequent development would have been vastly different had, for instance, Coe Fen or one of the four sites near the Botanic Garden been chosen for the station rather than the south side of the city: then west and north-west Cambridge would probably be what east and south Cambridge are now since development generally followed rail centres. The arrival of the railway with the decision to build the station well to the south of Cambridge led to huge residential and supporting development, and also transport infrastructure (including bridges over the river) to the east and south of Cambridge but not to the west.

Wide arcs of fields to the west of Cambridge, now North West and West Cambridge sites, and down to the Barton Road were enclosed in 1802, and a large area of land to the south of the heart of old Cambridge in 1807: in each case, within four years land ownership had been re-allocated.[23] However, Colleges and the University were notably slower than private landowners to develop land when enclosure made that possible. Until the twentieth century, development petered out around Newnham, the Backs and Castle Hill, giving Cambridge the imbalanced shape it has had until recent times.

So it was as late as 1923 that land acquisition in the area of West Cambridge began: 402 acres were bought from Trinity College, for £22,500, and then a further 32 acres bought from Professor Newall in 1931, and another 1,391 acres bought 1947–1949. The final land package for the current site, formerly Trinity College land, was bought from Cambridge Perfusion, in 1998: 395 acres for £524,000.[24]

Site development of the School of Veterinary Medicine began in 1954–1955. The School had been founded in 1949 with eight undergraduate students although its origins go back to 1909 when the University Department of Pathology set up an outstation to study diseases of large animals. In 1935 the University entered into an arrangement with the Royal College of Veterinary Surgeons whereby it ran a pre-clinical course and a postgraduate diploma in a building at West Cambridge. The fabric of the School grew markedly in the 1950s and 1960s, in part to accommodate increased student numbers. The 1980s saw further developments, abruptly halted in 1988 by a national review which recommended closing Glasgow and Cambridge clinical veterinary schools. This was one of the most politically unfortunate reports of the time as the Conservative front bench had many Cambridge alumni and the Party was desperately trying to staunch the loss of support in Scotland. Mrs Thatcher was said to have knocked it on the head for those reasons, which was easy to do in the face of a widespread campaign. A follow-up report took more account of the implications of the increasing animal checking legislation by the EU, and a better assessment of future needs for vets in UK, so the closure decisions were reversed, and, indeed, student intakes and built space requirements increased.

Across the Madingley Road, the Astronomy Department set itself up in 1965–1968, later to expand there enormously, especially when Royal Greenwich Observatory staff moved from Herstmonceaux to Cambridge. Also, University scientists, who during the war had been hugely involved in developing radar capacity, set about, with financial help from Mullard Ltd, building up a huge array of powerful telescopes on a site along the Barton Road to the south-west of Cambridge, beside the old LMS railway line (which opened in 1956 and ran to Bedford and Bletchley). How the astronomers who'd set up the early astronomy facility in 1822–1823 would have marvelled at those huge dishes.

The rather vague and meagre report (just over two pages) which had been in the report of the Council of the Senate in 1958 on how to meet the expansion needs of the science departments had achieved little or nothing.[25] In 1965, however, there came the 'Deer' Report[26] and that was a watershed in the development of the University estate. The 'Deer' report was initiated and drafted by four academic scientists and it changed policy; in particular it identified that the main thrust should be development to the west of Cambridge. A big step forward was taken in the early 1970s with the Swinnerton Dyer Report on the long-term development of the University:[27] the enclosed plan featured the famous Swinnerton Dyer ellipse showing where development of the University estate should go.[28] The area included an area from Lensfield Road, along King's Parade, out to the Observatory at Madingley Rise, and a small part of the West Cambridge site. Predicated on the concept of a 'steady-state university' with 'an upper limit to the University of Cambridge future growth' (14,000 total students), his report concluded that the University should not spread south (e.g. by purchase of the land off Brooklands Avenue given up by government agencies such as the Property Services Agency) but rather should spread west using the land mostly assembled before and during the Second World War: 'Future developments in physical sciences and engineering should normally take place on sites in West Cambridge.' The report was accepted. During the war there was a temporary Short Brothers aircraft factory at West Cambridge, some huts of which were used by the Physics Department for a while. (The company was nationalised between 1943 and 1948.) In 1971–1974 the Cavendish laboratory, which had outgrown its city centre site, moved to the closest end of West Cambridge and settled into its 30-year design-life CLASP-construction buildings: component-assembled accommodation which is flexible but, being designed before the oil crisis of the early 1970s, wasteful of energy.

Management of the twentieth-century estate

The Buildings Syndicate continued its work between the wars. Whereas 70 years later the Buildings Committee would typically have about 100 pages of report for each of its six annual meetings, seeking authority for about £100M of work, the annual report of the Buildings Committee on 30 October 1928 had just six short paragraphs: one dealt with treatment for death watch beetle in the oak roof of the South Room of the University Library. Another that 'A small room in the Arts School' had been fitted out with shelving; a third paragraph discussed four boilers.

Soon after the Anatomy building on the Downing Site was completed in 1938, the storm of war broke, and all matters relating to estates rightly took lower priority. Cambridge students of many disciplines disappeared from Cambridge; however, students of engineering, physics, chemistry and medicine stayed, doing a two-year degree as a longer-term call-up policy was taken by the government (unlike the drafting policy of the First World War which led to the disastrous loss of a generation of young men in those professions, with of course the loss of those who would have been their progeny). Also, thousands of students came up to Cambridge from London University; their domestic accommodation was provided by Colleges and teaching space by the University, provided in the sense of students having to squeeze up a bit, (defining the term 'squeeze up' by the generous standards of Cambridge).

After the War, estate management resumed much as before, led by academics on a piece-meal basis as capital funding was found, mostly by them. In 1958 Council set out in a two-page paper how the University should meet its future space needs; that was to be by developing current sites, with 'one or two buildings of more than 7 stories' on the New Museums Site, and for the 'acquisition of further central sites as opportunity may offer'. The report noted that these solutions might not be sufficient, but it offered no other options. From after the War until around 1975, the University Grants Committee paid relatively generously for university buildings. During those years the University should have thought and planned more cohesively, but there still was little motivation for central University planning; most people were still thinking just in terms of departments and colleges.

During the cuts of the 1980s the strategic direction of the University estate became not only more difficult to steer but also more crucial as huge advances in the sciences and engineering portended considerable expansion, and that led to most departments having ever-increasing and justified ambitions. Site strategy was overseen by the University's Long-Term Planning Committee until it was subsumed into committees described in Chapter 4. A seminal report was produced in 1989: this updated the priorities for the Sidgwick Site originally set out in a report in 1950,[29] and the equally seminal Deer Report;[30] specifically, this recommended that

> A development plan for the area of West Cambridge should be prepared in conjunction with St John's College [which owned land adjacent to the land bought earlier by the university] and Jesus College [which owned the 'Rifle Range' land, formerly used by the Officer Training Corps].

The question as to which science and related departments should move to West Cambridge was raised. The 1989 Report also recommended that the Downing Site should be earmarked for Biological Sciences, with Architecture and Anthropology, and Geography needing to remain there 'for at least many years'. It also recommended a study of the development of the Old Press Site: that need for development was studied and re-studied over the decades (until towards the end of the decade of this book a group under Pro-Vice-Chancellor Malcolm Grant came up with an excellent scheme which got good support from the city's planning department, but then was lost in 2005 because of changes to the membership of the group).

The management of space allocation in buildings, a fundamental in planning an estate, still rested on squatter's rights and tactical alliances.[31]

Funding

Benefactions to fund buildings continued throughout the Second World War, notably from 1941, and indeed for about another 20 years there was more capital funding by the University Grants Committee with the growing assumption in Britain (unlike the USA, with its liberal tax exemption for university benefaction) that it was the job of the state (the tax-payer) to fund universities, including capital funding. So expansion and new building work in universities in the UK between the 1960s and the 1990s was a measured, controlled and generally a 'public sector' matter, with strict costing criteria and capital procurement rules, generally sensible by the standards of that time, set by the UGC.

Historical build rates

An internal study was done in the Estates Department on the building rates of projects set out in this and subsequent chapters. The rate picked up steadily over the centuries until it slowed down in the 1950s onwards as buildings began to be much more heavily serviced, often over 30 per cent of the cost of a public sector

Table 1.1   Construction rates

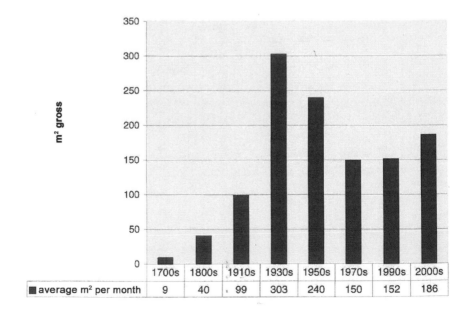

Average Speed of Construction (m² per month) on University Projects

average m² per month	1700s	1800s	1910s	1930s	1950s	1970s	1990s	2000s
■ average m² per month	9	40	99	303	240	150	152	186

building being in electrical and mechanical services. Then, as techniques were developed to speed up installation of those services, including a great increase in off-site manufacture, overall build rates speeded up again; construction in central Cambridge with such constricted streets never has been easy. See Table 1.1.

Conclusions

So, by 1996 where our review begins, the University estate was established with over 320 buildings. A line of buildings lay along the historic central axis of the University, founded on the gravel lens that runs through the soft clay, from the former Divinity School, past Senate House and the Old Schools, Mill Lane, the Pitt Building, and along Trumpington Street, with the major Downing and New Museums Sites nearby. Then along to the Engineering and Chemistry Department buildings and the Scott Polar building. To the south of Cambridge lay the large and fast-growing NHS/University medical campus on the 'New Addenbrooke's Site'. To the west, across the river, the Sidgwick Site was partly developed, and beyond that the University Library, and the land at West Cambridge with the Vet School, the Department of Physics and parts of the Engineering Department and the research buildings of Schlumberger and British Antarctic Survey on leased land at the west (further) end of the site.

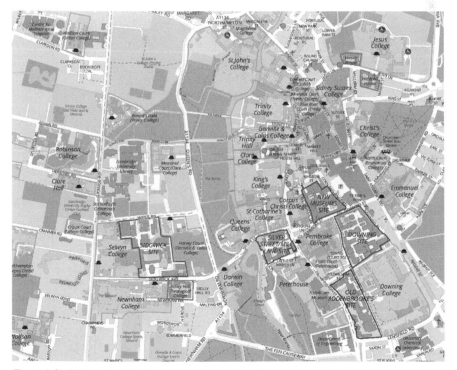

Figure 1.2   University site plan

By courtesy of the University of Cambridge

Most of the buildings on that estate were already getting cramped, and many had poor facilities. What would be essential for expansion and provision of new, high-quality buildings was to secure planning permissions and capital funding for an intense capital programme, and to secure government approval for release from the green belt and development of the huge site at North West Cambridge. In 1996 those were the greatest challenges in developing the estate satisfactorily: if they could be achieved, all might be well. If not, not.

Notes

1 There were also a couple of hospitals, one was a leper colony at Stourbridge; the other, for paupers, dedicated to St John, moved to the current St John's College site around 1200. After the University was started in Cambridge, there were no more universities founded in England for over 600 years, though four were set up in Scotland by 1583.

2 Dorothy Owen, *Cambridge University – A Classified List* (Cambridge University Press, 2011).

3 The first 'University Printer' was Thomas Thomas who was appointed in 1583. He set up a printing house on the site of what became the Senate-House lawn, near the current University Press Bookshop. At the time the Stationers' Company based in London, a Guild which regulated the profession, was very critical of the Cambridge operation, describing Thomas as 'vtterlie ignoraunte in printinge'.

4 The building was substantially extended in 1873 and more recently upgraded as an office and conference centre.

5 Then there were temporary syndicates, the word coming from the Greek for 'a delegacy'; later there were and are 'permanent' syndicates.

6 'Professions' relating to the construction industry as would be recognised nowadays were then rare: the Institution of Engineering started in January 1818 and the Royal Institute of British Architecture was founded in 1837. Other professional institutes followed.

7 Report of the Syndics of the Press on the Purchase of Sites, Cambridge University Reporter.

8 As late as 31 July 1952 a wonderful letter was sent from 'The Medical Research Council. Privy Council' to N. H. Hackworth Esq of the Applied Psychology Research Unit chiding him for purchase of 'the house in Chaucer Road' without permission – 'when the Council learned of the transaction they assumed the most portentous official frown!' Precisely the same happened, by the same Unit, with an implicit call on University utility services and maintenance money, in 2000, but the practice was stopped.

9 One signatory was J. N. Keynes – the father of the great J. M. Keynes, once Third Bursar of King's College, instigator of the Arts Theatre and later a director of the Bank of England; also a regrettably unheeded participant in, and strong critic of, the drafting of the disastrous Treaty of Versailles.

10 Reporter No. 4120. Vol 89, No.12, 26.11.58, pp.432–434.

11 Professor Fox was a bachelor Fellow of King's College, and apparently he commented that as there were six waiters on duty at breakfast, he was disappointed that they had to make the toast the night before.

12 Beacon Planning, Cambs, *Historic Environment Analysis New Museums Site* (September 2013).

13 W. Sindall got going in the 1860s, flourished and merged with the London company Morgan/Lovell in 1994, and now employ some 6,500 staff.

14 Organic and inorganic chemistry were then taught and researched in the same department; Professor Todd was head of 'Organic'.

15 Professor Mott was a Nobel Laureate physicist.

16 UGC's circular letter 4/86, 20.5.86, signed Swinnerton-Dyer, is printed in the Reporter

of 3 June 1986, p.547 and the text of the reply is printed as an Appendix to 'Planning for the late 1980s: notice' in the Reporter of 30 July 1986, pp.712–713.

17 Indeed, until undergraduates were allowed to borrow books, a surprisingly large proportion of undergraduates, and even of graduates, never made it to the University Library.

18 CUA was unfortunately later allowed to lapse and most of its crucial database of alumni was lost.

19 It was said that it was added so that it could be seen from the 'University' side of the river; another theory is that it was Rockefeller who wanted a tower. Also, it was at that time realised that not enough room for expansion had been specified.

20 The Hall was rented in 1860 by Queen Victoria for her son Edward, the future King Edward VII, to live in while he was an undergraduate at the University.

21 Great Eastern Railway from 1854.

22 An earlier plan had been to build the railway station on Coe Fen, to be built up out of the swampy land, but that plan was baulked by Peterhouse. At least seventeen different sites for a Cambridge railway station were studied between 1834 and 1864, principally to the south and east of the town centre. Worthy of passing note is that the railway station built in 1846 has lasted until the time of writing; it was due to be refurbished in 2014–2015. (Incidentally, at least four successive and unimpressive structures were re-built over the same period as Oxford's station.)

23 Enclosures around Huntingdon Road and Chesterton were not till 1838.

24 Robert Neild, *The Financial History of Cambridge University* (Thames River Press, 2012).

25 Reporter No. 4120. Vol 89, No.12, 26.11.58, pp.432–434.

26 Reporter 1965–1966, pp.546 et seq.

27 Reporter No 4884. Vol 105, Nos 13,17,12,74, pp.542–580.

28 Being before the days of computers, the ellipse was first drawn by the Assistant Registrary using two pins and a piece of string.

29 Reporter 28 June 1989 pp.791 et seq and Reporter 1949–1950 pp.831 et seq.

30 January 1965; see C. N. L. Brooke, *A History of the University of Cambridge* (Cambridge University Press, 1992).

31 A letter from J. N. Keynes, Registrary, dated 16 January 1913, read:

> Dear Grant,
> Want to move Economics Library out of Divinity Schools and into ½ bay in the Moral Science Library. But I daresay they have established a vested right to their present position.

2 Why did it all happen then, and why so fast?

Why did the sudden 'tsunami' of construction projects come about in British universities in the 1990s: why then, and why so suddenly, and why so massively? This building programme was the most intense in the history of the University of Cambridge, and so it was for many UK universities.

It was in mid-1998 that project activity really started to kick up in Cambridge, a bit later than in most universities: by August 1998 the Cambridge University Monthly Project Summary paper for building projects in design/construction totalled £214M: in June 2001 that was up to £427M[1] and at a peak of £635,638,890 in October 2003, with totals staying close to that level for several months. The gross built space of the University's operational estate increased by 33 per cent, from 499,000m^2 in 1996 to 662,000m^2 in 2006, after capital expenditure on major capital projects and the larger 'minor works contracts' of some £770M (as set out in Chapters 6–10).

So why and how did that happen?

Government funding

Universities suffered a series of funding cuts following the UK economic decline of the late 1960s. The system of quinquennial funding, whereby the UGC sent out cheques for the first year in the full expectation that the cheques, including inflation allowance, for years 2–5 would follow, gave universities an ability to plan ahead. Following the oil crisis from 1972, the UGC stopped adding inflation, and the quinquennial system was ditched altogether in 1973. A White Paper in December 1973 caused an immediate moratorium on all UGC building project starts until June 1974, and a 20 per cent cut on the capital budget thereafter.[2] Perhaps the most damage to Higher Education (HE) during the recession of the Thatcher Government was what was seen as an attack on universities' independence, and so universities' top priorities became to circle the wagons and defend academic freedom rather than think of expansion. (An epithet from a disgruntled academic was the hope that the iron lady would 'rust in peace'.)

As noted in the previous chapter, in its formal response to the University Grants Committee's Circular Letter 12/85, Cambridge University noted that the cuts 'had been extremely damaging. If it is now necessary to anticipate the reductions in funding used as the basis of your letter of 9 May 1985, the damage and major disruption to the work of the University cannot be over-emphasised'. In the final paragraph of the letter, the University also noted how the number of its historic nineteenth and early twentieth century buildings put a strain on maintenance budgets. It stated a desire 'to attempt to maintain its provision for building repairs in line with inflation'; initially that was to prove very difficult, and then too difficult. As the maintenance backlog increased, buildings became more run-down and in many cases they came close to or over the limits imposed by rapidly increasing regulations governing what came to be called 'Health and Safety' (or 'elfinsafety' as expressed by maintenance staff): the problems, for example the dangers of exposure to mercury, asbestos and radiation, were beginning to be recognised. (Much later, the Higher Education Funding Council for England (Hefce) was to embark on a series of funding rounds once the extent of the maintenance problem became widely apparent.)

One factor in the justification presented for budget cuts to universities followed the granting of university titles to the 'Polys': these came into the HE sector with lower unit costs and thereby gave ammunition to those driving down unit HE costs across the board. Ministers were of course happy to say that many more students were thus being educated 'more efficiently'.

Funding for individual, approved capital projects ended around 1990. Hence, by the 1990s, a series of funding cuts and uncertainties left universities' estates under-capitalised and under-maintained, and in government policies these were the early straws in a cold wind warning that the days of new buildings being routinely funded by government grants were numbered.

The Times 'Good University Guide' in May 1998 rated Cambridge 75th out of 96 universities for the allocation of expenditure on facilities, and top in Categories of Teaching, Research and Entry Standards: this contrast was stark (even allowing for the survey's exclusion of College student facilities support services, such as accommodation). To be fair, there had been some upgrades of premises and important infrastructural developments, as set out in Chapter 1, but the level and extent of those upgrades had nothing like kept up with increasing needs.

The operational stresses on increasingly shabby academic buildings were dire, and were starkly illustrated when the then Head of the Chemistry Department, a man of firm intent, later to be become Chief Science Advisor to the Blair Government, said on local TV that his labs would be in contravention of safety regulations unless the upgrade he was seeking was implemented. That clarion call set the cat among the University pigeons.[3]

Post-war increases in the numbers of students and staff

The three decades prior to 1996 were marked in Higher Education not just by the seemingly endless series of abrupt announcements of financial cuts to annual and capital budgets, but also by expansions of student numbers.

In particular, the 'Robbins Report by the Committee on Higher Education in England and Wales', accepted by the Government on 24 October 1963,[4] recommended, *inter alia*, immediate overall expansion of universities by some 10 per cent. Largely as a consequence of that report, the number of full-time university students rose in the UK from 197,000 in 1967–1968 to 217,000 in 1973–1974. (The report also concluded that university places 'should be available to all who were qualified for them by ability and attainment', and 'increased attention should be given to the problems of introducing men and women from families with scant background to the atmosphere of higher education', as well as recommending that all Colleges of Advanced Technology/Polytechics should be given the status of universities.)

Regarding the University estate at Cambridge, the increase in student numbers in many departments was accommodated by better (or rather, slightly less bad) space management. In departments already due for physical expansion, such as Chemistry, Medicine, Engineering (expansions spurred on perhaps by Harold Wilson's Labour Party Conference speech on 1 October 1963 about 'the Britain that is going to be forged in the white heat of this [technological] revolution'), the increased student and staff numbers were factored into new building plans. In 1974 it was noted that since the UGC policy now was to relate capital grants to student expansion and that most universities were expanding faster than Cambridge, 'the situation is clearly unpromising', and 'The situation in Cambridge is worse because of the large amount of obsolete accommodation on University sites'.[5]

After a period of rather depressed stability in Higher Education in the 1980s, the greatest and most unexpected expansion of student numbers came early in the 1990s when the then Conservative Government announced to a surprised public that the student population would go from one in five British 18-year-olds to one in three. When asked how that would be funded, the Chancellor of the Duchy of Lancaster, MP for Bristol West, Cabinet member William Waldegrave calmly said that it would take about three generations for such a social change to work through: it took 10 years, accelerated in part by the high levels of unemployment. It might be argued that this change in Higher Education policy was one of the most significant policy decisions of the latter part of that Conservative Government. Participation in Higher Education in the UK at school-leaving age grew from 3.4 per cent in 1950, to 8.4 per cent in 1970, 19.3 per cent in 1990 and 33 per cent in 2000.[6] See table 2.1.

Staff numbers were also growing apace in Cambridge: in 1926 there were about 600 members of the University of Cambridge Regent House, 3,300 in 1994. Total full time equivalent staff numbers increased from about 6,400 in 1997–1998 to over 7,500 in 2006 (and to nearly 9,500 by 2014). The numbers of unestablished research staff (mostly post-docs) rose from about 1,750 in 1997 to about 2,400 in 2006 (and to over 3,300 in 2014). In contrast, total established academic staff rose from about 1,340 in 1997 to about 1,490 in 2006 (and to 1,580 today). The number of established academic-related University Offices administrative staff was 118 total in 1996–1997; in 2005

Table 2.1   Numbers of students awarded first degrees in UK per year

Year	Number of students
1920	4,357
1930	9,129
1940	9,311
1950	17,337
1960	22,337
1970	51,189
1980	68,150
1990	77,163
2000	243,246
2010	330,720
2011	350,800

(Source: The Reporter)

Table 2.2   Students (total full-time undergraduate and postgraduate) in the University of Cambridge

Year	Number
1970/1971	10,720
1980/1981	11,479
1990/1991	13,553
2000/2001	16,519
2010/2011	18,484

(Source: Parliamentary Report – Education: Historical Statistics. 27 November 2012)

and 2014 there were 274 and 339 in the Unified Administrative Service/VC office.[7]

From the end of the Second World War, seven new colleges were established, all with full complement of academic and administrative staff; in the 1960s the number of Cambridge Colleges increased with New Hall (now Murray Edwards) and the graduate Colleges of Darwin and Wolfson. Clare Hall was set up in 1966 and only Robinson College was built after that (in 1981), originally planned to be a graduate college, but, partly in response to growing University undergraduate numbers, changed to become another undergraduate-based College. (Fitzwilliam changed status, and location, from Fitzwilliam House (near the Fitzwilliam Museum) to Fitzwilliam College.) So while the ratio of College Fellows to students increased between 1938 and 1964 (Fellows from 386 to 860, students from 5,457 to about 9,470, so increases of 123 per cent and 74 per cent respectively), from the 1980s that ratio started to decline, slowly at first then quickly, making it less easy to consider the University as being truly a 'Collegiate' university. In the late 1990s, the need to improve the proportion of academics with college affiliations was a main driver for the North West Cambridge development to have three new Colleges, two graduate and one undergraduate as then planned.

Against the general UK picture, Cambridge did not increase student numbers so dramatically as many universities: 6.4 per cent between 1989/1990 and 1994/1995 (10,190 to 10,600 undergraduates, 13,722 to 14,600 total students including research post graduates), and the relatively good staff:student ratios were defended as a very high priority. However, the number of graduate students continued to rise considerably from the mid-1990s as a matter of policy; the University's 1989/1990 Planning Statement, para 1.3 noted that 'First-degree work is increasingly unable to take students to the limits of their subjects'. (Increasing some first-degree course lengths did happen, but later.)

There was a growing number and proportion of students from overseas, partly because they bring in more fees. In 1996 there were 731 undergraduates and 1,522 postgraduate overseas students in the University (6.5 per cent and 32 per cent of totals respectively); by 2003 there were 966 overseas undergraduates and 2,233 overseas postgraduates (8.2 per cent and 35 per cent respectively); by 2012, there were 7,200 international students from 130 countries studying at Cambridge, including 870 from China and 800 from the USA. The growth of non-EU students was kick-started in the Thatcher era with the directive that they should pay 'full cost' fees. The number of post-docs on the University payroll at Cambridge doubled between 1998 and 2013, to 3,400, with several hundred more externally-funded academic staff.

With continued expansion of student numbers, and overall Higher Education budget cuts, it was not surprising that between 1979 and 1997 spending per UK student halved and UK university average staff:student ratios nearly halved,[8] libraries and other buildings decayed and the HE sector accrued deficits of £5bn in borrowing and maintenance backlog.

So in sum, there were coiled springs all over the University: poor and unsuitable working accommodation, caused in turn by constraints on the extent of space available for teaching and research as both increased in numbers and in demands, and degradation in the quality of premises through lack of capital and recurrent funding. These coiled springs were salient reasons for the suddenness and the extent of the surge of capital projects starting in the late 1990s.

University planning at Cambridge

On the very wet Friday of 10 December 1999, the VC's Senior Management Team (SMT) comprising the VC, 2 PVCs and six heads of administrative services set off, in many cases somewhat reluctantly, for an 'away-weekend' in a small Peak District conference centre, Riber Hall. It was cold and the rain beat down all weekend, and one member brought a vicious flu infection which most attendees had contracted by the Sunday. The first SMT 'away-weekend' in April 1997 had been mainly concerned with broad issues such as governance: the purpose and the outcome of the 1999 'away-weekend' discussions were more specifically crucial to the future shape of the University and to its capital programme. The VC, Sir Alec Broers, and the three officers who had been appointed by him were insistent that there should be a well-articulated outline plan for the future development of the University. It was clear that the University had unprecedented opportunities and, equally, that it was navigating in powerful currents such that drift would lead to falling standards of research, and hence University status in the increasingly competitive international Higher Education arena. Someone not so keen on change said that if the University didn't change it would still always get what it always got. Someone else disagreed: it wouldn't, because other universities would take away what Cambridge had always got. One crucial outcome of the 'away-weekend', apart from a flu-ridden SMT, was agreement to promulgate and, once approved by Council, to implement a strategy that annual undergraduate growth should be limited to 0.5 per cent, graduate growth to 2.5 per cent and academic research staffing growth to reflect current needs, at that time anticipated at up to 9–11 per cent annually. The drafting of an effective Estates Strategy had been on hold pending such policy agreement, and it began straight away; it articulated clearly specific needs for capital expansion. In subsequent Estate Plans, approved planning figures were respectively 0.5 per cent, 2 per cent and 5 per cent.[9] This clear enunciation of high compound expansion of the research base of the University was a crucial step to genuine planning and, crucially in this story of the development of the estate, showed the fundamental need for both a better quality of facilities and also a very considerable spatial expansion in premises; also much better use of space by better estate management at University and departmental levels.[10] Without such enhancement of the estate, the expansion of the research base just wouldn't happen: leading academics and industry researchers wouldn't be attracted by cramped or poor facilities, and indeed many of those already in post would move on. Demand for a large capital programme was thereby further stimulated by University-wide acceptance of the considerable research and graduate expansion policy.

Academic pro-activity

The considerable frustrations of funding and unit-capitation cuts, coupled with rising numbers of staff and students, coincided in Cambridge with a period of especially powerful academics. Some of those, like the Regius Professor of Physic (viz medicine) and the Heads of the Pure Maths and the Biochemistry Departments for example were at the top of their academic organisations: others were powerful academic research leaders who had less senior appointments. Many brilliant academics were joining the University. Running a university or university department is like a game of football, the players change but the game goes on much the same; but through the 1990s at Cambridge and many research-intensive universities very strong players were coming onto the pitch so the game became faster and tougher as the incentives, and competition, increased. (The Cambridge context was of 90 Nobel prize-winners having had affiliations to Cambridge; about one per year since the start of that award.)

Generally speaking, in the earlier years up to and during the expansion of universities' estates, the majority of the powerful academics who took up the challenge of getting more suitable academic built space were in those academic disciplines closer to business and industry: the sciences, medicine, engineering and maths. Prime Ministers Wilson, economist, and Thatcher, chemist, understood the importance of the sciences in universities. Many of the academics in the Arts, Social Sciences and Humanities were more prepared, surprisingly prepared, to keep on making the best of facilities they had; a nadir at Cambridge was perhaps when the Social and Political Sciences Faculty moved into the abandoned Cavendish Labs, until encouraged to go elsewhere by senior university academics. When in 2000 the VC was persuaded of the extent of potential perceptions of imbalance between the priorities of the University in fund-raising, he decided to emphasise in his annual address to the University that the academic areas in Arts, Humanities and Social Sciences had very high priority (specifically citing as an example the top-priority need for a new building for the Faculty of English) so that expansion of all academic disciplines was being supported.

Leadership at the top

'Cometh the hour, cometh the man'. One man saw the estate challenges clearly: Alec Nigel Broers, born 17 September 1938 in India, educated to first degree level in Melbourne, came to Cambridge to do a pathfinder doctorate in nanotechnology in 1965, then to depart, on the day of his viva, to USA to work for IBM.[11] He returned to UK in 1984, became Master of Churchill College in 1990, Head of the Engineering Department in 1992 and Vice-Chancellor in 1996, the first VC to give up being head of College on starting as VC. He took over as the first VC to be appointed at the start of his time for a full five years with an option of a further two. VCs through the centuries served for one year, then starting in the mid-nineteenth century, for two years[12] from the ranks of College heads (known as 'Heads of House'), a process sometimes described irreverently and unfairly as 'Buggins Turn': being in the job for just two years meant that no VC could conceive, champion and see through any capital building project, as these take 3–5 years to plan, design and build. Secondly, each knew that he, or in the single case of Dame Rosemary Murray, she, would soon be back to full-time work in the previous role, so there was good reason not to ruffle feathers by decisive management. Continuity and much policy, therefore, was achieved by administrators.

In 1987 there had begun a reform that was crucial to the ability of the University to attract funding for developing the University's estate: the Wass Report (see also Chapter 4). Sir David Harrison, VC of Keele and then of Exeter University, subsequently PVC at the University of Cambridge, and Master of Selwyn College recollects:

> In November 1987 the VC (Michael McCrum) was presented with a Memorial signed by 194 members of the Regent House requesting that

the Council set up a Syndicate to review the governance of the University, and this the Council agreed to do. The Syndicate's membership was announced in February 1988, with 4 external members (the Chairman – Sir Douglas Wass, a leading civil servant, Lord Kearton, Sir Geoffrey Allen and myself) and 6 internal members (Alan Cook, Gordon Johnson, Peter Mathias, Roger Needham, Anne Warburton and David Yale). The Syndicate presented its Report in May 1989. Its main recommendations were

- A full-time (5+2 year) VC.
- Regent House as the governing body of the University but some of its powers transferred to the central bodies, and to the Council in particular.
- Powers of the Senate to be restricted.
- Council recognised as the principal executive and policy-making body.
- A Board of Scrutiny to be established.
- General Board given additional powers, but formally accountable to the Council.
- Financial Board reconstituted as a committee of the Council.
- Council to have an Executive Committee.

It took at least four years to get these recommendations approved and new Statutes and Ordinances agreed. (There was a ballot of the Regent House that the term of office for the VC should be for either four years, as per Oxford, or for five years, or for five plus two further years if approved: to some surprise, the vote was for the third, and best, option).

The Regent House was of course happy to take the Board of Scrutiny but reluctant to give up its own powers. Nevertheless, the arrival of a full-time VC had a profound practical and psychological effect as time went by.

To support the post of VC, there were to be two PVCs and further office and domestic support.

The Planning and Resources Committee (PRC) also came along, arising from the Wass recommendation that the Council should have ultimate policy-making powers. A Consultative Committee was also recommended (I served on it), with a purpose to bring lay opinion to the Council. This was of course no longer required once it was agreed that there should be lay representation on Council.

Prior to implementation of the Wass report, the VC had one secretary and half of the 'Assistant to the VC', and no Pro-Vice-Chancellors (and no domestic support from the University such as was needed for official entertaining).

Implementation of these recommendations of the Wass Report was a crucial factor in getting the University into a position from which it could seek, receive and manage benefactions for new buildings, wisely and accountably, and that was another reason for the timing and speed of the surge of capital funding and projects.

PRC first met in January 1998, chaired by the VC with membership of two PVCs, and three members nominated from each of Council and the General Board, soon to be joined by all six 'Chairs of Councils of Schools' ('Deans' in normal parlance), and the University Librarian. PRC was served by a range of officers, including the Director of Estates, and the Director of Finance once that post was set up. The wise, and many would say overdue (having been discussed since at least 1974), decisions to appoint VCs up-front for a period of up to seven years and to appoint first one and then two PVCs, had a considerable and positive impact on the ability of the University to effect a busy capital programme and empower fund-raising.

These management decisions at Cambridge came at about the same time as the national factors already mentioned in this chapter. It is much to the credit of Council, and Sir David Williams as VC, that most of the Wass Report was implemented expeditiously by Cambridge standards (not at all fast by the standards of other universities, and at the speed of a striking sloth on a Bank Holiday by industry standards; universities won't sacrifice academic democracy for the sake of speed).

With reference to the implications of the expansion of student numbers and the simultaneous funding cuts, Sir Alec in his widely-acclaimed inaugural speech to the University on 1 October 1996, entitled 'A Vision of Bridges' said:

> We are witnessing the second biggest expansion in higher education ever … Yet in a world obsessed with league tables and gold medals, we are also facing the largest cut in capital funding that has been made to any other institution this year. No-one who has the guts to ask for the best can escape being accused of elitism. Egalitarianism is naturally desirable when votes are to be gained, but it snuffs out the flame of inspiration and is the executioner of the first division … If the funding schemes remain as they are, then, no matter how much we do to implement financial efficiencies in our administration, Cambridge will have to find alternative resources to maintain its excellence. And we will uphold our excellence … We need to invest in [increased numbers of graduate] students and will ensure we offer them the facilities they deserve – excellent technology, libraries and teaching. Antique architecture may attract our tourists, but antiquated facilities will eventually drive away our students. The only way to inspire these exceptional young people is with brilliant lecturers and professors. To attract brilliant academics, we will have to be innovative in their employment arrangements because, as things stand, we cannot pay them a world-competitive salary. What we can do, however, is to make sure that our research facilities are first-rate.

Visionary power sharpened by Australian directness.

Capital fund-raising

Sir Alec Broers was the first VC who in many ways was, and was seen as, 'coming from outside' the Oxbridge 'bubble', by virtue of his 19 years with IBM. He brought an experience of 'industry', and he was well networked with very senior and well-heeled industrialists in the USA and UK, many of whom were later to fund major capital building works in Cambridge. This was a time of the boom in stock markets in the USA, UK and elsewhere. The IT and later the 'dot-com' bubbles can be thought of as starting in 1993 with the launch of the Mosaic Browser (the first web browser from the National Centre of Super Computing) and, it could be said, reaching a peak on 10 March 2000. Many punters made fortunes, and, encouraged by kinder tax laws, some were inclined to look around for good universities on which to bestow benefactions. Buildings were rightly seen as good recipients for benevolence: they give very palpable benefits that last for decades, 'pet' areas of research and teaching can be favoured, and new buildings can bear prominently the name of the benefactor, thereby earning the benefactor a form of immortality. That source of capital funding came at a particularly good time to promote and feed a burst of desperately needed capital projects in HE in the UK. And, for a while until the practice was stopped, the benefactor could have the pleasant experience of personally selecting a famous and charming architect.

Industrial companies were also increasingly trusting and admiring universities as centres of applied research excellence. That of course had a positive feedback effect both on benefaction for and industrial investment in new facilities. University researchers seen to be successful, and helpful to local and national economies, attract more benefaction and that in turn leads to higher overall academic standing and benchmarking which in turn leads to more benefactions: success breeds success. To take just one example: in the energy industry, the researchers Greg Muttit and Chris Grimshaw found that in the year 2000, over 1,000 research contracts were being conducted in UK universities for oil/gas companies because of the excellence of university research, and that in turn the research got better still with the industry's investment.[13] As a local example, in 1999 BP gave Cambridge University £25M to fund research across five academic departments (another example of the rapidly increasing inter-disciplinarity of research and, to some extent, teaching) in a new high-tech building off the Madingley Road on the west side of Cambridge. All this brought further benefits as such success enhanced university departments' inter-university ranking, increasingly important in attracting capital funding, as well as sugaring the pill of the introduction of 'student fees'.

During and after the First World War many personal fortunes had been made on the back of scientific research and development, and some of those fortunes funded benefactions to universities. There was less interest in science than in other aspects of industry, such as motor engineering. From the Second World War, however, there was wider and more genuine interest by entrepreneurs and by large international science-dependent companies in the advancement of war-boosted science for commercial and for wider reasons, and hence there were many associated benefactions to scientists in universities.[14] Much of the exciting new science that came out of those wars was physics-related. The first half of the twentieth century has been described as 'the golden age of physics' and that brought funding to universities' physics departments both by benefaction and investment. The second half of the twentieth century was associated more with genetics, bio-technology computer sciences and nanoscience, all of which brought much money to fund facilities in those disciplines. By 2003/2004/2005 some 95 per cent of capital expenditure ('CapEx') in Cambridge University came from external sources, much of it from industry.[15] By 2011/2012, Oxbridge raised 45 per cent of the total funds secured through benefaction by British universities, and more than the rest of the 'Russell Group' of research universities put together.[16]

Cambridge University came late to highly professional and focused fund-raising through benefactors: other universities and many of the Cambridge Colleges already had well-developed alumni networks and 'development offices'. At Cambridge, the allegiance of most alumni was to College rather than to Department or to the University *per se*. Cambridge University Development Office (CUDO) was set up in 1988,[17] having existed previously as Cambridge University Association (set up in 1898) until that was scrapped (maybe under pressure from the Colleges). The Cambridge Foundation was set up in 1989 with some 20 leading international industrialists, all Cambridge alumni, and University academics. There was still some opposition from within the University as many academics believed that capital funding was, or at least should be, the responsibility of government rather than benefactors. The Foundation chairman was Sir Alastair Pilkington and the deputy chairman was Adrian Cadbury. CUDO serviced the Foundation; its Director, William Squire, and the Registrary and Treasurer or their deputies attended the quarterly meetings: three in London and one in Cambridge. Membership included Paul Judge, Martin Sorrell, Anthony Tennant and David Simon, and Alec Broers was a Foundation Trustee. Academic members included several Heads of House and leading academics from Law, Medicine and Biochemistry, Engineering, Management Studies, Mathematics and the University Library, those being the priorities decided by University Council for benefactions coming via the Foundation after assessments of bids from departments across the University. (Some bids were very ambitious, for large new buildings, some were for help more at the level of new photocopiers.) So potential benefactors were courted by strong academics, highly placed 'captains of industry' and efficient CUDO officers making benefaction for 'shiny-new' buildings, or overdue upgrades of existing buildings, satisfying and relatively pain-free. Martin Sorrell was persuasive in the argument that the University should start sending Alumni magazines outlining the benefits to the University and opportunities for benefactions. A target of raising £250M over 10 years was set; £200 million was to be by external benefaction and £50 million raised internally by the University and the colleges: for example, Trinity College contributed money to the construction of the Isaac Newton Institute, and St John's College the land. On 3 December 1991 an article in the Financial Times celebrated that the first £100M had been pledged and/or received. Capital projects that benefited included the Isaac Newton Institute, Faculty of Law and the Judge Business School.

Another benefit of the professionalisation of fund-raising for capital projects in the 1990s was being able to cope with the consultants increasingly hired by potential benefactors to assess the cost-effectiveness and the risk-profile of their proposed benefactions: was there likely to be waste, would there be hostile reaction from neighbours? And would the building be economical and 'green' to run? These issues were becoming increasingly important to potential benefactors. Close and well-integrated teamwork between CUDO and the capital project managers in the Estates Department provided comprehensible data and evidence of good practice in procurement, risk-management and sustainability. If those were found wanting, the benefactors would take their money elsewhere, or would seek to take direct charge of aspects of the project, making key decisions about design and construction themself, often, maybe even usually, to the disadvantage of future users and managers of the building; over the period of this study, all that was avoided.

In addition to the $210M benefaction for scholarships from the Gates Foundation, said to be the biggest benefaction ever to any university in Europe,[18] some of the most significant private and corporate benefactions for University buildings 1996–2006, included the following:

- £12.5M from W Gates Foundation for the Computer Lab
- £9.1M and £3.5M from The Atlantic Philanthropies Ltd for the School of Clinical Medicine
- £20.9M from The Wellcome Trust for the Wellcome Trust Institute of Cancer and Development Biology
- £3.6M from The Dill Faulkes Education Trust for Maths CMS
- £7.8M from Dr and Mrs Gordon Moore for Maths CMS
- £16.5M from Hutchison Whampoa (Europe) Ltd for the School of Clinical Medicine
- £13.8M from Cancer Research UK for the School of Clinical Medicine
- £5.6M from National Heritage Memorial Fund for the Fitzwilliam Museum
- £2.5M from The Wolfson Foundation for Cambridge Diabetes Centre
- £2.0M from The Wellcome Trust for Genetics
- £1.7M from The Wellcome Trust for Medicine
- £1.5M from The Wellcome Trust for Chemical Engineering
- £7.0M from The Atlantic Philanthropies Limited for the School of Clinical Medicine
- £3.0M from The Atlantic Philanthropies Limited for English
- £1.2M from Corfield for Astronomy
- £1.0M from The Wolfson Foundation for Physics

- £1.0M from Hong Kong Liaison Charitable Trust for the School of Clinical Medicine
- £1.0M from The Gatsby Charitable Foundation for Plant Sciences
- £1.0M from Garfield Weston Foundation for the School of Clinical Medicine
- £2.0M from The Wellcome Trust for Biological Anthropology
- £1.0M from The Wolfson Foundation for Chemistry
- £82.0M from The Gatsby Charitable Foundation for Plant Sciences
- £2.5M from The Wolfson Foundation for Centre for Physics of Medicine
- £25M from BPI, including £2.5M for a building
- A large donation from Dr and Mrs Gordon Moore for Maths CMS.

Impetus from high-tech industry and research institutions – the 'Cambridge Phenomenon'

An influence affecting the capital programme since the 1970s was that of the science parks being set up around Cambridge, the 'Cambridge Phenomenon' of spin-out companies growing out of research groups, and in turn stimulating investment in science, medical and other departments and benefitting those Colleges which invested in the Science Parks, most notably Trinity opening in 1973 (the first major science park in UK), St Johns and Peterhouse.[19] These supplemented the various research institutions that had earlier sprung up around Cambridge such as NIAB (National Institute of Agricultural Botany) on the Huntingdon Road, and the Milk Marketing Board in the Milton Road research centre.[20] These various research institutions round Cambridge also stimulated the University departments to expand, mainly by virtue of the research synergies. For example, the research base at Hinxton was established in 1948 and by the start of the period of this book was a thriving centre for cutting-edge life-sciences (largely genetic) research which was well integrated with research being done in the University's labs; at the time of writing, more drugs are being developed in the Cambridge region than in the rest of Europe, and there are 4,980 life-science companies in the UK. (Another much quoted 'statistic' is that there are more Nobel prize-winners 'from' within two miles of Great Saint Mary's, the University church, than in France.)

The Granta Backbone network, a network of underground ducts and cables providing an advanced communications infrastructure for the University and colleges of Cambridge, completed in March 1992, gave confidence in the communication infrastructure. The £3.6M cost of installing it was shared between the University and colleges, with contributions from the Cambridge Foundation and some non-University institutions. The project involved the excavation of some 30 kilometres of trenching linking 80 University and College sites stretching from Girton College in the north-west of the city through the central and western parts of Cambridge to Addenbrooke's Hospital in the south-east. Once the University had such a high-quality communication system, it became easier to attract benefactors to the University for high-tech research.

Management of building projects

To attract benefactions, and maintain confidence, the University capital programme had to be and be seen to be cost-effective. As is set out in Chapter 5, the 1990s saw reforms of the construction industry following two government-sponsored reviews, both of which had the strong support of industry leaders. The first was by Sir Michael Latham MA Cantab. and the second was led by the industrialist Sir John Egan, formerly of Imperial College and London Business School. Early adoption by the universities, particularly Cambridge,[21] of the improved procurement practices emerging from these reforms reduced the real-terms (inflation discounted) unit cost of construction (by about 15 per cent for non-labs at Cambridge; see Chapter 5); thus, benefaction and investment in capital projects was associated with more building per unit of real-terms funding, and with less planning controversy, and with an absence of legal disputes. (None of the Cambridge projects 1996–2006 went to court, nor anywhere near their steps). All that helped to secure external funding for buildings.

In Cambridge University, a dispersed organisation, the academic leaders and indeed administrators through the centuries saw their own local building requirements as individual and unrelated projects, as indeed in practice they were. By the latter-1990s, however, it was clear that taken together, the overall requirements for more and better built space would entail a very large programme of overlapping and interacting capital works. As a result, new building projects became part of a closely managed University programme with which benefactors and industrial investors could identify.

National and local government help in permitting expansion of the University

In 1998/1999 the Eastern England Regional Planning Unit was preparing to start, through Public Inquiry, the next review of how regional aspirations and government-imposed requirements for industrial growth and housing were to be accommodated across the region: such planning was crucial to getting subsequent planning permissions for strategic developments. (The Regional Authorities for all areas with universities were going through similar procedures around that time.) Cambridge University had a slot in the 1999 Public Inquiry to give evidence as to its case for growth. The case made (see Chapter 3) was accepted. The consequent report, Regional Planning Guidance 6 (RPG 6), was very helpful to the University; it nucleated the chain of Town and Country Planning reviews (as described in Chapter 3) down to Local Plans for Cambridge and for South Cambridgeshire District. Those Local Plans gave an in-principle 'presumption in favour' of appropriate planning applications for University development in general so that the availability of good and adjacent sites for new buildings could increasingly be assumed. That was very helpful for getting benefactions, and in eliminating the costs and time-delays implicit in resorting to legal defence of planning applications. And because of regional/ local government and University planning, Cambridge had, and has, unlike nearly all the 20 internationally highest-rated universities, a good land-bank immediately beside its existing campus; most other universities have major site problems finding land in between other buildings into which new buildings can be squirrelled (Yale and Oxford for example), or as in the case of Princeton, building sites are up against what in the UK would be called 'green belt' land with huge stresses from local communities when new buildings are proposed. In Cambridge at the start of the decade being studied, much of the Sidgwick and Addenbrooke's Sites and most of the West Cambridge and the Clarkson Road Sites were immediately available. Also, the 120-hectare site at North West Cambridge was expected to be available if a huge green belt release were to be achieved; that was achieved, by close involvement of the University with local government, over the 12 years before the Outline Planning Application was made. So site availability for new buildings was not a big issue at Cambridge.

Sudden increase in government and Research Council funding 1998 onwards

A final factor as to why the construction boom in research-intensive universities such as Cambridge came about from the late 1990s, and why it came so suddenly, arrived from an unexpected quarter. The years before the change of government in 1997 did not bring much capital funding to universities. Priorities were on funding the unexpectedly fast rise in student numbers described above, with an emphasis on recurrent grants rather than capital grants. What happened after 1997 was surprising. Not only did capital funding for universities increase after the election, and by much more than expected, but the incoming government adopted a policy whereby funding would not be handed out evenly, giving some to all but not a lot to any, nor would it be channelled to those universities in most dire need, as a Labour government might traditionally have done ('to each according to his need'). Rather, the unexpectedly large capital funding programme was to benefit mainly those universities which were already demonstrably successful. 'Reinforce success' was the cry. The Joint Infrastructure Fund (JIF) launched in 1998 ran for two years with five rounds of funding totaling £750M, funded by the Wellcome Trust, the Department for Trade and Industry Office of Science and Technology and the Higher Education Funding Council for England. The JIF programme was succeeded in 2000 by the Science Research Infrastructure Fund (SRIF); this funding allocated £1.6bn to research-rich universities on a formula basis, rather than on a bid basis as in the JIF programme. During the decade 1996–2006, Cambridge received JIF/SRIF funding of over £200M, nearly 6.3 per cent of the totals available, as shown in Table 2.3.

As well as JIF and SRIF, the seven UK Research Councils continued to fund research directly and indirectly at a good rate, with welcome emphasis on inter-disciplinarity, and on net benefits to be achieved. Also, Hefce launched

Table 2.3 JIF/SRIF funding of University of Cambridge projects

Funding	Sum
JIF from 1999	£30M
SRIF 02/08	£42M
SRIF 03/06	£60M
SRIF05/08	£69M
Total	£201M

two £30M funds for 'Improving Poor Estates' (the University put in five bids, all were rejected) and for 'Refurbishing Research Laboratories' (the same five bids were accepted).

Summary of factors

To sum up: by 1996, the estate of Cambridge University, as in many other universities, was in poor condition after successive rounds of Higher Education funding cuts, and it was too small to cope with the increased numbers and demands of students and staff. Researchers were creating ever-more exciting research opportunities but were hampered by the deficiencies in buildings' extent, facilities and condition, especially where heavier burdens were being imposed by the health and safety industry. There was a particularly powerful population of highly successful, confident and demanding academics fed up with years of austerity and with government 'interference'. A lot of academics were like simmering kettles with all the pressures caused by poor facilities; but more money was becoming available especially from UK and US benefactors as huge gains were made from stock exchanges, as tax legislation made benefaction to universities more attractive and as government capital funding was suddenly made available. In Cambridge there were the first VCs who knew that they would be in post long enough to champion capital projects and see them through, and they were supported by an increasingly effective committee structure and a fund-raising office which could provide evidence of efficient and more collaborative procurement of increasingly sustainable buildings. There was a growing acceptance of integrated estate-wide planning, with programme and project procurement and decision-making, achieving cost-effectiveness and adequate risk management. As the decade progressed, there was growing evidence of regional and local support for University expansion, even for expanding into the green belt, and good building sites were available for suggested projects. And, above all perhaps, the University articulated clearly a bold academic plan that rested on expansion and provision of better facilities especially for research: physical expansion had to stimulate and support and attract brilliant academics.

The stage thus was set. Three years into the decade being studied the VC was to report that the University of Cambridge was 'the fifth largest building site in the world'. He may have been right. He usually was.

The cranes were flying over Cambridge.

Notes

1 Construction cost of projects in design/construction   £358,844,474
   Plus total planning/design/management   £47,302,372
   Add VAT   £21,323,293
   Total   £427,470,139

2 Michael Shattock, *Making Policy in British Higher Education* (Oxford University Press, 2012).

3 A report by Department of Trade and Industry Secretary Patricia Hewitt in 2002 (reported in the *Daily Telegraph,* 13 October 2004) found that:

   ● The general higher education research infrastructure in the UK needed £2.7bn to remove a worsening maintenance backlog.

   ● Over the past 10 years research income and student numbers had risen by 70 per cent compared with an increase in physical space of just 26.5 per cent … Many of the 750 higher education sites covering 5,000 hectares in total are 'land-locked' with little or no space for further expansion.

   ● Given the perceived choice of reducing activity and making staff redundant … or cut back on addressing maintenance back logs, the latter nearly always has been the preferred option.

   Universities' premises costs have long been second only to staff costs, and therefore an attractive and easy, and often misguided, target for financial cuts in struggling universities. As a result of all these cuts, together with expansion in student numbers, by the start of the period of this study, buildings in universities were getting not only under-maintained but also under-spaced, exacerbated in Cambridge and some other universities by particularly bad space management with crowding in some areas at some times, with other spaces empty much of the time. At Cambridge, the weekly teaching schedule was typically 7 hours for 4 days and 4 hours for one day, for 24 weeks per year.

4 HMSO Cmnd.2154.

5 Reporter 17 December 1974.

6 House of Commons Education Statistics; Paul Bolton November 2012. Standard Note: SN/SG/4252, House of Commons Library.

7 Cambridge Reporters 28 May 1998, 14 June 2000, 19 June 2002, 31 May 2007, 14 May 2014. Note that some definitions are not self-consistent, and therefore need cautious interpretation. A survey of UK universities by *The Sunday Times* (4 Nov 2011) cited the largest and smallest proportion of academics in universities as a proportion of total university staff, at 58.9 per cent and 33.1 per cent. These statistics are of general interest but do not have relevance to Oxbridge where there are a great number of college teaching staff who have no formal locus in the university.

8 *THE*, 11 May 2001.

9 University of Cambridge Estate Plans 2000, 2003 and 2005.

10 There was around this time a great deal of good work done on space management by the Association of University Directors of Estate (AUDE, founded at a conference in Cambridge, Easter 1990), and by the Committee of VCs and Principals (CVCP – then a powerful and effective body which diminished as it was increasingly upstaged by the growth of the 'Russell' group of universities: it converted into UUK on 1 December 2001).

11 The story is told that on the day of his PhD viva he went to Heathrow and, rushing to check in, was stopped by a young woman doing a 'Brain Drain' survey of adults flying to USA and asked if he was going there to work, and if so, if in a science-based occupation. He said that he was. The next question was 'For how long have you been qualified' (expecting a reply in years). The answer, after reference to his watch, was 'Oh, about 3 hours'.

12 Strictly speaking, election to VC was 'annually, with a two year limit.' (No VC was bumped off after one year; one died between election and start date). The VC prior to Sir Alec Broers, Sir David Williams, of legal discipline, and President of Wolfson

College, was appointed in 1989 for two years; that term was then extended as a consequence of the Wass Report to three years, and then for another four. So, in practice, although he served as VC for a total of seven years (until 1996), only for a short time (after his final re-appointment) did he know that he would be in post for sufficiently long to see through a major capital building project.

13 Greg Mutti and Chris Grimshaw, *Captive State* (Macmillan, 2013).

14 As one example, the Cambridge Chemistry Department benefitted hugely from investments from Shell and Unilever. In many further examples, help from industrial companies was secured for departments developing at West Cambridge.

15 University CapEx was running at about 16 per cent of OpEx (operating expenditure).

16 Richard Adams in *The Guardian*, 10 April 2013.

17 The first, and very successful, head of CUDO had been ambassador to Israel: it was perhaps felt by the appointment panel that an ambassador to Israel would be well able to handle the pressures from Colleges' development departments.

18 The Gates benefaction was courted for years and when, during a wet winter Monday morning VC's senior management team meeting, success was reported, it was decided to wait for a week before making the announcement so as to let Oxford prepare for anticipated press commenting because Gates' would be bigger than Oxford's Rhodes fund. The VC and one staff member at Oxford were briefed very confidentially. One of those leaked, and the *Sunday Times* gave it front-page coverage. All discussions had been with Bill Gates senior ('Daddy Gates') and not with the Bill Gates junior who made the money; he was going to be briefed before the official announcement was made. It was decided that the VC should call Bill Gates junior at start of his work at about 6.30am Seattle time, it being usual to tell people if they were giving $210M. Apparently Bill Gates just asked if 'Mom and Pop' knew, and being reassured on that, immediately turned the conversation to other research matters. Maybe $210M wasn't much of an issue. The outcome of the phone call could have been otherwise.

19 By summer 2013 the 'Cambridge Cluster,' as it was importantly designated by government (see Chapter 3), had led to twelve $1bn companies, and in total 1,540 technology-based companies with a combined turnover of £12.3bn, and employing over 54,000 people. £1bn of follow-on funding had been raised by University spin-out companies. 25.9 per cent of people in work in 2013 in Cambridge worked in the knowledge-intensive industries compared with 12.3 per cent for England generally. Cambridge by then had more patents per head of population than the next six UK cities combined: facts which may be useful for a pub quiz. If the pub is in Cambridge.

20 The Milton Road Site had also included a research unit whose work included 'bleeding out' and burying horses (some of which some people thought might have had anthrax). It later included the previous University athletics ground. The site was sold off for housing, after some rather heated investigations about the possible anthrax; a more basic reason for opposition to development of the land came from local people who walked their dogs there, a typical, and understandable, example of the Town and Country planning challenges common in University estate development.

21 Its Director of Estates during the period of this study was involved in both those industry reviews and in their implementation.

3 Planning the estate – regional and local land planning

The national system of town and country planning

No building can be put up until the required land is owned or leased, enough money is arranged to pay for construction and, since the first national Housing and Town Planning Act (1909), you have to submit building plans for any building of any meaningful size to the Local Planning Authority (LPA) before work can start. In fact there have been local bye-laws limiting house building ever since the plagues and the Great Fire in London, but the first national Town and Country Planning laws came into force because of the problems caused by vehicles whizzing along at 20 mph on gravel roads through ribbon housing developments before tar (later bitumen) macadam was invented. The 'Town and Country Planning Act 1947' consolidated previous legislation, and green belts were added in 1955 by Government Circular in an effective move to stop 'urban sprawl' and ribbon development.[1]

To help central and local government policy-makers encourage growth where they want it and prevent it where they don't want it, a 'plan-led' system has been the cornerstone of the planning process in England since it was consolidated by the Planning and Compensation Act 1991 (which followed section 54a of the Town and Country Planning Act 1990: now effected through Section 38(6) of the Planning and Compulsory Purchase Act 2004). The plan-led system followed many other countries which had by then long been 'plan-led'. There is a good deal of political and social logic in a plan–led system which cascades policy from central government downwards to local communities in meeting the important, emotive and never-ending challenge of balancing the perceived good of development to a local community against the wider good of others. For example, a university may want to provide a new building for residential accommodation, or to alleviate academic space deficits for departments researching medical advances, while some neighbours may be inconvenienced during and after construction and so oppose the application despite the potential benefits to many. An example of how this can work in practice was the application for lighting of the University's athletics ground at Adams Road, and that is set out in the Annex to this chapter. It is important as a lesson in how badly things can go in the planning world even for a small project, and as a start-point in the University's re-assessment of how to conduct local relationships for capital development. Another example of a big planning issue is given in Chapter 9.

During the period studied in this book and until 2011, the plan-led system was based on a linear succession of policies, starting with national guidance at government level on what generally is required for the national building stock, and what are the acceptable limits on how that may be achieved. Then there is 'Regional Guidance' on environmental issues and the outline allocations of land for such purposes as housing, employment, public sector social and physical infrastructure (including education at all levels). The regional level was cut out in 2010; new business-driven Local Enterprise Partnerships (LEPs) were set up soon afterwards to articulate and promote development needs in self-selected areas, such as the local Cambridge/Peterborough LEP. The next level was a Structure Plan at county level defining allocations for the various types of land-use. The 2003 County Structure Plan was the last of its type and will not in the foreseeable future be replaced as the counties no longer have an infrastructural plan-making role other than for transport; there is now a 'duty to co-operate' imposed on Local Planning Authorities who must consider cross-boundary issues.

Finally comes the crucial Local Plan which sets out in great detail what can be built where, and how. When an applicant applies for Outline or Detailed Planning Permission to build, there is a presumption for or against that application depending whether or not it complies with the Local Plan.[2]

How the national system worked around Cambridge

In 1950, Cambridge City Council formulated a policy based on the Holford-Wright Report for Cambridgeshire County Council;[3] accepted by Government, promulgated in 1954, it led to the Cambridge Town Map adopted in 1965. This policy, which set out to limit Cambridge's population to 100,000 as a good place to live, strangled growth by achieving a green belt tightly drawn around the city, the reason for this being avoidance of the effects of 'industry'. The 'City Fathers' looked west to Oxford and its heavy industry (such as Cowley car works) and railed against 'industry' that would spoil Cambridge as they knew and loved it. (Indeed, Oxford University was sometimes jokingly referred to as 'the Latin quarter of Cowley'; in response, Oxford dons would refer to Cambridge as 'the Polytechnic of the Fens', or 'that place beyond Milton Keynes'.) What seems extraordinary is the conflation of heavy industry such as seen around Oxford with the high-tech industry trying to come into Cambridge which is largely carried out in offices in a manner, and with infrastructural needs, very similar to those of an academic office. High-tech industry development proposals which would have brought a lot of employment and taxes to Cambridge were refused planning permission in the 1950s and 1960s. For example, IBM wanted to set up its European Research and Development HQ in Cambridge in the redundant Cattle Yard near the railway station, wanting to be close to the University's Mathematical Laboratory (later the Computer Laboratory) under Professor Maurice Wilkes; the IBM application was refused, losing a huge amount of employment and prosperity. (The Cattle Market stayed largely derelict for 40 years, used for non-commercial purposes like children bringing their dogs for 'dog-training', or vice versa.) An excellent description of the history of industrial growth in Cambridge is set out in *The Cambridge Phenomenon: 50 Years of Innovation and Enterprise.*[4]

Employment in Cambridge grew as best it could, with levels of employment and housing accommodation becoming ever more out of sync: housing and support services were having to jump over the green belt to the 'necklace villages' (whose early expansion was associated with academics without University appointment who were until 1882 not allowed to be married, so kept their mistresses/wives far, but not too far, from the censorious eyes of University authorities).[5] At the start of the 1800s the population of Cambridge was less than 10,000; in 1963 it was 96,000, and in 2011 it was 124,000. In 1951 about 8,000 people worked in Cambridge but lived elsewhere with 39,000 living and working in the City; by 1991, there were 36,000 working in the City but living elsewhere, in large part because of lack of city land for houses, and just 33,000 living and working in Cambridge.

A key issue was and is that the attractiveness of Cambridge, especially the historic centre, where most of the University then lay, had to be preserved (not least to safeguard tourism income: 3.5M tourists in 1997, an almost three-fold growth since 1972). The public was wary, having seen what could happen if planning policies became too lenient; for example, the wanton destruction of Petty Cury in the Lion Yard development of 1977–1979, and the development of shopping and housing to replace what was considered a slum area around the district known as the 'Kite' (from its shape on the map). Those social impacts had to be somehow balanced against the long-term benefit of the shopping and housing created.

Car use was rocketing among the citizens of Cambridge and among the staff of the University, although students under the age of 25 were and are forbidden to have or keep cars within six miles of Great St Mary's University church (except for a few concessions) and so the relative lack of parking provision in the City caused a lot of the surface area of University sites to be given over to car-parking; that provision was cherished, indeed fought for,[6] so that for many decades the layout of most of the University's major sites was seriously constrained by parking-space provision and retention. City policy was for many years to insist on high parking provision in Planning Applications: about the turn of the Millennium, that was quickly reversed to a policy of opposing more than a stated minimum.

In the 1960s, following Harold Wilson's speech about the 'white heat of the technological revolution', and an anodyne two-page inconclusive report by the University Council about the need for growth, Cambridge University set up a sub-committee of Council in 1967 to consider the planning aspects of the relationship between the University and science-based industry. The City Council supported this; it was at the time feeling frustrated by the County Council's policies on housing and employment expansion (and hence income from the rates). Local employers also supported it as they faced serious recruitment problems due to the lack of housing. A group of four University scientists led by Prof Sir Nevill Mott wrote a positive and clear report in 1969 recommending careful relaxation of restrictive policies, clearly setting out the need for physical expansion of the University, both in contiguous development (thus showing a need for development at West Cambridge) and in particular re the establishment of a science park on the edge of Cambridge. The committee's recommendations were accepted in the early 1970s and Trinity College founded the country's first significant Science Park in 1970, with the Knapp laboratories as anchor-tenant. (In fact, the main work in that lab then was not research but the production of the first multi-layer pills to obviate taking several doses of pills per day.)

The Swinnerton-Dyer Report in 1974 importantly concluded that the University should spread west using the land mostly assembled before and during the Second World War. The 1996 Cambridge Local Plan was the culmination of a lot of discussion about the precise shape of the University's westwards expansion. The University had wanted to fill in the space between the existing edge of West Cambridge development and the proposed by-pass 'New West Road' (NWR) as the main expansion site, but because of heated local opposition to the NWR (there was a general move away from building new roads) the focus for expansion shifted to the area of the current West Cambridge site instead, and the Rifle Range and Grange Farm were brought within the green belt in the 1996 Local Plan to protect them from future development and effectively scupper the NWR in perpetuity. At the same time, land was taken out of the green belt at Gravel Hill Farm in NW Cambridge and the 19 Acre Field for University or College development, providing useful compensation.

By the 1990s, half of all those who worked in Cambridge were living elsewhere and were teeming down the feeder roads from the necklace villages into Cambridge, causing and suffering from delay and pollution. So the emotive matter of the green belt became a bigger and bigger issue for the University and all Cambridge, including such organisations as the Cambridge Preservation Society, CPS, now known as Cambridge Past Present and Future. There are 14 green belts in England, protecting 1.6M hectares covering some 12 per cent of England, roughly the same percentage as the total built-up area in England; the Cambridge Green Belt currently is 26,340 hectares, but despite its relatively small extent it has always been totemic in the debate on the quality of the environment. That is despite its relatively low visual quality and general inaccessibility: 74 per cent of it is agricultural/horticultural, 13 per cent is semi-natural grassland, 5 per cent wooded and 8 per cent 'miscellaneous'. Further, there are relatively few views through the green belt to see and savour Cambridge. It has, nonetheless, much appeal to those opposing 'urban sprawl'.

During the post-War period relationships between the University ('the Gown') and 'the Town' regarding 'Planning' were often tense, sometimes antagonistic. From the 1990s, a small group consisting of a leading Bursar, a leading academic well versed in planning matters, the Registrary and the

Director of Estates met bi-annually with the City Council represented by a leading Councillor, its Chief Executive and its Director of Planning; that conduit gained focus and purpose and was generally helpful in anticipating and preventing, or at least minimising, friction about development plans. At the end of the 1990s, this group was succeeded by a similar and more effective group with the addition of the Vice-Chancellor, and also County Council representatives mainly because of the county's responsibility for transport, an increasingly important aspect of nearly all development as congestion got worse and sustainability got more important; prior to 1999, liaison with the County Council on infrastructural matters had been sporadic. Planning sensitivities were usually about opposition to developments seen as intrusive or as adding to traffic problems, or, to varying degrees, just 'Nimby-ism' as is common everywhere. (The more recent and absolute relation of 'Nimby' is 'Banana'—'Build absolutely nothing anywhere near anyone'.) Traffic congestion was a major component in South Cambridgeshire District Council's general opposition to development, notably for development proposed for the Addenbrooke's hospital area. (Opposition to moving Papworth Hospital to the Addenbrooke's site was an even sharper issue, responding to resistive campaigns by locals.)

When planning permission had to be obtained, University buildings were generally classified under Planning Law as either Use Class B1b ('Research and Development') or D1 (non-residential institutions; here Education), or, occasionally Use Class C1 or C3 for residential (including student accommodation). University development was seen by most citizens as less threatening than commercial or industrial development, and there was an increasing realisation that unless the University's future viability and development were safeguarded as it sought to respond to new and competitive demands and opportunities, much prosperity would be lost to Cambridge and its inhabitants, many of whom were living in Cambridge directly or indirectly because of all that the University brought.

Regional planning guidance

As set out in the previous chapter, the University of Cambridge in 1999 agreed expansion targets for research and teaching: 0.5 per cent annual growth for undergraduates, 2.5 per cent for graduate students and up to 9–11 per cent annual growth in post-docs. (University student numbers have roughly doubled on a rolling average of every 40 years since 1760.) Fortuitously, 1999 was when the East of England Region was required by government to set out how it planned employment, accommodation, social and physical infrastructure, sustainability and other requirements across the Region, and to set out its priorities until 2016 in a Regional Planning Guidance plan.

In this and subsequent Public Inquiries down the plan-led system, all organisations of any size had to fight their corners to get the policies, land-allocation and priorities they wanted: the stakes for developers and institutions were very high, with several large rival proposals each representing over £1bn of development. The University saw that a successful outcome would be crucial in the development battles to come in the years ahead. Nearly all Cambridge's competitor universities around the world were getting increasingly constrained by a lack of contiguous land on which to build new offices and labs for growing research. Cambridge did own contiguous land, and, crucially, it would be in a position to pull ahead of competitors in provision of facilities if it won its planning battles for the expansion it wanted.

The VC, Sir (now Lord) Alec Broers was due to give evidence over two days during the month-long 'Public Examination before Inspectors' for the Regional Plan in the Malthouse in Ely in February 1999, but at the last minute he was called by another Government Inspector to give evidence in the Appeal Against Refusal to allow development of genetic research facilities at Hinxton. So University estates officers were mustered to argue the University's case for future expansion; there were no planning barristers briefed and the planning consultants, also thought to be a bit weak and very expensive, had been given a more limited role.

The atmosphere in Ely was superficially congenial but all knew what was at stake, and were ruthless in seeking to gain advantage in how scarce land and economic resources would be prioritised. Cross-examination was sharp. The University traced the trajectory of research and noted how much of that could be provided by sites that the University owned or could reasonably acquire, in particular how much could be provided at West Cambridge. Although spare site capacity was considerable, extrapolation of the trends of research, and, to a lesser degree, graduate teaching, over the following decades showed that the University would need much more space for development than was available, and that that should be at North West Cambridge, an area perfectly suited by proximity to the University and hence more sustainable. (This was the first time the sustainability issue was raised in this public debate, and it was used to the advantage of the University.) More especially, analysis showed how the University desperately needed accommodation for its young members of staff and 'soft-money' researchers. Some of that could be provided at West Cambridge, but not nearly enough. The issue here again was the huge deficit of housing in Cambridge because of the tightness of the green belt boundary, University people being forced to live away from Cambridge, notably in the 'necklace' villages and market towns (such as Waterbeach and Chatteris) surrounding the city. The wasted time, and the pollution thus caused, made for a very strong case to develop housing and employment at NW Cambridge, within and at the edge of the city, and in other edge developments by going into the green belt. This was analysed and articulated by a group known as 'Cambridge Futures', one of whose founder members was the Vice-Chancellor, Sir Alec Broers; the leading member was Professor Echenique, who ran his own transport consultancy company, did consultancy for Cambridge City Council regarding other possible development sites and was a professor in the Architecture Faculty of the University; Councillor John Durrent and Michael Marshall were other prominent members. The rather startled attendees of the Regional Planning Guidance Examination in Public were treated to an unexpected and sudden academic presentation, complete with PowerPoint, by 'Cambridge Futures' which wanted, inter alia, development of land around Cambridge, with imaginative solutions to the urban traffic problems. The presentation had a strong impact, and reinforced the case that had been made by the University for green belt release at North West Cambridge. As Peter Studdert, former Director of Planning at Cambridge City Council, recalls,

> the 1996 Local Plan had not been particularly radical, and the more radical thinking about the future growth of Cambridge was really kicked off by the Cambridge Futures initiative in 1998. In retrospect this was an incredibly successful collaboration between Town and Gown which made a significant impact at the RPG plan Examination in 1999 leading to the adoption of an ambitious growth strategy for the Cambridge sub-region in RPG6 in 2000. Local authorities developed the strategy, in particular with Buchanans, Llewelyn Davies and DEGW between 2000 and 2003 to evaluate possible green belt changes and to identify the best site for the new settlement, and this was subsequently embedded in the 2003 County Structure Plan.

Later, the options set out by Cambridge Futures were to be prominently set out in a series of editions of the *Cambridge Evening News*.

The outcome of the expansion was promulgated in 'Regional Planning Guidance (RPG) 6 of November 2000' by the then Dept of Transport, Environment and the Regions: 82 crisply worded pages. The report was welcomed by the University. For example, Policy 26 of RPG6 required that:

> Development Plans should continue to include policies for the selective management of development within the area close to Cambridge, discriminating in favour of uses that have an essential need for a Cambridge location'

The key aspects affecting the University were:

- that priority should be given to 'Extension of the clusters of research and technology-based industries (Policy 7)
- that Local Authorities with [others] should develop a vision and a framework for Cambridge that would allow it to develop further as a world leader in research and technology based industries and the fields of Higher Education and research (Policy 21), and, crucially,
- that there was to be review of the green belt.

Those conclusions laid the foundations for the University to get enough building land for expansion for a further 70 years at targeted growth rates; if University input into a green belt review could result in a considerable green belt release, then the crucial development of North-west Cambridge was a question of 'how' and 'when', not 'whether'.

As soon as green belt release looked more likely, the University bought further land west of the M11 to secure expansion beyond 2080, the very, very approximate date when on current trends all currently owned sites might be full. (Some wag suggested that would be at 9.25am on 14 May 2080.)

Cambridgeshire County Structure Plan

The next and crucial stage was the Structure Plan for Cambridgeshire County; a Deposit Draft Plan was issued early in 2002, with then an Examination in Public (EiP) which opened for evidence with three government-appointed Inspectors. The stakes were even higher for the University because by now battle-lines were being drawn up between rival teams pushing their areas for achieving the growth in employment, housing and infrastructure being imposed by central government and/or adopted by Cambridgeshire County Council as matters of policy: policy that would shape the all-important Local Plan and hence whether building development could or could not proceed. Rival development proposals included Waterbeach, East Cambridge around Teversham and Fulbourn and the airport, South Cambridge around Trumpington and Clay Farm, and NIAB (National Institute for Agricultural Botany) land off Huntingdon Road, as well as the University site of North West Cambridge. Clearly, only some of those would get through. Developers whose multi-billion-pound schemes did get through stood to earn hundreds of millions of pounds profit over the years, and those that didn't would lose profit potential, and their considerable preparation costs.

The first day of the Public Inquiry regarding University evidence started rather oddly. The previous evening a senior county councillor had sent an email to a leading city official suggesting how the next day a trap could be laid to stitch up the University. His transmission coincided with a computer virus whereby emails were sent out randomly to large chunks of address books: that included the University officer due to lead the next day. The officer cheerfully emailed back welcoming the newly-open approach of the councillor, and then made sure that by the dawn most people knew about the rogue email. An apparently mild comment from the University near the start of the formal proceedings next morning brought great mirth around the chamber. The councillor did not appear that day.

Again, cross-examination was sharp, and even at times dramatic, such as when a letter was borne into the chamber to the Inspectors saying that the newly formed Network Rail did not undertake to honour commitments made by its predecessor body, Railtrack, and that therefore there was no assurance at all that a new station could be provided at Waterbeach if that development were to proceed: collapse of stout party. The University case was introduced by the Vice-Chancellor, and supported at the first session by Pro Vice-Chancellor, Professor Malcolm Grant; then for the next two days the University's case was put by the Director of Estates[7]. This Inquiry was, even more than at the

Regional Planning Guidance Public Inquiry, highly competitive between rival plans for expansion: the City Council was pressing for expansion to the east, and several commercial development companies pushed hard for their schemes, such as at Waterbeach, around the south of Cambridge, and to the west around the Barton Road. Also, and entirely reasonably, there was a lobby which sought to defend the green belt, and hence oppose its release for any development.

The main outcomes for the University set out in the consequent Structure Plan (Cambridgeshire and Peterborough Structure Plan 2003) were favourable, and crucial to University development: notably confirmation that development for University use could take place at North-west Cambridge if academic need was confirmed.

A report in October 2004 (RSS14 for the East of England: report of the Strategic Environmental Assessment) noted that the East of England Regional Assembly had 'cleared the way for 478,000 homes to be built' and warned of 'chronic shortages of piped water, the risk of floods, harm to the environment and wildlife, a slump in the region's quality of life, and a battle for land between the public sector and private enterprise'. Clearly, a green belt release on the huge scale of North-west Cambridge was going to be emotive; indeed it already was.

Cambridge and South Cambridgeshire District Local Plans

The earlier 1996 Cambridge Local Plan was initially placed on deposit as draft in June/July 1992, and the Public Inquiry into objections was held in March 1993 and again in April 1994. The Inspector's report was considered in December 1995 and further modifications were placed on deposit in January–March and July–August 1996 before final adoption in November 1996. This Plan built on the spirit of Holford, but recognised the changed local, regional and national contexts, such as commuting and car ownership, connectivity to London and sustainability issues; those required a relaxation of the constraint on growth, while seeking to ensure that Cambridge remained a fine place to live, work, study, play, visit and invest in.

The final major planning steps for the decade up to 2006 were the next rounds of Local Plans for Cambridge City and for South Cambridgeshire District Council; these would set out in detail the context for taking planning decisions, including land-use for the University estate (which is split between these two Local Planning Authorities) and ruling on specific detailed planning applications, for new buildings or Listed Building Consents; around that time the University was making about 50–60 planning/Listed Building applications per year.

The University's case regarding the Cambridge City Local Plan was examined in front of a government-appointed Inspector in a Public Inquiry October–November 2005. The University put its case with conviction, most notably about expansion space needed in the north-west for the continued, rapid rise in research activity, and to address the shortage of affordable housing available for young academics and staff as key workers and hence within the definition of Social Housing provision (The campaign by the University for this definition had started a few years earlier.)[8] The matter of 'key worker' provision was crucial to the nature and financial viability of development at North West Cambridge and is discussed further below and in Chapter 11.

There was a good level of accord between the University and Cambridge City Council, and the University had wide support generally; also, as noted above, the joint University/City/Business organisation called 'Cambridge Futures' had articulated persuasively the general case for releasing green belt land around Cambridge. In 2006 the University learned that the Inspector's Report supported all its proposals made at the Public Inquiry, giving very strong support not just for removal of North West Cambridge from the green belt (while reinforcing the importance of the 'green fingers' that run into Cambridge mainly from the south) but also giving clear support for the proposed master plan for development at North-west Cambridge: that was crucial as about half of the plan lay in the area of South Cambridgeshire, and the Inspector had said, very informally, that he might well be also appointed as Inspector for the South Cambridgeshire Local Plan. He was indeed so appointed, so the odds were already good for getting that master plan through all the way. The North West development scheme which finally got planning permission late in 2012 was very similar to that which was proposed at the Examination in Public in November 2005. (See Chapter 10.)

As Peter Studdert, former Director of Planning at Cambridge City Council, notes,

> Although the 2006 City Local Plan removed land within the City from the Green Belt in NW Cambridge, the precise boundary for the whole site including the part of the site within South Cambridgeshire was not defined until the adoption of the Joint Area Action Plan (JAAP) in 2009. The JAAP was a helpful process in steering the development forward and setting out the main parameters and quality standards for the development, and the University had got a very generous hearing from the Inspector on the question of the quantum of development that could be allowed within South Cambridgeshire!

Regarding the JAAP, by law, a sustainability Appraisal must be carried out for Local Authority spatial plans and EU Directive 2001/42/EC requires Strategic Environmental Assessment for major developments, so those were done for North-west Cambridge prior to 2009.

More generally, the adopted Local Plan set out a strategy for growth of the City with extensions on north and south sides, and a multi-modal transport plan: it was widely regarded as an extremely good planning document, and in the way of the plan-led system, it had the backing of central government, and was in consonance both with the Regional Plan (RPG 6) and with the County

Structure Plan. There should be 12,500 new homes over 17 years (representing about 20 per cent increase in the population of the city) linked to a strong low-carbon strategy delivering over 60 per cent of all trips in the new development from public transport, cycling and walking. A summary of the new vision was of a compact, dynamic city with a thriving historic core surrounded by attractive and accessible green spaces, the city continuing to develop as a world leader in the fields of higher education and research, and fostering the dynamism, prosperity and further expansion of the knowledge-based economy. The Local Plan was recognised as providing a forward-thinking, environmentally-led spatial strategy.

The detailed planning of the various proposed developments around Cambridge was co-ordinated by a new organisation called Cambridgeshire Horizons which lobbied central government and was controlled by the local councils: some have argued that Horizons should have had the power of an Urban Development Corporation (and that probably would have prevented the planning problems at Cambourne). Horizons was wound up in September 2011; it had done a lot of successful lobbying of government and others, and successful work on implementing the Local Plans in conjunction with the Local Planning Authorities and the University; its planning principles were adopted formally within a Quality Charter.[9]

One of the wider implications of high property values in the city, average house prices being nine times lower-quartile earnings, was the impact on recruitment of academics, and that was a vital part of the argument for releasing North-west Cambridge from the green belt and agreeing to provision of 50 per cent key worker rather than a mix of social and key-worker housing: provision of much-needed and long-awaited affordable accommodation for academics near the University would relieve pressures for others in other parts of the city. Some of the University's estate planning staff had a concern that some citizens might worry that concentration of University accommodation would be seen as 'ghetto-isation' but such concern was not heard. Officers planned that this accommodation would be for support staff as well as for academics. In 2014, the subsequent round of District and Cambridge Local Plans were being examined; between them they envisage a further 33,000 new homes and 44,000 new jobs by 2031.

Two other different but important planning matters: firstly, when the University was in the final stages of negotiating the Outline Planning Permission for the development of West Cambridge and the associated S106 Agreement[10] (which set out the requirements for associated works to be done, such as road and sewer upgrade), the City Council agreed to a clause whereby every two years there should be a mutual review of the Outline Planning Permission for West Cambridge with a view to updating it in the light of changing requirements for the site. It is thought that this agreement may be unique, at least it is unusual: it certainly proved to be of immense value to the University, and indeed to the local community as the way the site was developed could be improved rather than be bound by an aging Planning Permission.

Secondly a planning issue that was thought rather silly was that of Greater Crested Newts (GCN). GCN were an issue both at West and North West Cambridge, although managed sensibly and at moderate cost. According to EU legislation, GCN are an endangered species in Europe and therefore, reasonably, must be protected. However, in England GCN are relatively plentiful; indeed many if not most major development sites have them, and if they don't, it is said, whimsically perhaps, that for a modest fee of a few thousand pounds sterling at least a pair will mysteriously appear on a proposed site, causing huge delay and extra cost in their protection, and providing an argument for not allowing development.

A useful tool in getting a clean and flexible Permission for West Cambridge was the use of digital modelling, an under-used approach in those days. A very good unit in the Architecture Department of the University was developing ground and building digital software so that proposed development could be shown in 3-D fly-through with accurate settings of ground and other environmental data, and councillors and planning officers could 'play with' the model on their screens, seeing how the development would look from any angle and in any condition of light and weather.[11] Nowadays with Building Information Modelling (BIM) such digital modelling is a common aid in gaining planning permissions.

It is important also to note that the land purchases (as set out in Chapter 1) gave the University the land it needed for expansion to at least the middle of the twenty-first century and was mostly bought before the price of land around Cambridge rocketed when the green belt corset was drawn tightly around Cambridge, causing city development to be to be expensively squeezed in. The balance of land acquisition needed for the University's expansion came after the University had won the above planning battles to have expansion land designated in local councils' statutory plans as 'for University use' so that those land-costs became very much lower than they would have been on an open market without such restriction. The average cost of land bought at West Cambridge was £54k per acre,[12] compared with the price of land south of Cambridge bought for development of a 'Medi-Park' rumoured to have cost about nearly £2M per acre. By 2013 arable land prices around Cambridge had risen to about £10k per acre, compared with the national average of £8.5k.[13]

That the University was thus able to achieve a favourable planning environment, and acquire plenty of affordable and adjacent expansion land, unlike nearly all of its international competitors, is one big reason why there was such an unprecedented and sudden wave of capital projects in the University from the late 1990s.

Annex: Planning Application for lighting at Adams Road

At the University's athletic ground at the bottom of Adams Road, after an unsuccessful Appeal against refusal of planning permission in 1992, and then a successful, amended application, a multi-purpose sports pavilion was erected.

A planning application for floodlighting the athletics track and the hockey pitches was submitted by the University in April 1994, and that was opposed by the University's Institute of Astronomy. (It does happen in the dispersed community of scholars that is called a University that one part of the University opposes a planning application by the University.) Following objections by a small number of local residents, the Local Planning Authority commissioned an independent audit of the proposed lighting scheme, and the floodlighting planning application was further deferred in November 1994. The University appointed W. S. Atkins as lighting consultants. Discussions continued in a rather bad-tempered manner until the University withdrew its application in October 1996. A replacement application for floodlighting only the athletics track was submitted in December 1997, supported by the Institute of Astronomy. The plan proposed was based on a lighting system with eight columns, each 16m high. The objectors communicated by letters written on expensive, headed notepaper under the title of the West Cambridge Preservation Society. Their leaders asked for balloons to be 'erected' in the position of the proposed heads of the front lighting columns so as to carry out night-time simulation. The University, suspecting that this exercise would give the local residents an opportunity to stage-manage a media event to the detriment of the planning application, would not agree. The ill-feeling caused by this proposed flood-lighting scheme was damaging the ability of the University to establish a good relationship with the Local Planning Authority, so there was a change in tactics in the autumn of 1998. A new round of meetings with the objectors was opened, mainly with the purpose of getting a better understanding of just how much support the leading objectors really had, and what really were their 'red lines'. It was agreed that the University and representatives of the objectors should meet at what was called a Development Control Forum, which was open to the public. The University invited supporters from local secondary schools which would also be using the athletics track: that made a difference. It emerged that the West Cambridge Preservation Society was a very small organisation of a few individuals. The planning application was approved at the September 1999 meeting of the City Council Planning subcommittee, subject to investigation of telescopic, hydraulically raised columns being suggested by Dr Pellegrino of the Engineering Department. On 4 October 2000 another Planning Application was made by the University and on 13 December 2000 Planning Permission was finally granted, by one vote. After some six years of battling with the local planning authority, and bad public relations, the gaining of the Planning Permission at the end of 2000 was however a Pyrrhic victory as it emerged that the scheme could not in fact be afforded.

As a learning experience for the University, the saga was crucially important.

Notes

1 There are exceptions. The Town and Country Planning (General Permitted Development) Order 1995 (as amended) does allow certain buildings to be erected

as 'permitted development' in specified circumstances. See www.legislation.gov.uk/uksi/1995/418/schedule/2/made for further information and specifically Part 32 for Universities.

2 Following publication of the National Planning Policy Framework in March 2012, if an application meets a high-priority local need and is deemed to be 'sustainable development' it may be permitted even if not in accordance with the Local Plan; at the time of writing, it is far from clear how this is working out in practice. See paragraph 11 onwards of the NPPF: www.gov.uk/government/uploads/system/uploads/attachment_data/file/6077/2116950.pdf

3 William Holford and H. Myles Wright, *A Report to the Town and Country Planning Committee of the Cambridgeshire County Council*, 1950.

4 Kate Kirk and Charles Cotton, *The Cambridge Phenomenon: 50 Years of Innovation and Enterprise* (Third Millennium Publishing, 2012).

5 The ordinance prohibiting Dons from marrying was enacted by Elizabeth I. It was relaxed in 1882, in rather a piece-meal sort of way: in King's College, for example, relaxation initially came only for those who had had their fellowships for at least ten years. When the ban was lifted, many dons around the University maintained to skeptical authorities that the first time they had set eyes on their wives was when they were coming off the first subsequent train from London.

6 Above the exit of the main University car-park at UCLA is the homily: 'A University is a loose collection of academics bound together only by their common opposition to the University's car-parking policy'.

7 Professor Grant was a planning lawyer by background: subsequently when he was Provost of UCL he hosted the launch of the UrbanBuzz Report on sustainability in UK construction (see David Adamson and Peter Shewry, 'Towards Sustainable Construction': a 2-year study with seminars around the country', in *UrbanBuzz: The Challenge of Sustainability of Cities* (UCL, 2007)). A very senior civil servant who didn't know him asked: 'What do you do?' Malcolm replied, 'I am an academic planning barrister.' 'Ah', said the civil servant 'then in one person you encapsulate the three things I most hate in life'. Professor Grant enjoyed recounting this story; as a brilliant Provost (VC) his position was very secure.

8 The earlier definition of affordable housing as being social housing did not suit the University as it wanted to maximise key worker housing for academic and support staff. It therefore argued that affordable housing should include key worker as well as social housing. Once that was achieved, the University offered that a high percentage of the total housing provision should be affordable housing.

9 See www.cambridge.gov.uk/sites/www.cambridge.gov.uk/files/documents/cambridgeshire_quality.

10 Known as Section 106 planning agreements with reference to that section of the Town and Country Planning Act which empowers a Local Planning Authority to require work to be done and/or financial contributions to be made for the provision of support facilities or infrastructure associated with the development. (The House of Lords test case *Tesco v SSE* in 1995 drew a line in the sand requiring greater relevance of imposed S106 works to the site being developed.)

11 Some of the software bore a resemblance to Russian software which had been 'derived' from the US Tomahawk Cruise missile software developed during the Cold War.

12 Robert Neild, *The Financial History of Cambridge University* (Thames River Press, 2012).

13 Savills, *Market Survey of Agricultural Land*. (This is an annual survey, and can be directly accessed by Google search.)

4   University management of capital expenditure

The physical shape of the University of Cambridge at the start of the 1990s was the culmination of about 800 years of a steady and successful evolution, of which one of its famous sons would have been proud: the fittest, the most brilliant, had survived and flourished. But the 'managerial systems' of the University, however, hadn't evolved much since Darwin was sailing around the world. The question can be asked, and indeed it was asked by administrators in the mid-1990s: is the University so very successful because of or in spite of the way things are done? Answer came there none. Indeed, the very term 'managerial system' was despised by some academics, rather as a matter of principle or of academic honour.

As in most sensible institutions, decision-making at University level rests on two managerial systems which sometimes overlap, and sometimes underlap, leaving gaps unattended. Firstly, chronologically and in university priority, there is the system of management by committee and, secondly, there is the line-management system. This chapter tells a story of how the committee management and how the line-management of the University both had to cope with rapidly increasing pressure and rapidly increasing opportunity, and how both had to grow in authority without jeopardising the academic excellence and spirit that made Cambridge what it is, 'an academics' co-operative': that tension between management and academic independence was to be seen in nearly every capital project into the 1990s, decreasingly so from then.

Management by committee

In a widely-acclaimed speech at the start of his vice-chancellorship in October 1996 entitled 'A Vision of Bridges', the VC, Alec Broers, (knighted in 1998, made a life peer in 2004) said:

> For the last decade or so, universities have been constantly urged to model themselves on business structures. Executive authority, however, in this University, unlike in the majority of industry and business, is exercised through chairmanship. The VC is chairman of the University Council (properly, the Council of the Senate, set up in 1856), and of the General Board of the Faculties (set up in 1926; from 1856 it existed as the 'General Board of Studies'), as well as of Statutory Committees of these two major

bodies; he was also chairman or member of innumerable other boards and committees, on which he was represented by deputies in most instances. This may seem a bizarre interpretation of executive power, but it does work, drawing together a diverse and strong-minded group of academics.

The committee system at Cambridge was based on the primacy of the Regent House whose membership, of over 3,000 by the mid 1990s, comes from the academic and the more senior administrative staff of the University and the Colleges. Motions ('Graces') are put to the Regent House; generally these come from the University Council which is the University's principal decision-making and executive committee; direct management by Regent House could be described as 'crowd-sourcing'. The power of Council was strengthened by the implementation of the Wass Report published in May 1987 (as set out in Chapter 2). In the mid-1990s Council comprised the VC, four members from among the Heads of Houses (Colleges), four senior academics, eight other members of the Regent House and three students. Council meets regularly in term time, but not in August or September (many other key committees didn't meet between early June and early October quite regardless of activity levels; not helpful for capital projects needing decisions week by week). Until around 1999/2000, the Council agenda, prepared by the Registrary, related more to issues that had been around for a while rather than to planning how the University should meet future challenges and create new visions.

By the end of the Vice-Chancellorship of Sir Alec Broers in October 2003, agreement had been reached on going from two to five Pro-Vice-Chancellors (PVCs), having two lay members of Council, and requiring 25 rather than just 10 members of the Regent House to sign a petition requiring a particular issue to be voted on by the Regent House before proceeding; such votes took a huge amount of time and caused great delay. These were important steps forward, if not as radical as most of the University wanted.

The other main committee, the 'General Board of the Faculties', advises the University on educational policy, with a secondary role 'to control the resources necessary for the proper implementation of that policy'. The General Board is chaired by the VC and the main members are the academic leaders of the six 'Councils of Schools' (which are what in most Western universities are known as 'Faculties' headed by 'Deans' who are professional academic administrators appointed to the post of Dean, or senior academics elected by their peers for a limited tenure); the structure of the Councils of the Schools was developed by Dr Ian Nicol, Secretary General of the Faculties in the mid 1970s. The balance of responsibility between the General Board and Council shifted a bit after a Council paper on 26 April 2004 suggested that

> The General Board, which is accountable to the Council for its conduct of academic business, should in its annual report to the Council deal primarily with this accountability, concentrating on its own direct conduct

of University business, rather than producing a commentary on University matters more generally.

The same Council paper noted that, 'officers with managerial responsibilities often say that they feel that they do not know what the major items of University business under consideration are'. This, some felt, resulted from transitional issues relating to the widening of power of PVCs, then five in number.

In the autumn of 1996, Council voted to disband the Planning Committee and the Resources Committee in favour of a new consolidated committee, the Planning and Resources Committee (PRC), established in January 1998, reporting executively directly to Council, and reporting to the General Board. This was a powerful and well-balanced committee chaired until October 2003 by the VC, and with a membership which included the PVCs, the academic heads of the six 'Councils of Schools', a representative Head of House (i.e. a College head) representing Council; officers channelling papers to the committee were heads of administrative services including the Registrary and the Directors of Estates and Finance. That membership, and chairmanship at VC level, made PRC an excellent group to decide, among many core matters, significant capital and estates issues. PRC worked well, except in matters of prioritisation: the University had a good record of agreeing to things, and quite a good record in saying 'no' to some things, but not a good record in deciding what should be sacrificed to achieve what was to go ahead. In January 2001, a Resources Management Committee was set up to prioritise/approve allocation of resources, reporting to PRC and replacing the General Board's Needs Committee and Council's Allocations Committee.

The quality of output of PRC gave authority to Council as an executive body, and its establishment was a crucial factor in managing the rush of capital projects that was about to burst over the University, and in overseeing external and internal policy matters to which the management of the University estate had to respond, and often had to anticipate if building procurement time-scales were to be met.

Answering to PRC and to Council from 1998 was the newly established Buildings Committee (BC). Previously there had been Building Committees in each of the six Councils of the Schools (some chaired by College Bursars), plus a 'Sites and Town Planning Committee' (chaired by College Bursars) and a plethora of other working groups, syndicates and working parties elsewhere in the University as described in Chapter 1. Happily, the establishment of that one central Buildings Committee (BC) in 1998 came just in time to cope with the rapidly growing agenda.[1] Management of the BC soon, logically, passed from a generalist officer in the Treasurer's Department with no procurement background, to the Estates Department. Chairmanship passed in 1999 until 2004 to an academic, Dr Derek Nichols, who had experience of some aspects of industry and of local government, and was on Finance Committee for 15 years. He successfully chaired the BC for the five peak years of the University's capital programme; the University owes him a great deal, even if it doesn't know that. His ability to steer adroitly nearly every BC meeting, even those with decisions relating to hundreds of millions of pounds, to a satisfactory conclusion within ten minutes of the scheduled end-time endeared him to members and officers alike.

The BC became significantly more authoritative and therefore more powerful in 1999 when its membership of the leading academics of the University, and the Head of the Architecture Dept, was augmented by a lay membership of five leading representatives from the construction industries: from architectural and engineering design, construction/contracting, property and local government backgrounds. There was only moderate opposition, and that was not from academics worried about a threat to their authority,[2] but from some senior administrators; their opposition was expressed by delaying the change, but in this case, however, delay by the mantra 'the best way to prevent is to delay', usually a highly effective tactic in the University, was easily overcome. (Prevention by delay was sometimes expressed as the 'principle of unripe time': very 'Cambridge'.) The knowledge and experience of and the generous amount of time given by these lay members greatly enhanced the level and depth of scrutiny and advice afforded to Estates staff, and also gave assurance to academics; most academics understandably had little knowledge or comprehension of the complexity of the fragmented and inefficient construction industry that worked at high volume–low profit levels in ways baffling to highly intelligent academics not used to commercial businesses. Lay members are rather like non-executive directors, but were not paid (beyond expenses), and there were few perks beyond a couple of dinners a year.

During the decade being studied, the main functions of the BC in its role of overseeing and developing the University's estate were to address two streams of decision-making, the first planned and fundamentally logical, the other event-driven.

Firstly, the logical stream: asset management. Until the late 1990s, the University had neither a consolidated schedule of property it owned or leased, nor a comprehensive schedule of building condition with a strategic plan for estate development.[3] An attempt in the early 1990s to get a consolidated list of University property had foundered: it began hopefully by going through the University telephone book seeking addresses, which nowadays seems rather bizarre. In 2001 a property committee was set up to enhance the planning and the use-value of the considerable property portfolio. Many buildings had had condition surveys, some had not, and very few had any assessment of fitness for current purpose, far less fitness for expected future purposes as is required as a base-line for any decent planning. The Higher Education Funding Council for England (Hefce) required all universities by the mid-1990s to have an effective estate management plan, and, nominally, all had. However, the rigour, extent and usefulness of these plans varied considerably. Some, a dwindling number, were merely a bland expression of what the estate was in area, with some views on condition and the extent of 'backlog maintenance' (which typically in a large civic university at that time was up to about £50M). At Cambridge, the estate plan at that stage was superficial and of little practical use to the University or anyone else. It didn't go through any committee; in fact, there was no suitable committee for it to go through. For example, a major gap in the strategic management of the estate right into the 1990s was a lack of assessment of how much of existing and planned research and teaching should move to West Cambridge. A master plan for West Cambridge, produced by the architectural practice MJP in 1998, was regarded by many as satisfactory in scale and potential, but lacking the vision of what the University stood for and wanted to become academically. (Surprisingly, this was the first University site master plan since the abandoned and costly master plan of the Downing Site by the architect Denys Lasdun in the early 1960s.)

After a lot of scrambling and hard work, and 'politicking' to get enough academic support, there was produced for the BC by 2000 a credible and useful plan for the whole University of Cambridge estate. It was rather late for the surge in the capital programme starting in 1999–2000, but the logical analysis that led up to it, and the emerging decisions about estate planning were starting to work through by mid 1999, hence allowing, even stimulating, better-informed decision-making and planning processes.

The basic estate planning process was straightforward. First was to set out the University's current estate, the buildings, land and facilities, by location, quantity (amount of useable ('assignable') space, and gross space as had to be maintained) and quality (in terms of the standard of the premises and the fitness for the purposes for building users). Next was to set out the agreed needs for buildings and support functions out to five, and, with less precision, ten years ahead. It was this key link between academic aspiration and estate capability that was forged when the University specifically decided in 1999 to plan its future annual growth: 0.5 per cent for undergraduates, 2.5 per cent for graduates and the large figure of up to 9–11 per cent annual expansion for post-docs and research assistants/associates (later set at 5 per cent). The practice previously was to assess and extrapolate current numbers and trends, with subsequent planning for building upgrade or expansion once needs were clearly evident, a practice of 'predict and provide' rather than the more pro-active 'plan and provide'. Individual Schools and Departments could decide on their own growth plans subject to agreement at Council of School level, but once the University estate plan had been argued through to consensus, and then agreed by BC, PRC and other committees, and then Council, they had to plan within that University Estate Plan.

The next and key step was the 'gap analysis' which compared current quantity and quality against the quantity and quality of built space and facilities that were agreed at University level as justified to meet the planned future five and ten years ahead. From that comparison of 'present' and 'justifiably required', came a schedule, the 'gap analysis', of what was actually needed. The building deficits from that gap analysis were then costed, prioritised and assessed for fundability and practicability by PRC, advised by Buildings and

Finance Committees. Finally, from that assessment came a widely agreed programme of who had what facilities, who needed what in terms of capital requirement in broad priority groupings and, as far as possible, how those should be resourced. It was promulgated and made openly available to all once cleared by PRC and Council. Each Estate Plan (a new plan was produced about every two years from 2000, with annual internal updates) was preceded by an overview of the environmental context of the University in its local and statutory contexts; increasingly that included matters of sustainability.

To take one significant example of how policy was brought in via the Estate Plan: by 2002, aspects of sustainability, and therefore energy reduction, were becoming higher priority than hitherto, and so a policy of managing design and procurement on a whole-life basis was adopted so as to optimise the total cost of a new building over ten years of its life, rather than just to minimise the initial capital cost and carry higher running (e.g. energy) costs. From then on, planning new building projects was done on a whole-life, costed and holistic basis; this was agreed as an underlying task of the Buildings Committee. The introduction of the requirement to provide a clear ten-year costing plan to PRC for all major project proposals prior to approval was only initially opposed by a few academics, mostly because of their concern that projects they already had well in train would be derailed by the need to produce a sound ten-year plan.

This estate planning process was managed consciously and proactively by the University committee system as advised by its officers, and was timely in coping with the rush of capital funding.

The second challenge to the Buildings Committee was less planned and less pro-active than the process of developing and implementing a practical estate plan: it was to respond to the suddenly increasing level of funding that was starting to flow for the reasons set out in Chapter 2. The Committee worked through papers sent out by the Estates department at least a week in advance of the seven meetings annually, each paper marked up as being for either 'Decision', 'Comment' or 'Information' regarding procurement on all current projects. For example, for the meeting on 8 November 2000, there were five Decision papers relating to a total of £54M of capital projects, plus approval of the maintenance/minor works budget of £14.5M, and approval of an environmental plan for the estate; one paper for discussion (utilities policy) and one Information paper with progress reports on 31 current capital projects. The ability to handle a flow of projects which was both much greater and more complex than ever in the University's history, with (correctly) far greater requirements for accountability and consultation than previously, rested on the extent of experience, judgement, teamwork and capability of the Buildings Committee and its officers. The extent of their successes and failures will be assessed in subsequent chapters.

One committee that did not change with the flow of capital works was the Finance Committee (FC), which according to University Ordinances was responsible for capital works. It was clear, however, that that was being looked after by the Buildings Committee and PRC, so the FC was relatively content as long as it was kept informed and retained responsibility for the Property sub-committee.

Over most of the decade under discussion there was a joint committee between the University and the NHS Trust at Addenbrooke's, co-chaired by the VC and the Trust chairman. Small in size, positive in tenor and stream-lined in bureaucracy, it coordinated capital projects on the Addenbrooke's site (even coping well with a joint project which the Trust was required, as ever, to procure using Public Finance Initiative (PFI) and with the University opting, correctly, to procure not by PFI but with an initial bullet-payment to cover design, construction and through-life maintenance).

Almost invariably the academic members of committees were public-spirited and made considerable sacrifice of time and effort to volunteer, or agree, to serve on committees so as to improve University work and life. One challenge, however, in any committee system is bringing to the surface any conflicts of interest, so that members' input can be discounted if a personal interest is being followed. Indeed, personal interests may require that a member may not speak or influence discussion, or even remain present during a discussion. In a research-based university this challenge is considerable since research bringing personal gain may very well be a legitimate and welcome activity even though much of the research may in practice have been done by funded assistants, and in University-funded premises; the ingredients for the success of that research may well be affected by changes of university policy. The challenge is greater if a committee member not declaring an interest is powerful in research reputation: still more of a problem if that member has a degree of Asperger's syndrome and is less aware of the feelings and responses of fellow-members of the committee. All universities, and all academic disci-plines, possibly even more than some other professions, have those smitten with that ailment. Cambridge was no exception to these management issues, and that very occasionally weakened committee performance regarding estates projects; over the period examined in this book successful measures were brought in to safeguard disinterested committee management.

Another problem tough to overcome was that of achieving what VC Alison Richard (VC 2003–2010) called 'continuity of business'. In an institution which rightly prides itself on academic freedom and intellectual rigour, there was a tendency for individuals to review and reason *ab initio* whenever an issue came before a committee, even though that issue had previously been considered and certain decisions already taken. This was compounded if new members of a committee felt little or no responsibility for previous decisions, so that momentum and continuity of management were both lost. Sometimes a committee member would argue in different directions in different committees when 'wearing different hats'. Lack of continuity of business is a sort of 'institutional Alzheimers'. Continuity of business improved from 2004, but opportunities and time were lost on several important occasions by lack of 'continuity of business'; for example, in 2005 an opportunity to demolish the Arup Tower in favour of a much better new building was lost.

Committee over-sight of capital procurement and advice by lay members of BC were especially important as the pace picked up, and procurement officers became increasingly stretched in their daily work. As will be discussed in Chapter 5, capital projects were especially risky when there were a lot of them at the same time. If projects had started going wrong, then the intense capital programme could have imploded as scarce resources of time and effort were diverted: it was through the authority and the active support of PRC and BC and the re-ordering of project procurement systems, and good fortune, that it didn't. The then-chairman of the Bursars' Committee said, 'if the early projects had come unstuck, the wheels would have come off the whole University strategy of expansion and research growth' (except, as a former contractor, he put it more bluntly). Had development of the committee system not been achieved in time, and it was touch and go, there would have been little chance that the ensuing stream of capital projects could have been properly managed as a co-ordinated programme.

Still on committees, it is worth noting that by the early 'noughties' there was an effective committee to ensure that there should be maximum access into and around buildings and the estate generally by those with any form of disability. The pressure for access for wheel-chair users was the most powerful, as it is nationally, even although in a university ratios between different disabilities are different from national averages, because of the age-range: lower proportion of ambulatory disabled among young people for example. The committee had the benefit of work developed to provide access for Professor Stephen Hawking through the complex new building for the mathematicians.

Whether the standards of committee management and decision-making really were good enough, only time will tell, but by virtue of all that was done as set out above, it is concluded that the committee aspect of capital management in the University had in fact got its act together just about in time. One might argue that it took a year or so before the BC became fully professional, with its sound balance of senior academic leaders and a balanced group of industry professionals drawn from the top echelons of their professions. And one might say that it really was a further year or even two before the PRC fully found its feet and was understood by decision-makers through the University for what it truly was, *inter alia*, the real filter that stopped, delayed, refined or approved strategic capital planning and the flow of individual capital projects within that. One might also note that it took several years until the committee system developed and implemented management of capital projects on a ten-year 'life-cycle' business-case basis, including the ethical and financial aspects of sustainable construction, notably the production of CO_2. Those criticisms would be fair, but it would also be fair to accept that the University did come to such process ahead of most public-sector clients of the construction industry. And that a few years earlier it had been widely regarded as a poor manager of construction procurement with projects suffering from overspend, delay and legal dispute.

So it is a fair conclusion that by 1996, when the wave of capital projects arrived, the re-ordered committee system was being planned, and that it was sufficiently in place by early 1999 when the tempo of capital works really kicked up.

Line management

That's the story of management through institutional committee management: what of the other thread of management: line management?

In parallel with the committee system, the line management system ran throughout the University much as it had done for centuries, until the arrival in 1996 of Alec Broers as VC, and as noted in Chapter 2, from the implementation of the Wass Report. Sir Douglas Wass was appointed in November 1987 to consider, in the light of changing circumstances, the government of the University with particular reference to:

a   the tenure, powers and duties of the vice-chancellor
b   the functions of the central bodies, and the relationship between them
c   the relationship between the central bodies, the Councils of the Schools, and the Faculty Boards and other bodies
d   the role of the Regent House
e   the inter-relationship between University and College policy.

This made the VC's position as the head of the line of management of the University much stronger than previously, being appointed for five–seven years, having two, and later five, pro-VCs, and a more sensible level of office and domestic support. Sir Alec from the start led and encouraged and achieved change so as to respond to the rapidly growing challenges in HE, although hampered and delayed by the efforts of a small minority of academics (referred to by some as 'the backwoodsmen') suspicious of modernisation of management; indeed, 'managerialism' was seen by some academics, including some who were gifted and well-meaning, as being mutually exclusive to academic freedom of expression. (Around that time there were jokes going round about 'how many xxx does it take to change a light-bulb? The version going round Cambridge was: 'How many Cambridge academics does it take to change a light-bulb?' ... 'Change? What do you mean "change"?!')

Certainly, the success of the University down the centuries has rested on independence of thought, relatively unfettered by managerial systems, but ever-increasing accountability and nationally and internationally competitive performance review meant that some reform of the very *laissez-faire* culture was necessary. The changes came gradually, and sometimes painfully, but more and better reforms were achieved at that time in Cambridge than at Oxford, although less change than Sir Alec and many, probably the large majority, of the University wished. The nervousness about throwing babies out with bathwater still permeated deep into the psyche of some. As the then-Registrary noted, Sir Alec had a clear view of the changes and reforms he wanted, and he was good at taking advice from those who had the capability to effect those.

(A much earlier VC said jokingly 'my indecision is final'. Another was said to be 'a man of iron whim'.)

Taking office, after being Head of the Engineering Department and Master of Churchill College, (the first VC to give up being head of a College to become VC), on 1 October 1996, Alec Broers came with considerable experience of both academia and industry, and the clear vision which he set out in his 'Visions' speech. Moving quickly to implement that vision under his leadership, the University soon appointed a new Registrary ('Registrar' in other universities), a Director of Estates in 1998 and in 1999/2000 new Directors of Personnel and of Research Services. Of the two Pro-Vice-Chancellors (PVCs) at that time, one was for 'external' matters and the other 'internal', and they with four heads of administrative services including the Director of Estates, plus the head of the Development Office, formed his Senior Management Team (SMT) which met to inform, discuss and decide on actions early each Monday morning. (Sir Alec's penchant for earlier 'breakfast meetings' wound down after a couple of years of polite but resolute unenthusiasm among his colleagues.) That SMT mechanism worked well (although there were sometimes 'streams of consciousness' meetings which were less productive) mainly because membership, while small, represented sufficiently the range of University activity and planning. On more specific aspects of the emerging capital programme, the VC had bilateral meetings with the Director of Estates.

From the establishment of the post of 'Registrary' in 1506 the administrative services of the University were run by him, then along with a Treasurer as from 1926, and then from 1934 also with the 'Secretary General of the Faculties', a quaint term which suggested comparison with the League of Nations, or conjured up grand visions of the Empire. (Indeed, many earlier leading administrators in Cambridge and some other universities came from the colonial, military or diplomatic services.) These three were referred to as the 'Principal Officers'. The Registrary administered Council and its business, the Secretary-General of the Faculties administered the General Board and the Treasurer through the Finance Board looked after financial accounting. This triangular arrangement of the three 'principal officers' made sense when VCs were coming into office for just one year or two years as they generally had done since 1587. After their two years, most VCs reverted to their previous roles, or retired, and therefore didn't want to cause changes that would 'frighten the horses', so a good measure of administrative continuity, professionalism and leadership was vital to make up for the lack of long-termism at the academic top. So for centuries there was a small administration which in many ways actually led policy. However, from the time of Sir David Williams (appointed for two years, then another five altogether) and Sir Alec Broers (appointed at the outset for five–seven years) there was at last enough continuity for a VC to envisage, understand and implement capital projects.

According to the Statutes of the University, the Director of Estates was line-managed by the Treasurer, but that became impractical when there was a fast-moving capital programme with its range of contractual risks, particularly since the Director needed to make daily decisions directly, quickly and with enough knowledge of the construction industry with all its traps. The incoming Director of Estates took on the job in 1998 on the basis that in practice he would report to the VC in line management and to the Buildings Committee for executive authority, as he had in his previous university. That said, the Treasurer was a source of excellent advice about how best to manage the important dimension of assessing and securing sufficient support for the accelerating capital procurement programme from the diverse academic community.

Throughout the decade there were of course some ups and downs, but there never were serious or long-running clashes between those managing the capital programme and the academics, as there were clashes in some other areas of administration. This was in part because the capital programme was not seen by most academics as affecting their professional lives as much as other aspects of administration such as accounting or personnel management, or seen as a threat to the preservation of academic freedom and democratic decision-making through committees and the Regent House; or seen to be incompetent enough to warrant close interest. A member of the Regent House for many years took a close and very critical interest in all administrative matters and made frequent and eloquent (if repetitive and unhelpful) challenges through what were euphemistically called Regent House 'Discussions' but were in fact usually a series of individuals' prepared speeches with little or no discussion. Her challenges to senior staff and policies were pursued even via the legal system, and were very expensive in staff and managerial time, and emotional reserves, not least for the VC, but her focus was on issues other than estates matters, on which she was broadly supportive, deeming the *modus operandi* as being 'open'.

By the spring of 2000, the administration started to become grouped under eight Divisions, each line-managed by its director or equivalent. Only one of those services (Estate Management and Building Services) was formally established in the Statutes and Ordinances which regulated the governance of the University: an odd situation for a University that was so bound by conservative adherence to regulation. It was a useful coincidence that the huge challenge to the University administration posed by the wave of capital projects was in that one area whose basis was legitimised in the Statutes and Ordinances.

In September 2000 the University Council 'established a Working Party to review the roles of the Principal Officers'. Many British universities had gone onto a system whereby the Registrar took the role of head of administration, and the current Registrary at Cambridge was determined that Cambridge should go that way. To achieve that, the post of the Secretary General was mutated into an Academic Secretary (with the post-holder of that unit directly appointed internally) in charge of one of the eight divisions of administration. The post of Treasurer was considerably changed with the addition of, and then substitution by, a new post of Finance Director who would be a professional accountant and head of the Finance Division.[4] So two of the three 'principal officers' posts came to an end. Those processes took several years to complete, and the paths towards those changes were not easy, inevitably leading to tensions which were for some years a diversion. And at times, some administrative issues became consuming across the University. For example, the new recurrent and capital accounting system 'CAPSA',[5] pushed through in 2000, soon failed amidst acrimony and at a cost stated, probably much over-stated, to be £9M. In fact, CAPSA had much in it to be commended, and it came from Oracle, a world-leader in such software, but it was said to have been insufficiently road-tested in trials with staff not well enough trained. A lot of people, especially in academic departments, eschewed the threat, real and perceived, to their accounting independence. An external review was demanded, and it was decided that that should be led by Anthony Finkelstein, Professor of Software Engineering at UCL, and the former Registrar of Warwick University, Michael Shattock, author of several books on university management. Their report, widely-publicised within and beyond the University (front page of the Daily Telegraph), spoke of the University suffering from 'a culture of amateurism' with 'tortuous decision-making through a network of academic-led committees'.[6] The then-Finance Director left soon after. Her successor was the first to be professionally qualified for that role; directors of the other administrative services had long been professionally qualified. Quite soon the accountancy function stabilised, and started to improve levels of confidence. However unpleasant the Finkelstein/Shattock report was, it did provide ammunition for those seeking reform. In the wake of the report that followed the collapse of 'CAPSA', a *Times Higher* article (27 April 2001) 'seriously doubted if Cambridge could keep up': not only did it keep up but within a few years it became one of the world's four leading universities according to some of the league-tables.

Even after the rationalisation of the line-management of the administration following the move from a triumvirate to a service functionally based on eight Divisions, there still was no formal statement as to which officer had actual responsibility for the capital building programme: the laws of the University in its Statutes and Ordinances consciously shied away from clarifying the issue, and most senior administrators and academic leaders wanted to keep it that way. It was however perfectly clear that it would be the Director of Estates who would have to resign if things went badly wrong. (At that time, the highest turn-over among senior university administrators in British universities was among directors of estates, reflecting high exposure, risks associated with dependency on a flawed industry and the risks from changing academic requirements after legal contracts were signed.) The capital programme at Cambridge was moving so quickly that its professional in-house managers in the estates department were best-placed to manage the procurement programme, and the VC was content for that to be so, implicitly leaving responsibility with the Director of Estates reporting to him, and to committees.

Deliberately, estates officers largely kept well away from internal re-organisational tensions 'at the centre'. When the Unified Administrative Service

(UAS) with the directors of the eight administrative services, chaired by the Registrary, found its feet with its members meeting and discussing interface issues, and, occasionally, matters of wider policy, there was negligible discussion of capital estate matters. One problem of the UAS was that the Registrary's own department was so thinly staffed that he had to rely directly on administrative divisions to do what should have been a co-ordinating role in his Office: an example was that in March 2003 co-ordination and management of overall University resources was set up as a new unit in the academic services division rather than in the Registrary's Office. That could have led to organisational distortions common when co-ordinating functions are done by one of many departments rather than the umbrella organisation/department. However, all concerned were able to make the system work albeit somewhat against the grain of good management principles in some aspects. Most of the organisational challenges faced by the University management were those commonly taught on MBA courses. (Only one officer at director level at the time had had such formal management training, but most had during their careers acquired excellent managerial and organisational skills.)

In terms of line management of the estates divison, it was universally agreed that there was over-riding merit in keeping together under one department all aspects of management of the estate, from conceptual planning, through land assembly, land management and Town and Country Planning, master-planning, planning applications, design and procurement of new buildings and refurbishment projects, maintenance and running of buildings throughout their lives, through to demolition or disposal. That was and is the common practice in other universities, and the logic is obvious: the Director of Estates could brief and could take direction and responsibility for all these overlapping and inter-dependent aspects of the estate, with the estates department taking an integrated, collaborative and holistic view so as to minimise the incidence of glitches arising between the different aspects of estate management. If some of these aspects are split off, disjunctions soon emerge; for example, land planning doesn't foresee the subsequent issues in building planning; maintenance staff do not inform designers well enough about required technical systems; or buildings arise that are difficult or overly expensive to look after. People who design and put up buildings don't feel full responsibility to those who maintain or service the buildings, and commonly rivalries spring up between them. Sir Alec and the PVC Malcolm Grant, soon to go off to be Provost (VC) of UCL, were especially clear that it was vital to keep all aspects of estate management under one department. During the decade being here considered and previously, that system worked well. (Subsequently and counter-productively, the functions were taken apart from 2007–2013 before being put back together again, after the loss of morale and many good staff.)

An issue that was increasingly affecting University management generally, and the management of the University's capital construction in particular, was the problem of recruiting enough good managerial staff. One reason was that as the nation went into the boom of the mid-2000s, there was nearly full employment among trained construction and accounting professionals, and company salaries and 'perks' rose to match. The University, for understandable reasons, pegged administrative salaries to what was seen as being comparable to academic salary scales. This had the direct effect in boom-times of reducing the capacity of the University to recruit high-quality staff to manage its projects well. Also, there were some consequences on morale of existing staff in income comparisons. Many, though certainly not all, academics rightly had the benefit of being able to earn income from other activities, some academic (such as taking supervisions), some external (such as consultancy), whereas some senior administrators who were, of necessity or through professional pride, working extremely long hours were without such opportunities and, anyway, 'moonlighting' by administrative staff was, correctly, disallowed, or at the least, strongly discouraged. Also, academics rightly get sabbatical leave, a bonus not afforded to any administrators.[7]

Salaries at Cambridge were generally lower and housing costs higher than for the same administrative jobs at many other good UK universities, and some senior administrators came into post at Cambridge with salaries lower than they had had in the posts they had left in other universities: that said, there were of course many offsetting benefits of working in Cambridge, such as professional job-satisfaction and the wonderful cultural and architectural life of Cambridge. Also, there were compensations for working in the public sector on a lower salary than on a higher salary in industry, such as better security and pension. Recruiting of administrators whose professions did not map onto non-university jobs continued without too much difficulty; however, recruiting those who could find good jobs outside the university sector did become very difficult in the 'boom times', especially 2002 onwards, and especially so in areas of capital project management. Some job adverts for project managers and more senior posts secured just a tiny number of barely qualified applicants; one senior job in estates management attracted just two even remotely qualified. The much-delayed introduction of a new and interim 'pay-spine', finally starting in a very tentative way in 2002, did in the end help to reward good performance to a limited extent and, to a tiny but important extent, to restrain pay increases for those performing badly. Later, a full review in 2005/2006 made for more pay enhancement flexibility, with some degree of 'merit' pay for administrators, and some flexibility in pay offered that could be offered when recruiting in expensive professions; known as 'market supplement'.

In 2001 Council noted that the cumbersome committees that made new staff appointments were causing such delays that many good candidates were lost. That position was somewhat improved when heads of administrative divisions were allowed to make junior appointments in their own divisions, but for appointments which in industry would be done quickly by senior professionals, University administrative heads still had to spend huge amounts of time to achieve even middle-grade professional appointments relating to work, the nature of which the academics running appointments committees, through no fault, had no knowledge and sometimes no great interest.

When the decision was made in 2003 to expand the number of PVCs to five covering most aspects of the administration, their role was to oversee but not manage; however, by taking key committee roles they also took, to varying degrees, some direct line-managerial responsibility. This change was welcomed by many academics but it made some decision-making by senior administrators more diffuse and sometimes more difficult. Increasingly, the selection of PVCs included senior academics who also had experience of administrative and/ or business matters, but in the early days there were some situations in which PVCs took over some decision-making on professional matters of which they had little or no experience. The professionals who were running procurement of the capital programme, aware of the complexities and the risks involved, were determined to avoid any confusion that might arise from over-detailed PVC intervention; that took a lot of effort but largely worked during the decade.

As elsewhere throughout the developed world, the nature of line-management mutated over the decade as personal contact was reducing because of the rapidly increasing reliance on digital rather than personal communication. The effects of this rapidly became significant. As in other aspects of working (and personal) life, management through meeting and conversation with colleagues was increasingly giving way to email and telephone, or indeed voicemail as people (extensively in some areas of the University) avoided even answering telephone calls, other than recording them and then responding at a time convenient to the called, if at all. The ease of distributing information digitally, exacerbated by 'Reply to All' overuse, was hugely increasing the volume of information being brought to managers so that decision-making was getting better informed for those capable of rapidly filtering out what was relevant and credible, discarding the rest; for those who couldn't or didn't do that, decision-making was not in fact 'better' informed. Further, because of the permeation of digital communication means, pressure on decision-making time resulted in decisions being made more quickly.[8] Whether the quality of decision-making got better or worse with increased information, much less personal contact, and decreased decision-making time, is a crucial but seemingly under-studied matter. Certainly, during the decade those managers and administrators most capable of maintaining good personal contact, and also achieving good information management (by working much longer hours, and/or having good support staff), performed demonstrably better than others. In management of the capital programme in particular, those who conducted more business through personal contact ('Walking the Talk' or 'Management by Walking About') were observably more successful in their buildings projects than those who relied more on email and voicemail; but again, no research or authoritative data to substantiate or quantitate this observation has been found.

Conclusion

From the late 1990s huge changes were made in line management capacity and capability as complementary to and in parallel with effective changes of the committee system of management. For the all-important committee system it is concluded that just about enough of those changes did come just about in time. Committee re-organisation came mainly through the establishment of effective Planning and Resources and Buildings Committees and their sub-committees. Some of the line-management changes didn't come in time, and some worked imperfectly, but the 'direction of travel' was right: line-management moved first to a mixed academic/administrative senior management team comprising the effective VC's SMT, to 'oversight but not management' through five PVCs, and through the re-ordering of support administration from a triumvirate of 'Principal Officers' to a relatively unified administration based on eight divisions co-ordinated by the Registrary. And if seen as painful and very lengthy, these reforms to committee and line-management systems were in fact quite speedy in the context of the long and slow history of the University. As the Cambridge tortoise said, 'I am glad to be moving slowly because I may be moving in the wrong direction'.

So the University was just about able to deal in a business-like manner with the unprecedented estate expansion programme starting in the late 1990s. Such a programme would probably have collapsed had it come earlier, or had reform been later, or slower, or more effectively opposed.

Notes

1 The Buildings Committee started life under the name of Major Projects Sub-committee, then became the Buildings Sub-committee (under the assumption that it then reported to Finance Committee (FC): in reality it reported to PRC but internal politics delayed recognition of that, without causing problems). The 'Estates Committee' concerned with land needs continued, reporting to the FC.
2 Academic opposition couple of times stopped such lay membership at Oxford, then probably the only UK university still not to have such lay membership on its Buildings Committee. In Ivy League universities in the USA, the equivalent of a UK Buildings Committee is comprised largely of lay members.
3 An attempt in the early 1990s to get a consolidated schedule of University property foundered: it began hopefully by going through the University telephone book seeking addresses. In 2001 a property sub-committee of Buildings Committee was set up to enhance the planning and the use of the considerable property portfolio.
4 The Finance Officer was an accounting officer, rather than a finance director as most organisations would understand that term. In effect, the PVC for Resources became close to what most organisations would recognise as a finance director: he was a lot of other things as well, and retained some of his academic work.
5 CAPSA was/is the name of an accounting system: literally (CAPSULA, CAPSELLA), the box for holding books and scrolls among the Romans.
6 Reporter 2 November 2001.
7 In fact, statutory administrative officers were not precluded from applying for sabbatical leave, but none did.
8 In the late 1980s it would take about five days after receipt of a contractor's letter to draft a reply, with carbon copies, post it and get a reply. A decade later, five email exchanges in a day was not uncommon.

5 Management of building projects in the University

Building procurement in the UK

The bad and the ugly: the industry as it was

Infrastructure is a large and expensive part of national investment, so its procurement matters a lot to policy makers and to tax payers. For decades, the construction industry, the largest industry in the UK, was generally the least efficient, the most fragmented and the most litigious, and to some extent that is still true. In 2006 the size of the construction industry was thought to be £114bn including about £10bn in 'the grey economy'.[1] The industry represented 8 per cent of GDP then, employing some 2.1 million people of whom 785,000 were working in 'one-man' businesses and a further 223,000 in registered firms employing between one and four employees. Construction turnover in 2013 was 6.7 per cent of UK GDP. The public procurement proportion of the total varies from year to year, typically in the range 25–45 per cent, as governments turn the public sector construction tap up and down. (Most governments use construction activity as a lever on economic activity levels.)

Just as the procurement of infrastructure is a substantial slice of national investment and annual budgets, so it is in universities: in 2002 for example, of the total recurrent expenditure of the University of Cambridge of £450M, the cost of premises was £49M, the third largest expenditure after staff costs, and research grants and contracts. Investment in UK university estates in 2012/2013 was £2.4bn on buildings and £0.7bn on equipment funded increasingly from internal University sources rather than from government.[2] Most of universities' capital expenditure is on buildings.

The industry has long been the most fragmented industry in the UK: in the mid 1990s in evidence to the Latham review (see below) it was estimated that there were on average eight levels of sub-contractual interfaces on a typical £12M building project: from the client, through the designers and main contractor and through the various layers of sub-contracting. At each interface there is a different contractual arrangement, with different people with different attitudes, motivations and understanding of the work, and at each interface there are transactional costs in money and often in commitment. The fragmentation in the industry largely relates back to the introduction of Selective

Employment Tax in 1966, and to the growth of Health and Safety legislation which caused employers to cut down their staff levels so that someone else would carry the responsibilities for tax and for safety. (Safety rules now are different and don't allow that loop-hole.) There is certainly a need for some compartmentalisation in the industry so that construction companies can react best to rapidly changing economic conditions, but the level of fragmentation is much too great. In recent years, some large and progressive companies have been increasing the size and range of their workforce.

Typically, contractors' net profit margins were about 1.6 per cent at the start of the decade under review, rose to about 2.8 per cent in 2007, then fell back again.[3] So they rely on high volume at low profit margin, and because of the thin profit margins, projects became litigious even from an early stage; being denied even quite small payments has a surprisingly large effect on a contractor: clients needed to understand that. Designers were taking less risk and, conversely, often making more than twice as much profit margin as contractors. Construction lawyers had and have higher profit margins and less risk: probably the case in most industries.

The poor performance of the construction industry generally became a national byword, and a national scandal. In the early 1990s the situation was so bad that the government was, finally, persuaded to set up a national review of the industry; it was led by Sir Michael Latham MP (who had taken a first-class degree in history while at King's College, Cambridge). The specific prompt for the review came when, in the recession of the early 1990s, four self-appointed representatives of the industry, from the 'Group of Eight' pressure group, led by a small architect with a large bow tie, secured an hour's meeting with Mrs Thatcher. Witnesses said that they spent most of the hour demanding that the government provide more work for the UK construction industry or else it would be taken over by foreign companies, and that she then gave them one minute, explaining, in surprisingly expressive language, that if the industry couldn't get its act together and raise its standards then British people would be better off if foreign companies did just that.[4]

One of the most important bits of evidence for the Latham review came from HM Treasury: a high correlation between projects that went badly over budget, and those for which appointments had been made at cheapest stated price.[5] Partly because of the long-established practice in the industry of bidding low and then finding ways of claiming extra payment and time, the client, or tax-payer, often ended up, on overall average, paying more if they accepted the lowest tender for the initial capital cost of the building without due reference to assessment of quality or ability to meet specification than if they had taken a tender at higher price. This was because, as the Latham Report put it, 'Lowest priced tenders may well contain no profit margin for the contractor, whose commercial response is then to try to claw back the margin through variations, claims and 'Dutch auctioning' of sub-contractors and suppliers'.

John Ruskin, 1819–1900, noted, overstating somewhat, the dangers of getting the quality/price balance wrong:

It is unwise to pay too much, but worse to pay too little; when you pay too much, you lose a little money, when you pay too little you sometimes lose everything, because the thing you bought was incapable of doing the things it was bought to do. The common law of business balance prohibits paying a little and getting a lot. It can't be done. If you deal with the lowest bidder, it is as well to add something for the risk you run ... There is hardly anything in the world that someone can't make a little worse and sell a little cheaper – and people who consider price alone are this man's lawful prey.[6]

Another part of the evidence in the Latham review was the observation that showed that for the construction industry, legal process costs (that is, excluding the actual settlements) exceeded the total spent on all the research and innovation in the industry.

The conclusions of the Latham report set out a brave vision of an industry led by informed and active clients working directly with not only designers but also with contractors, since they, better than other parties, know the practical aspects of how buildings actually get built, and what the different processes cost. As the industry was so seriously fragmented, Latham set out methods of achieving integrated supply chains, with appointment by the client of a selection of designers and early involvement of the contractor, all on a quality/price balance basis. Following publication of his report, Sir Michael set up, with pro-active and very able Civil Service support,[7] 12 working groups, each with senior professionals drawn nationally from clients, designers, managers, constructors, suppliers; their task was to develop methods of achieving the recommended way of procuring new buildings. Some better clients soon started to implement the more open, less confrontational ways of working, and some of these ways of working were brought into draft legislation in 1997. The effect of the Latham Report on the construction industry was like the detonation of a cluster bomb: deep penetration and widespread individual explosions of change, with further bursts delayed but powerful.

Another report, *Rethinking Construction*, was commissioned by the incoming Labour Government and published in 1998 in the name of Sir John Egan, (formerly of Imperial College and London Business School and later chief executive of Jaguar Cars).[8] It reinforced some aspects of the Latham review, and added the importance of measuring all aspects of design/construction performance, notably the extent to which pre-agreed criteria were or were not met, so as to improve efficiency. Much of the inspiration for this review came from the Japanese car industry which relentlessly drove out waste by 'lean' production and deep use of metrics, and that was applied to the construction industry under the banner of 'Lean Construction'. As the regent said, 'if you can't measure it you can't manage it': not strictly true, but generally helpful guidance. However, the Egan Report sought to bring the clients into bilateral relationship with only constructors, leaving designers just to work for the contractors with very much less direct relationship with the clients: that mattered a lot. That trend continued, and currently more than half of the architects in the UK work for contractors.

One reason why many architects resisted for so long the introduction of contractors during the detailed design process was a fear that they would 'dumb-down' architects' designs, and that constructors would not understand the work of architects. As Renzo Piano (who wanted to work for the University during the decade being reviewed but was thought unsuitable for what was then wanted) said of architects' work, 'As an architect you have to be a sociologist, a poet, and to learn all those things takes time'. Another reason was that early contractor involvement (ECI) was seen as a threat to what many architects saw, and see, with some justification, as a very special relationship with the client. Denys Lasdun, who designed the first building at Fitzwilliam College, had said that an architect should not provide what the client wants, but what the architect sees that he wants, and then will delight him. The old apocryphal story of an architect said to have a policy that 'a client appoints me, keeps out of the way except to pay design fees, and then is invited to the opening ceremony' was from a bygone age, but there was still some nervousness about early client/contractor relationship.[9]

Changing the ways and improving the standards of a largely disaggregated and dysfunctional industry is a huge undertaking at any time, and construction reform was being attempted over a change of government, with rivalries between supporters of the Conservative-led Latham review and the New Labour-backed Egan review, exacerbated by personality issues.[10]

In pursuit of fairness – and cost-effectiveness

An Act of Parliament (the first ever with a large part specifically for the construction industry) was passed in 1997, drafted under the Conservative Government and carried through under the New Labour Government.[11] This Act brought reform in the concept of 'fairness': if a contract wasn't fair, it was void. More especially, that dispute resolution in the first instance was to be by adjudication rather than by court resolution or by the worst of all systems, arbitration (which combines huge cost, with many lawyers and expert witnesses, and often leads to a poor outcome after long proceedings). With that Act of Parliament as 'backbone', the many recommendations for more business-like, more open and more collaborative and measured procurement from the two national reports were encouraged across the construction industry and its clients.

Outcomes of the Latham and Egan reviews of the construction industry

Although given huge support by the Labour Government, the bilateral client/contractor emphasis of the Egan Report became somewhat displaced by the client/designer/contractor teamwork approach recommended by Latham when economic times were good; when times were hard, the less collaborative, bi-lateral approach of Egan became more popular again. A further report for government by Sir Peter Levene recommended,[12] *inter alia*, better communication with the construction industry to reduce disputes, but by that time industrial and client leaders, after expenditure of millions of pounds, and hundreds of thousands of hours given, were 'initiatived-out'.[13]

Although turning round a huge industry usually takes decades, there were many early improvements. In the early 1990s there were estimates in HM Treasury that some 90 per cent of UK public sector projects came in at least 10 per cent over budget, and 10 per cent were 90 per cent over budget. That did improve following implementation of the Latham and Egan Reports, and the work of the Movement for Innovation (M4I) set up directly after the 1997 election by the incoming New Labour Government:[14] in 1999, 34 per cent of public sector projects were on time, 25 per cent were on budget; in 2005, 63 per cent were on time and 55 per cent were on budget (average overspend 4.1 per cent). And budgeting got better: averaged over 2000–2004, setting the costs of public sector projects had an average error of 33 per cent; that error fell to 9 per cent averaged over the following four years.[15] The seminal NAO report of March 2005 noted that £2.6bn could be saved annually by reducing waste in public sector construction projects. (A leading consulting partner of NAO, Concerto, put the waste figure at £7.8bn.)[16]

An initiative which came out of implementation of the Latham review which has stood the test of time is the Considerate Contractors' Scheme (CCS; also known as the Considerate Constructors' Scheme), a national programme started in 1997 by the Construction Industry Board (CIB) whereby builders commit themselves to reducing the adverse impacts of building projects on the local community by better control of noise, with constrained early-morning, late-evening and week-end work, control of dust and workers' behaviour (no more 'builders' bum cleavage' or blaring radios), arrangements for visits by adults and by schoolchildren, and a large display of what the building should look like when finished; also viewing ports at child and adult eye-levels allowing people to see the building works, and a number to call out-of-hours in the case of complaint, had to be made public. Construction sites and companies voluntarily register with the Scheme and agree to abide by the Code of Considerate Practice which is designed to encourage performance beyond statutory requirements.

As early as 1996 a few of the better-informed and pro-active clients, including a few universities, were reducing risk and cost, and improving building quality, and the morale of construction professionals, by more open and collaborative procurement as recommended in the follow-up to Latham, even although some of those progressive clients were constrained by consultants and designers, particularly by some architects and quantity surveyors, who were reluctant to learn new methods and new contract forms, and even in some cases were thriving on disputes: a lot of legal firms had spent much time and money trying to stop reform legislation going through Parliament. By and large, engineers were more open to the new system, partly because the new, managerial contract form recommended by Latham and later by HM Treasury, the New Engineering Contract (NEC), in particular its derivative the Engineering and Construction Contract (ECC), had come from the Institution of Civil Engineers, written by people sick of the existing confrontational contract forms, notably the then current form of the Joint Contracts Tribunal (JCT) contract which was closely associated with high levels of litigation.

So, many who had not previously worked outside their comfort zones were reluctant to learn and adopt new collaborative contract systems, and contracts written in plain English rather than in 'lawyer'. And those clients, including the many public sector clients (especially those who had contracted out much of their procurement, and especially in local government where so often procurement was under-informed), who did not trouble to keep themselves informed about progressive changes and new 'best practice' remained unaware of the changes afoot: the very concept of 'clientship' was foreign to them. Many clients felt more comfortable with the long-established system of working just with designers and consultants who spoke the way they did, rather than bringing contractors into the process. This was to a degree a 'class' issue: many clients still subconsciously felt that designers and consultants, 'professionals', came from backgrounds similar to their own whereas contractors and specialist contractors, 'builders', came from different schools and hadn't been to university. That had long ceased to be true: the majority of contractor and supply managers had long had good education up to university level with many universities running undergraduate and post-graduate courses for management of construction projects.[17] However, perhaps contrary to old-fashioned stereotypes, it is worth noting that regarding working hours in the UK (which on average are higher than in most developed countries), two of the professions with highest average hours worked per week are architecture and project management; full-time architects work on average over 50 hours per week, partly because of intense work as deadlines approach.[18]

Safety and productivity

The safety record of the UK construction industry is and long has been very good, though of course, every accident is a matter of huge regret.[19] The United Kingdom accident rate is the second lowest within the European Union and considerably less than the average. In 1982/3 there were 6 fatal accidents per year per 100,000 construction workers; in 2012/13 there were 150 worker fatalities in the UK, corresponding to a rate of 0.51 deaths per 100 000 workers.[20]

While the organisation of the industry and procurement management had improved, worker-productivity hadn't: when HM Treasury internally reviewed productivity in the building industry in 2006, it found that unlike in all other major UK industries, it was not improving. Noting that safety standards were better than in most countries, it concluded that much safety legislation brought in during the late 1990s was marginal in effect and had reduced productivity.

Improvements to the industry came mostly from the industry itself, with some pushing by government, although that was muted by lack of actual support from central HM Treasury and the increased politicisation of parts of the civil service which meant that Ministers were notably more responsive to short-term issues rather than to the longer-term underlying challenges in reforming the construction industry.

Overall, however, while it is fair to say at the time of writing that British capital procurement at its worst is little better than it was in the early 1990s, at its best it leads the world: compare London 2012 procurement to that of Athens or Beijing Olympics venues, both of which were hugely expensive with bad time over-runs: the last buildings of London 2012 were handed over 365 days before the first event, and there was an underspend currently estimated to be in the range £377 to £760M depending on accounting definitions under its £8.1bn budget (with other definitions of budget scope, budget is quoted at £9.3bn). Sustainability was very much better in 2012. There were no fatalities or serious reportable injuries, compared with 14 reported deaths in Athens 2004, and a number not known in Beijing 2008: six were reported but that is probably an under-reporting.

University procurement pre-1998

The traditional and long-standing system in Cambridge University had been that an academic department would identify a need for new, bigger or improved built space, and would then enlist the support of the Vice-Chancellor and those who would help find the money to cover the capital cost; the University centrally at that time paid all building running costs. When there was enough confidence and money to pay for initial studies of design and cost, the project would go into the University's committee system, starting with one of the various buildings committees or syndicates. By the end of the 1990s, the changes in committee structure as set out in Chapter 4 made it possible to improve the way that the University went about getting new buildings. In summary, a project would go through the unified Buildings Committee as advised by the University's Estate Management and Building Services (EMBS) department which had responsibility for planning and managing the estate,[21] including responsibility for managing capital projects, and then, after a whole-life assessment, on to Planning Resources Committee and Council for final authority (apart from the formal and sometimes time-consuming process of clearance by Regent House). Many committees met just six times per year, and a sequence through the hierarchy of committees and their sub-committees had to be maintained, and that made management of procurement more difficult, time-consuming and at times frustrating.

Until 1998, the drawings and specification (which the designers assured were complete) for an approved project would go to as many as six pre-selected contractors who had just a few weeks to consult the top layers of the supply chain and submit a costed tender. Almost invariably, the cheapest was selected, the contractor would be told to proceed, and legal contracts would be drawn up for signature in due course (which was often when the construction work was at least half-finished; it is not unknown for contracts to be signed during an opening ceremony). The contractor, having been kept away from the design process, had little idea about the detail of the design, and often no idea about the reasoning behind the design, or why better solutions, in his view, had not been used. The rush to get together a tender price (against competitors) from the long and fragmented supply chain within the usual 3–6 weeks meant that the tendered price was based on limited knowledge, very limited understanding and a judgement as to the extent to what profit margins down the supply chain could be squeezed after the price had been agreed.

Procurement until 1998 always started with design rather than management and hence with the appointment of an architect. Architecture was, and at the time of writing largely still is, the only profession in the industry not to teach management *per se* in universities at undergraduate level; there is an assumption that management can be learned 'on the job' from practising architects, who themselves were never specifically trained in management. This was as issue long recognised by some members of the Royal Institute of British Architects (RIBA),[22] and picked up during the Latham review by that outstanding RIBA President, Dr Frank Duffy, who also tried to change this. (More progress is being achieved in many US architecture schools.)

As the architect and construction lawyers were appointed at about the same time and ahead of other appointments, each project set off grounded in an ethos set by architects and lawyers, rather than by a team which included professional project managers as well as designers and soon supplemented by those who would actually manage the building procurement once the outline design was approved, and the contractors, or better named the 'constructors'.

Changes in the University's procurement policies

At the start of the decade being considered, the general situation in the UK construction industry was one of poor but improving performance. In the University of Cambridge fundamental change in procurement was clearly necessary since neither the draft legislation affecting building procurement nor the Latham Report recommendations were known. Following a series of major projects which had ended up significantly over the original budget, beyond agreed time schedules and with defect issues that then needed employment of claims consultants and lawyers to resolve over years, a guide to construction procurement was issued in April 1998. It was sent to Sir Michael Latham for comment; he declined to reply substantively, saying privately that in fact the guide illustrated most of the reasons why the government had called for a review of the industry in 1994. On 5 October 1998, he came to address an open forum of many academics and procurement staff in the Judge School of Management; that talk was useful in setting out to academics and to procurement staff the urgent need for a radically different way of building procurement. The client-led team-approach to procurement came as a surprise to those present, academics and administrative staff alike. The principles of teamwork procurement as led by a pro-active and informed client were then set out and discussed in detail; from that meeting on the agenda was clear.

The April 1998 guide to procurement was superseded later that year by a guide that reflected the recommendations that had emerged from the 12 professional working groups set up to implement the Latham review.[23] It was based on the general rule that success in capital project management comes from alignment of people, and alignment of peoples' motives, as well as good design, management and construction. It required early appointment of all involved (designers and contractors included), by a pre-considered, promulgated and quantitated balance of quality and cost assessments; the basic principles of all appointments in procurement were agreed to be: assured quality, value for actual out-turn cost, compliance with fair trading law (both UK and EU), accountability and transparency. 'Cost' as assessed in quality/price balance was noted to be initial capital (i.e. tender/offer) cost, but it was noted that expected whole-life cost was a factor in quality evaluation. Resolution of any disputes was to be through adjudication (a process not carried out by lawyers and expert witnesses but by senior, trained members of the construction professions, and within a couple of months). It was subsequently agreed that internal project managers were to carry out early value for money analyses, and then recommend a procurement strategy, and to undertake a formal exercise to define, achieve and compare key performance indicators.

The new University guide also required all contracts over £2M to use the more recent and managerial, less litigious, form of contract, the Engineering and Construction Contract (ECC) as that had a far better track record for teamwork and the avoidance of legal problems than the previously used then-adversarial JCT form of contract.[24] It is worth noting that during the decade being considered only one project, Biochemistry, went to adjudication (plus one tiny West Cambridge project), and none of the over 100 projects during the decade came anywhere near going to court, to arbitration or to any other legal-based conflict-resolution.[25]

In 1999, following full interviews of several legal practices, scored on a quality/price balance, the University switched to a firm of construction lawyers more open to modern contract forms, more adapted to collaborative ways of working and more determined to concentrate on pre-contract legal work rather than post-contract.

On 18 November 1998, these first changes in procurement were formally approved by the University's Buildings Committee.[26] It was agreed that the University should move to a two-stage procurement process bringing contractors into the procurement team at the start of, rather than after, detailed design, being clear that the designers would continue to develop designs to RIBA Stage 3 (formerly Stage C) using their unique skills to set the outline design, and then, only then, for constructors to join the team in the development of that design, bringing their practical skills: the procurement strategy of 'develop and construct'. The University did not want either full design before constructors got involved, or to take the risk of loss of quality potential by resorting to the easier 'design–build' form of procurement; rather the in-between policy of 'develop/construct' that proved more successful.

It was also agreed that design of building services (such as temperature and quality of air, noise levels, reduced needs for energy with reduction of costs and production of pollutant gases) should have a higher priority than previously, and that all appointments were to be made on a quality/price balance rather than cheapest quoted price. Formal criteria for making appointments of designers and contractors had not hitherto been set out (rather, the reputation and charisma of designers had tended to be the main factor); it was agreed that selection of designers should be based on their ability to provide, in priority order:

- more and usually better space for the users of the building than the users had before,
- a more stimulating, efficient and pleasant environment for the building users
- enhancement of the environment of the neighbourhood
- enhancement locally and more widely of the environment, prestige and success of the University generally, and of Cambridge itself
- an opportunity to enhance the prestige and afford professional satisfaction to all involved in the programme.

In all, nine pages of procurement policy statements were agreed at that pivotal Buildings Committee meeting.

Changes in the management of project procurement

Once the Buildings Committee (BC) had assessed proposed projects at an early stage, and the Planning and Resources Committee (PRC) had approved the academic case and the analyses of capital and whole-life costs, projects were then specifically approved stage by stage as design and costing developed until the new building (or building refurbishment) plan was designed in outline (generally to about RIBA Stage 3). The new committee system stimulated procurement officers to report personally on their projects to the Buildings Committee with its five new lay members who were all very experienced and senior in their professions (see Chapter 4). Committee papers were drafted by project management staff themselves rather than by consultants and that made for procurement procedures that were better researched and considered. Project report papers were always circulated one week before meetings, in good time for them to be read by all committee members in advance. Before the start of BC meetings, project managers, with mounted drawings, took detailed questions by members directly, and informally. Each summer, there was a special 'away-day' meeting of the BC at which there were overview presentations and discussions about Town and Country Planning and community relations issues, analyses of recent and current projects' compliance with budgets and programmes, value for money in terms of unit capital costing, planned whole-life costing, and sustainability and other quality issues. Each term, there was a bus tour of BC members (with local councillors who wished to attend) round current projects, with the contractors present to conduct site tours and answer questions.

Representation of building users

The introduction of a mechanism of appointing a Representative User (RU) for new building projects was pivotal; in his farewell speech to the Estates Department the VC, Sir Alec (now Lord) Broers, said that that was the most significant change in recent University procurement. During the inevitable ups and downs of building procurement it was crucial to have one clear voice (albeit a voice that rested on consensus within the Department) in the key appointments of designers and contractors, through the design processes and also through the vital processes of 'value engineering' (VE), the series of reviews carried out at various key stages to ensure that all that is being designed is necessary, that the project is held down to capital budget and that the projected whole-life performance and cost are acceptable. Perhaps most important of all, the RU co-signed with the Director of Estates the post-completion project report and the Post-Occupancy Evaluation about two years later, both reports going into the University committee system (and the latter going into the public arena for the good of others in the industry). So everyone knew that the RU had considerable power and, jointly with the Director of EMBS, 'the last say'. Getting an RU to understand how the industry works, including its weaknesses and strengths was a good start. A single-sheet list of 'roles and duties of RUs' was given to and discussed with all RUs to ensure that all was reasonably well understood. (A 28-page document setting out procurement policy with more detail was issued in 2005.) One adjunct was the practice often recommended and sometimes adopted whereby the RU employed a professional 'specifier' who had a good knowledge of the academic discipline and also of the procurement and construction processes, usually a graduate of and practitioner in the academic discipline concerned: such a specifier could assure bridges between the client and designers who would be unaware of some technical aspects of the needs of building users, and academics who knew but assumed that provision would be automatic without being specified. The nightmare scenario was handing over a building which seemed fine but did not in fact do what the users needed in some vital aspect; remedial work at that stage is very expensive and leads to disagreements as to who is to blame and therefore who is liable. No building is perfect for its users, but all need to be generally acceptable: the following chapters on projects set out how suitable new buildings were found to be.

Setting out how much space was needed in a new building

One of the basic steps in procurement is 'writing the brief', and one of the basic aspects of that is: how big should the new building be? Optimisation of existing built space to reduce needs for new space is important at this early stage: space management of existing buildings affects the capital programme considerably because a new building should only be built/acquired if existing and planned activities can't be sensibly carried out in existing premises if people could sensibly squeeze up a bit, and share space and/or facilities such as lecture theatres when reasonably possible. Many academics fought long and hard against space management by those outside their departments (and many didn't like much departmental space-management either). They felt that their buildings were just that, their buildings, even though the University centrally was paying their premises costs, which could be a high six-figure sum annually for some departments. In 2003/2004, Cambridge space allocation per student was about double the average for Russell Group universities.[27] The dawning realisation of what built space costs to run, and the view that built space should be paid for through devolved departmental budgets, began to persuade academic leaders that space should not be wasted, and that for new buildings, space requirements should be limited to that which could be justified by realistically estimated research/teaching needs, costs and income assessments, and the ability to pay for all the new space during the building's lifetime. Space planning and space-cost attribution to Departments followed, with an allowance assessment based on reasonable requirements.[28] Implementation of such space charging as part of devolved budgeting, Resource Allocation Method (RAM), took a long time and was patchier than in other universities. Getting better use of space led to more realistic space demands in new projects once budgets were devolved. Policy for and allocation of space by a committee led by a senior academic, usually a Pro Vice-Chancellor, had been common in most universities since the mid-1990s: it started in the University of Cambridge in 2004/5, as a 'Space Management Advisory Group' (abbreviated as SMAG; unhappily misprinted once as SMAUG, per the dragon in 'Lord of the Rings'), chaired by one of the deputy directors of estates. There was a joint group set up by the Colleges and PRC to investigate the concept of the 'Principal Place of Work' to see if rooms and/or lecture theatres could be better used through closer liaison between Colleges and University in space-use: less was achieved than expected.

Selecting the designers, project managers and constructors in practice

The new criteria for selecting designers and constructors were noted above, in particular, how they were to be selected on a quality/cost balance basis. The procedure of 'design competitions' was banned because it was thought better to make a good assessment of the capabilities of the field, and then to select one practice from a shortlist on the basis of track-record, suitability and initial ideas, rather than get practices to work up (usually at their own expense) a design in the limited time before interview, and then being stuck with that, as often happens in design competitions. In the UK for large/medium size projects, just 3 per cent of architect appointments were by design competition; 25 per cent were appointed without any competition, but much of that was for design of serial retail or office projects; the RIBA's dedicated team ran just 17 design competitions in 2013.[29]

Although there was some resistance to these changes in University procurement from designers and consultants, and indeed, for a while from a few University staff, change was made easier by the increasing promulgation of examples of really good and really bad procurement from around the country from the body set up to implement Sir John Egan's review of the industry, the Movement for Innovation (M4I), for example, comparison between procurement of the Wembley and Emirates football stadiums: the former, designed by Foster et al. and constructed by an Australian contractor, procured in a combative and 'silo' manner ran very late, hugely over budget and with massive legal costs, while the latter, procured on a teamwork basis, was the converse; the success in the procurement of HS1 rail project, as compared with such projects as the British Library with its repeated design changes. Closer to 'home', the Higher Education Funding Council for England (Hefce) was beginning to make it clear that it took a dim view of universities which procured low-performance buildings or over-spent budgets on Hefce-funded projects, and they sent their small and dwindling number of staff around those universities which were attracting much capital funding, notably Cambridge, Oxford and some of the London universities, to monitor procurement performance. And it is fair to note that within the University there was some resentment against the previous overspending of budgets and over-running of project programmes, and hence pressure to change and do better.

There was another issue. In many leading universities in the UK and USA, there was a habit whereby the benefactor of a building would make the gift conditional on appointment of 'their' architect, or at least would put pressure on the university in choice of designer. (A watershed in USA came at Princeton University where a benefaction was turned down when staff and students were especially averse to the proposed design demanded by the benefactor's favourite architect.) Such a scenario was prevented at Cambridge once the relevant committees approved the above ranking of designer appointment priorities (see p.83) and an open procedure for appointing designers; imposition of a benefactor's architect didn't happen during the period 1996–2006.

In Cambridge, managing capital projects with an external professional project manager working alongside an internal project manager and the designers, worked well during the decade, although sometimes the external project manager tended to create a distance between the designers and the client, inserting themselves too much as intermediary. Occasionally, Estates Department project managers would hear that siren voice and be tempted into a less pro-active role making it more difficult for the architect and engineer to get a perceptive understanding of the users' needs and for the RU to get a full vision of what could be made possible. All appointments required compliance with the University's 'Design Guide' (e.g. DG + CB updated Sept 06) which set out many detailed requirements for University Buildings.

Contractors appreciated early involvement. The regional manager of the national contractor Willmott Dixon, Chris Tredget said,

> Cambridge University is a highly prized client within the construction industry with the prestige reflecting upon those involved with it. It was therefore vitally important to us when an opportunity arose to tender for its Plant Growth facility on the Botanic Garden in Cambridge that we secure the contract with the intention of securing a long-term relationship. The project was procured using the University's 'Develop and Construct' procurement route. Our own business had engaged Sir Michael Latham as a non-executive director and as a result had adopted collaborative working and a team approach which married well with the University's desire on this project. The early engagement gave us a greater input but also a greater responsibility to make the project a success. We worked closely and diligently with the designers, supply chain and the University project management to deliver a project that was truly excellent. The project was of good quality, was value for money and was well coordinated and constructed due to the high level of engagement of all parties involved. The project put greater demands on us in terms of the engagement required but this was all worthwhile due to the smoothness of construction. The team demonstrated a strong cross discipline bond that overcame the technical challenges with speed and efficiency. Having learnt from this first scheme we successfully delivered subsequent schemes for the University; many of the team's relationships have stood the test of time, as has the procurement route with quality of outcomes.

In 2000 there was an illustration of the benefits of early contractor appointment during a visit by the Minister for Construction Nick Raynsford MP (MA Cantab): when he asked a particularly blunt-speaking contractor manager what he thought about 'all this new-fangled procurement', the reply (those present stiffened, fearing what might be a string of expletives) was 'Well, when I started this job I knew a lot more about it than I've known when I've finished most jobs'.

The University followed voluntarily the rules for appointment laid down by the EU, as set out in the Official Journal of the EU, OJEU. In 1997–1999 the University sought to establish in EU courts that it was exempt from EU tendering rules on the basis that it was more than 50 per cent privately funded and therefore not in the public sector. Finally after much cost and effort the case was won, but that outcome was not to be relied on in capital procurement because it just needed a dip in any one year below the 50 per cent level and the University would become liable; the contractual liability chain in construction projects in the pipeline would be nigh impossible to unscramble in time, leaving the University open to legal claims of unfair competition. In any case, there were clear advantages in tendering widely for design and construction, especially as there was a suspicion that there might be 'ringing': a tacit or overt agreement between designers and/or contractors to keep cost rates above average. The Director of Estates therefore decided to continue to follow EU procedures, despite the time and cost of wading through the large numbers of applicants. From the large number of applications from contractors for a job, through OJEU, about 20 or so were sifted out, and then about four applicants were interviewed, and an appointment made on the quality/price balance described above.

Post-Egan the use of 'Framework Agreements' became more popular among clients: after assessing possible designers or contractors, an agreement is drawn up whereby companies, typically three or four, are put on the framework; thereafter clients do not need to go through EU procedures but can tender directly. Some universities chose the framework route; Cambridge decided after much consideration not to set up a framework because the benefits did not justify the costs as the University had set up a good list of trusted designers and constructors who could tender competitively. Nearly all practices/companies appointed were British-based. Recent mergers and take-overs have resulted in an increasing proportion of UK companies being foreign-owned. Davis Langdon, for example, is now part of the US giant AECOM which has over 100,000 employees world-wide. This will accelerate as Chinese start taking over construction as they have for utilities such as nuclear energy.

Quality and fitness for purpose achieved

Regarding the quality of designers appointed during the decade, Richard Saxon CBE, member of the Buildings Committee since 2004 and former Chairman of the architect-engineer practice BDP, commented,

> Apart from their functionality on opening day, the value achieved by the buildings built over the last two decades lies in their flexibility and brand value. The buildings are for the most part spatially generic and able to adapt to changing requirements over time. Their character is on the other hand sufficiently attractive to create loyalty from users and to attract donors willing to fund further new buildings.
>
> Whilst the low density and separateness of the recent building stock reduces opportunities to share space, it increases the clarity of identity of each department and of the University as a whole. On a scale of 1–5, the suitability of buildings in the University during this decade would score about 4, and they are used well by their users (despite low sharing rates). It is fair to say that by the standards of HE buildings, those designed and erected 1996–2006 represented suitable quality and fair value for money. Designers were made to work hard for the University as a demanding

client but were fairly paid and shielded from user-driven changes. Novation to the contractors at the second stage of design–build did not inhibit relations between designers and client as the professionals expected to be re-engaged by the University if they did well. The University believes in the value of good design, if you define a building design which achieves its various goals as a good design.

That said, it was difficult to show correlation between money spent and quality achieved on the one hand, and actual performance of the new building on the other. It is relatively easy to correlate the level of capital cost against hard operational costs such as utilities and maintenance (through the life–cycle analysis described above), but it is extremely difficult to correlate 'soft', less quantifiable performance such as staff productivity, attractiveness to potential staff being sought retained, academic results of students, etc. Chapter 9 describes how a successful bid, fronted by Cambridge, to Hefce funded a study, carried out by the University of the West of England, to seek such correlation, but the results were insubstantial. Since then, further research relating to schools is set out in the article 'Added value of good design' by Dr Sebastian Macmillan.[30]

- There is strong, consistent evidence for the effect of basic physical variables (air quality, temperature, noise) on learning, but that once minimal standard are attained, evidence of the effect of changing basic physical variables is less significant.
- There are forceful opinions on the effects of lighting and colour, but the evidence is conflicting.
- Other physical characteristics affect student perceptions and behaviour, but it is difficult to draw definite, general conclusions.
- The interactions of different elements are as important as the consideration of single elements.

For the higher education sector, CABE's *Design for Distinction: The Value of Good Building Design in Higher Education* reports that there are at least links between building design and the recruitment, retention and performance of staff and students.[31] Fifty articles are reviewed, and five new case studies are reported.

The matter was of concern to the Estates Department because the questions always have to be asked: would we have got just about as good a building if we'd spent less money (and used that for some other building); would we have got a much better building if we'd spent just a bit more? That was an issue taken up by HM Treasury/NAO from 2006, especially in the procurement of Olympic venues (whose procurement principles, as set out by government, was similar to those [that] developed during the CMS project at Cambridge; see Chapter 9).

Capital budgeting and cost control

Until 1998, projects had proceeded by agreement of committees on the basis of a 'cost-estimate' which was updated (in practice, increased) as the project progressed. Further funding was found (a process known as 'goat-bag funding' from the old Greek habit of burying goat-bags of money for hard times) as design changes and additional requirements were brought in by the designer or by the future building users. The word 'change' suggests a lack of foresight or lack of clarity, or early design error, and sometimes that was the case; however, change was more often the result of changes in the research that was to be carried out, especially in the fast-moving sciences or when, because of the new building project or otherwise, researchers from elsewhere suddenly became available.

There had been little use of metrics setting out unit cost (such as capital or running cost per unit area of built space provided), as the Egan Report had correctly recommended, and hence there was little benchmarking against similar projects elsewhere in the UK or overseas. There being little sanction against rising costs, procurement staff, designers and contractors, knew that costs could rise and be met, so a local design/construction economy grew on that basis. From 1998 onwards only in the early design stages did the project progress by 'cost-estimate': as soon as concept/outline design was agreed, then a budget was set, and generally remained as the maximum project cost allowed, and there was an emphasis on compliance with budgets. From 2004, every project was benchmarked, £/m², at an early stage, even at the stage of appointing designers (and sometimes the out-turn unit cost expected was set out in the designers' contract of employment as a guide); this benchmark was measured both in terms of £ per square metre of gross and useable ('assignable') built space provided because assignable space was what mattered most to building users while it was the gross area that has to be maintained and heated. A metric-based benchmarking system was later adopted by the Cabinet Office as one of the main criteria for approving public sector projects.

One of the factors in budget-management is professional pride and morale: there is professional stigma in coming in over budget, so therefore always a temptation to set higher budgets so as to preserve morale and avoid criticism from the committees. There is an old adage that 'any fool can come in within budget if a bigger fool gives him enough budget'. A favourite habit was to dribble in a measure of contingency into many elements of expenditure and then also apply an overall contingency to the bottom-line figure: it took a lot of time to identify and eliminate that habit. A technique picked up from an American university was to set benchmarked budgets tight enough such that from the risk-analyses there was a statistical probability that, say, 20 per cent of projects would come in over budget. That drives down actual unit costs. There had to be sufficiently good management and leadership so as to be able to maintain pressure on project managers to stay within budget while also accepting that such a percentage will on statistical average come in over budget. (It is interesting to note how many projects came in 'smack on budget': this is generally because projects coming in under budget were liable to professors or departmental secretaries quite reasonably wanting extra/nicer furniture as 'the money was there to be spent').

There was a feeling in the late 1990s that there might not be full and open competition between rival bidders for work around Cambridge. There was no specific evidence, and certainly there was never during the decade any suspicion or allegation of any graft or corruption; just the feeling that prices seemed higher than one might expect compared with national prices even allowing for higher standards at Cambridge in the University and Colleges. There was informal notification relating to that to the Office of Fair Trading; subsequently, in 2008/2009, and probably unrelatedly, there was a series of prosecutions across the UK construction industry for 'ringing' between contractors, and some large fines were imposed.

By 2003 there was growing acceptance that there needed to be a high level authorisation system to ensure that there was a full and viable business plan out to at least ten years before the project could be authorised; the Capital Procurement Process was established and taken on by the Planning and Resources Committee. This significantly improved sustainability.

In 2005 an internal study was carried out on total project cost (including tax) per square metre with reference to construction inflation figures from the Building Cost Information Service of the Royal Institution of Chartered Surveyors (BCIS; RICS). Throughout most of the period studied, 1995–2004, construction inflation was running much higher than RPI. Discounting compounded construction inflation, the cost of new non-lab buildings (less VAT, furnishings/fittings and off-site infrastructure) fell in real terms from £4,020/m^2 in 1995 to £2,950/m^2 at first quarter 2005 prices according to a regression analysis carried out through the Maths Department. Prices on the same basis levelled out 2005–2006 and over the decade the overall real-terms reduction was about 15 per cent.

The unit cost figures in Figure 5.1 are in 'real terms', that is, taking out national, official construction inflation figures; BCIS figures were used for construction inflation, which over the 'boom period' 1985 to 2005 was up to 2.5 times national inflation (RPI).[32] For lab buildings, it was found to be impossible to do better than keep real-terms project costs flat because of the ever-increasing Home Office legislation requirements for enhanced 'Health and Safety', some of it as promoted by the UK Health and Safety industry, much of it resulting from (and usually exceeding) EU requirements (much, but not all of it, sensible). The procurement management systems adopted reduced waste; the programme during the decade was achieved within a fraction of 1 per cent of the total of set budgets. Budgets set became steadily tighter in real terms with closer bench marking, greater team-work through 'develop and construct' procurement, much lower legal costs and higher assignable/ gross area ratios. Considerable economies in fee bills came from bringing much more work 'in-house'.

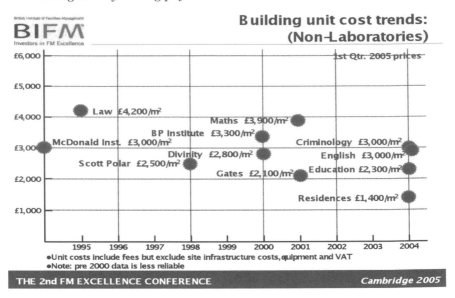

Figure 5.1   Unit project costs per m² gross at normalised 1Q 2005 prices

As well as the amount of payment, there was the issue of payment cash-flow down the long supply chains: how fast the University paid its bills and how fast sub-contractors got paid. As a matter of policy, it was agreed in contracts that invoices should be paid within 30 days, partly because that was a fair way to do business with companies that needed predictability, partly because it helped to get slightly cheaper prices and partly because late payment down the supply chain is one of the biggest sources of disputes. By setting the example and by monitoring onward payment down the supply chain, late payments never became a problem for the University. In the late 1990s there were trials elsewhere of project bank accounts (pba) whereby client payments go into a bank account with note of how much is due for each of the sub-contractors: those payments are made automatically unless intermediate payers can show that payment is not in fact justified. This was popular with sub-contractors, but by no means with all members of the industry; the University did not trial pba, though maybe it should have. The Highways Agency, for example, embraced and improved pba. (In the procurement document '2012 Commitments', which set the context for procurement of the London Olympics, pba were set to be used unless there were reasons not to do so: in practice, they weren't).

During the 1980s VAT came to be applied to Higher Education capital projects although that was never intended by the government. So, late in the 1980s it was agreed, first at the University of Bristol, that universities could set up 'holding companies' (that in Cambridge was set up as Lynxvale) which nominally ran major contracts; the directors of these companies had the duties of directors so they did hold meetings to 'oversee' procurement of the larger projects even though most directors were earlier involved in the decisions being overseen. That added a layer of bureaucracy and sometimes caused irritation, but was necessary.

The system of keeping projects under cost and programme control included monthly project reports with the details of all current projects, including size, key dates, total and unit costs, details of designers and contractors, progress and prediction. These reports went to members of the BC. At the end of each calendar year there was an analysis of unit and total costs against budgets for all projects completed in that year; this was also circulated and briefed to BC.

Net budget compliance for projects stayed on marginal under-spend over the years reviewed in this book: for the £750M spent on just over 100 projects documented in this book for the period 1996–2006, over the decade, there was a net fraction of a percent (0.1 per cent) under-spend: within that, some projects were under and some over, and as it was rarely possible to transfer money from one to the other, in practical financial terms the picture was less good than was the overall balance, though greatly better than the national average. There were a few projects that hit serious snags, and the need to resolve cost disputes; four had notable problems. In all those cases the users of the completed buildings judged them to be well fit for purpose, but all four were behind contracted programme (though in some cases ready in time for planned use having used up all the 'float' time reserve) and were above budget, by far the worst being 15.1 per cent over budget (Faculty of English);[33] it is notable that contrary to general University practice adopted in 1999, each of those four projects used forms of contract other than the NEC/ECC. Chapter 12 will discuss whether there is in fact a causal and significant correlation between choice of contract form and probability of success of a capital project. What was and is generally agreed is that the early involvement of the contractor during the design and package tendering process generally resulted in better and more cost-effective procurement. Details of budget/cost for all projects are set out in the project chapters, and value for money, in terms of capital and whole-life costs, is explored in Chapter 12.

Risk management

How the set budget and the actual out-turn cost correlate depends largely on how well risks are assessed, quantified and managed: the evaluation of the long list of things that can go wrong or simply turn out worse than expected (typically, underground conditions, the weather, sub-contractors going bust) and who should carry the various risks, and how much they should be paid for taking them. (One of the common failings in Local Government procurement is not to recognize and pay for risk). This takes some skill and experience and the public sector is often bad at it, in the worst cases leading to naive contracts and then legal disputes which are hugely expensive to the tax-payer or rates/buyer rate-payer, and rarely with blame being trailed back to those civil servants or councillors responsible. (Humorous cynics say that any project in the public sector goes through six stages: enthusiasm, disillusionment, panic, search for the guilty, punishment of the innocent, praise and honours for the uninvolved.)

One important issue in risk management was to assess how to set a sensible, total contingency sum (typically 6 per cent upwards depending on the risk profile, plus about 0.5–2 per cent written into the contract), consolidating all the 'goat-bags' of spare money previously embedded all over the place in project cost-estimates into one initially set and maintained contingency. Risks include sub-contractor collapse, necessary design change, human error, unexpected ground or material condition, or exceptionally severe weather. The risk register, regularly updated, set out the amount risked for various eventualities, who was to carry that risk, and the percentage probability of it happening (assessed by straight analysis or by statistical 'Monte Carlo' probability assessment). The level of contingency set reflected the risk register, on the project risk profile, and also on the risk to be accepted of going over budget.[35]

Risk was substantially reduced in the 'develop and construct' form of procurement, not just because the constructors were involved in design and there was a better team-work approach. A key matter was that sub-contract packages of work were being tendered as design proceeded, and a clear picture of the total contracted cost built up. The decision to proceed to second-stage, the construction contract, was made when the total of works package tenders reached about 85–90 per cent of the total works; that is, there was about 85–90 per cent cost-certainty on the main-contractor contract.

A Hefce grant in 2003 was won by Cambridge University to develop a process of risk-sharing between the University and its consultants by under-writing design mistakes: no one is infallible, so some mistakes are inevitable; that simple, statistical fact should be faced and paid for through a risk evalu-ation which is costed on the basis of attributing risk to those best placed to get the matter right and allocating money for the expected consequence times the probability of it happening. A main recommendation was that there should be a 3 per cent 'error allowance' on projects to underwrite errors, with unspent money being distributed at the end of the project between those involved: professional reputation protection and peer pressure meant that the fund was rarely spent and it did considerably reduce the blame culture long endemic in the industry, and hence increase team-work and mutual respect. The contin-gency percentage could be reduced to take account of the 'error allowance'. Later, the practice slipped away. (The move to digitally based building infor-mation management, BIM, systems has reduced the incidence of human design errors).

A new aspect of risk management came to the University in the late 1990s in that there was such a large number of simultaneous projects that they had to be managed as a 'programme' rather than as a series of individual projects being managed separately. A programme can be defined as a collection of inter-linked, complex, sometimes geographically separated, projects with the same overall objective and generally the same end-client. A programme has to be managed in an integrated way because the risk profile is quite different, not least because if one project in a programme goes pear-shaped it sucks in already tightly-drawn resources and expertise, and that jeopardises the other projects which in turn go wrong. Because there were no critical building procurement crises during the decade, most senior people in the University made a general assumption that all would go well, not realising that some of the complex set of risks in each project are down to chance or matters outside control of the project managers or designers, and how the overall programme could implode because of one project in serious trouble. As it happened, it didn't.

One aspect of risk is the amount of exposure the client and others in the design and construction team are willing to take to achieve innovation. Most clients in the UK construction business are singularly averse to innovation if, as it usually does, it involves some risk. The University made it clear that it would accept some innovation risk, though that commitment was often weakened as projects were pared down to meet capital project budgets by value engineering. Again, a remarkable example of UK clients and construction at their best is found in London 2012 Olympics; while the project came well under budget as noted above, there was a huge amount of innovation especially in sustainability measures of materials and logistics. The lesson had long been clear: contractors should be employed intelligently around a sound risk analysis and not just against base-line capital cost and schedule. (Local government councils often got that wrong).

Some elements of risk could be and were reduced by a switch towards off-site pre-manufacture, as in the construction of the Computer Lab, for example. The industry was shifting, very slowly, from a site-based to a factory-based industry. This was accomplished by some progress towards standardisation of components, though the UK still lags behind the USA and most countries in standardisation; at a recent conference a speaker admitted that his estate had 472 different types of light bulb, an extreme case. The move, albeit slow, to better standardisation also helped reduce maintenance costs and hassles.

A main principle here is that mistakes are to be learned from, not hanged for.

Project management after project completion

Soft Landings

Traditionally, after the construction period, when the architect (and from 1999, the client and RU also) agreed that the work on site was sufficiently completed, the department would move in and the last workers leave the site. As further inspections continue, defects are identified for the contractor to return and remedy, and any disputes about demands for extra payment get settled. One of several initiatives which emerged in the University in this decade was 'Soft Landings' (SL). SL was developed in 2000–2003 by the University with an architect of the practice RMJM, Mark Wade. At its simplest, it is agreed from the very start of a project that after building hand-over representatives of the designers, generally a mid-career architect and someone from the contractor company, stay on in an office in the building for a month or two, working with the building users to show them how best to optimise how the building works and sorting out problems (and otherwise getting on with their normal work). SL led to a significantly higher incidence of problems being sorted out by those two, without recourse to contractual remedies and arguments as to who is to blame, a common feature of construction projects. This arrangement also leads to a better comparison between building performance as designed, and how well it actually performs in practice, a key and long under-addressed aspect of UK building procurement, recently picked up by the Cabinet Office in its guide/requirements for better-value public procurement.[35] The two who stayed with the building under SL learned a lot themselves, and all members of the team tended to work better from the start because they knew that there would be some responsibility for building performance after it was handed over; the usual practice had been, and often still is, to thin down staff to zero by hand-over and then for every one working on the project to get straight into other projects. Furthermore, and importantly, SL was shown to be cheaper for all concerned than the usual practice of dragging back designers and builders when it is thought that there is under-performance.

The system was first developed after the first phase of the Maths project in time for the new building for the Computer Laboratory, designed by practice RMJM who had tried on-site building-optimisation on a previous job. The University Estates Department got financial support funded by a successful bid to a Hefce fund for the spreading of innovation in good practice and, with professional time and £8k given by invited consultants, design companies and construction lawyers to work up the scheme into a form that could easily be agreed within and at the start of a contract. The University website described Soft Landings as a means of getting some of 'the benefits of PFI thinking for non-PFI projects': designers and constructors thinking about how the building will actually operate in practice after handed over to the building users. (The University managed to avoid using PFI procurement except for one project for which the NHS was the senior partner so the University could not avoid PFI, and that experience was as bad as expected. In the mid-1990s all universities came within an ace of having to use PFI across the board, as NHS had to: it took an emergency delegation from the Association of University Directors of Estates to Westminster to stop that when there was a leak of news about universities having to go PFI.)

'Soft Landings' procedure spread from Cambridge through the UK public sector. The idea became more popular after its promulgation by HM Treasury Office of Government Commerce and various best practice organisations. It became official UK Government policy in some Departments around 2010, in a form made longer and more comprehensive, and more expensive, after it was taken over and developed (some say over-developed) by the Building Services and Research and Information Association,[36] and spread across all government departments by Cabinet Office requirement from 2012. Soft Landings has spread across many other countries such as USA, reaching Australia in 2010 with a launch which was advertised as follows:

> Soft Landings asks us to think about design and delivery of sustainable buildings, but throughout their lives, not just when they are built. This concept of on-going care of a building's sustainability is gaining traction across the developed world.[37]

Cambridge University never tried to control or take any intellectual property rights over Soft Landings; some say it should. As the University had not protected the name, it never got wide credit for inventing and developing Soft Landings, but that matters little. As former US President Harry Truman said, 'it is amazing what you can achieve if you do not care who gets the credit'.

By 2013, some leading companies were developing SL into even better systems for optimising the use of a building to reduce running costs and CO_2 emissions. In May 2013, the RIBA Plan of Work, which sets out for all across the industry the main stages in the work of an architect through a project, was updated and re-issued, the main change being the introduction of 'Stage 7 which embraces further duties arising from post-completion and post-occupancy evaluation activities', specifically Soft Landings. It took a long time, but the professional practice of 'doing' a building and then swiftly moving on to the next, which many saw as the main obstacle to industry improvement, was to a degree reformed by the need to optimise building performance, and thereby to develop knowledge and capability of designers and builders and clients and building-users.

Feedback of lessons learnt: post-occupancy evaluations

There is still a very long way to go in learning by experience in the construction industry. It is often said that one of the main weakness in the design and construction industry is the reluctance of practitioners to assess and promulgate and learn lessons from construction projects, and that is true; the UK construction industry is notably less good than those of many developed countries in picking up lessons, and also in picking up opportunities from academic and industry research. In an effort to remedy that, a system of post-occupancy evaluation (poe) was brought in by a few progressive clients through the early 1990s. The Faculty of Divinity project was the first Cambridge project to be subject to a poe. Poes are done a couple of years after occupation of the building so that the building has had a chance to 'settle in' and be run efficiently. The poe as used in the University was based on four half-day sessions: the first, attended by the client, the Representative User (RU), planners, neighbours and users, considered whether a new building was needed at all, whether it was in the right place and of about the right scale, and the effect on the local environment. The second, attended by the client, RU and the leading designers, consider the design process: how was it done and how could it have been done better? The third session is concerned with the procurement and construction phase, so is attended by the client and RU, and the constructor, and the leading designers. Finally, the last half-day considers how well the building was functioning operationally for the users and how efficiently in terms of energy/water/maintenance requirements. This system was developed by a joint national committee, known as the Higher Education Design Quality Forum, which had been set up by Dr (and US Prof) Frank Duffy, President of RIBA, and the then Bursar of the University of Bristol as a joint committee of RIBA and the Association of University Directors of Estates (AUDE), and supported by Higher Education Funding Council for England (Hefce). This form of poe was recommended by Hefce, and for a while its use was a condition of Hefce funding, until unfortunately that ceased to be mandatory, a casualty of a campaign to achieve what was stated to be 'a light touch policy approach by Hefce'; that was a step backwards for sustainability in universities.

Poes were carried out in Cambridge 1999–2006 on all capital projects over £1M. The reports, which went to the University committee system via the Buildings Committee, were drafted by the Estate Management and Building Services department project manager in consultation with all attendees of the four sessions, and, crucially, were co-signed by the Director of Estates as the officer personally held responsible to the University for the project, and the RU. Crucially, all poes for the decade were publicly available so as to allow all to learn, within and beyond the University.

Management of transport needs

Each year the University developed a travel plan. The central aim was to encourage and help those who used bicycles or public transport, or walked to and from work. The Citi 4 Bus service was set up and largely funded, and the amount of on-site car parking was steadily reduced accordingly, and as building land was needed. In 2005, there were 1,914 parking spaces on the city centre sites/West Cambridge, plus 129 at Addenbrooke's. The average one-way commute distance was 5.5 miles. The modal split was as shown in Table 5.1. Bicycle use was, and is, increasing at about 1 per cent per year on average.

Maintenance

By 2005, the University had an estate of some 650,000 m^2 gross built area, 350,000m^2 net assignable, in its operational estate, valued at £1.2bn replacement cost, plus a 'non-operational' estate (including 338 staff residential units) valued at £41.5M achieving an income of £1.5M *per annum*.[38] The operational cost of running the estate in 2005 was £24.4M, including the cost of utilities of £4.9M. The backlog maintenance cost was assessed as being £15M which was low compared with many research universities. The

Table 5.1   Transport to work pattern 2005

Mode of transport	%
Cycle	38
Drive alone	27
Car share	10
Walk	9
Public bus	8
Staff bus	Not yet recorded
Train	4
Motorbike	1
Telework	2
Unrecorded	1

assessment of building condition for the operational estate, carried out externally using the RICS scale and the operational condition assessed by academic and administrative staff were as set out in Tables 5.2 and 5.3.

For 2003/2004, 81 per cent of University buildings were rated as being in Condition A or B according to the RICS definitions; for the Russell Group that figure was 74 per cent.[39]

The Hefce-recommended level of maintenance funding was 1.5 per cent of the depreciated value of the estate. Generally, at Cambridge less was spent: 1.33 per cent in 2001–2002 for example. However, in some refurbishment projects, heavy maintenance was picked up as part of the project and hence capitalised. Early in the decade maintenance budget was around £24M with a further £4M for minor works.

It is worth noting changes during this decade in how the estate was maintained. Until 1999, the various trade groups, plumbing, electrical and the traditional building skills such as carpentry, brick and stone work etc. (the

Table 5.2   RICS building condition 2005

Condition	%
A. New or highly satisfactory	14
B. Satisfactory	64
C. Operational; some work required within 2–3 years	18
D. Substantial work required	0
Not classified	4

Table 5.3   Assessment of operational condition 2005

Condition	%
A. Excellent	20
B. Very good	70
C. Satisfactory	10
D. Unsatisfactory	0

name sometimes given to these trades is 'the Biblical' trades) were separately located, had different conditions of employment such as different rates of pay, holidays and working hours, and each followed the long-held traditions of their respective trades unions. Lack of co-ordination and communication led to lack of collaboration and was a considerable source of irritation around the University. In 1999, all the maintenance trades were brought together under one departmental head in much better accommodation on the west and then the south-west side of Cambridge, and with unified conditions of service and pay rates; it was not an easy process to initiate but it made a huge difference to overheads and to response times.

As a check on value for money, every few years maintenance work was tendered externally and compared with internal costings; it proved to be cheaper and better to keep the service in-house, largely because VAT was chargeable if contracted out, and the enormous cost of getting companies to guarantee swift responses in times of emergency out of normal working hours. In 1999 SLA (Service Level Agreements) were introduced after the Department negotiated them individually with all those throughout the University with whom it worked, setting out agreed duties, ways of working and standards to be achieved by all parties; that cleared the air. SLAs were generally popular with departments, and certainly clarified what level of service should be expected. There was some fear in other administrative departments that they might be pressured into developing SLA; that fear was soon justified.

Running-cost budgets; design for reduction for energy-use and carbon dioxide reduction

When the term 'budget' is used, most people reasonably take that to mean 'capital budget'; setting and holding to a capital budget is a basic and crucial management device used since time immemorial. Running costs, including utilities such as energy and water, plus maintenance, cleaning, grounds maintenance, insurances, are less apparent. Knowledge of the ratio of running to capital costs, 2–3/1 over about 20 years for university buildings (less for highly serviced buildings),[40] came as a surprise around University staff, who considered buildings to be 'theirs' if they had been instrumental in raising capital funding. This ratio was even more disturbing to central University administrators charged with running the University's finances while watching a fast-growing estate. As soon as the cost and ethical consequences of emissions such as CO_2 became realised, there was increasing emphasis on 'whole-life costs'. From 2000 onwards, the University started moving to a requirement for total costing over 10 years as a condition for project approval. In a speech on 25 September 2006, Sir John Bourne Auditor General and head of the National Audit Office commended the work of the Office of Government Commerce (see *Achieving Excellence Procurement Guides*);[41] he also noted that the recent NAO report 'set out the key dimensions of construction good practice including investing in well thought-through design and sound decision-making processes based on whole life value of the project rather than just the initial costs.' The subsequent press release said that, 'he will look for evidence that this is the case in Government construction': a fundamental change in public procurement.[42]

After the government set a monetary charge, albeit at a low rate, for production of carbon dioxide by larger users of energy, whole-life project cost included the cost of CO_2 emission. As a result of that, and of steadily increasing awareness of the social responsibilities of a university in minimising environmental damage, there grew an imperative in the procurement and the running of buildings to reduce the 'carbon footprint' of the building. In the 2003 Estate Plan there was a commitment that all new buildings should be at least rated as 'Very Good' in the Building Research Establishment Energy Assessment Method (BREEAM), and in December 2004 a policy to achieve 'Excellent' rating when sensibly possible was agreed by the Buildings Committee. BREEAM was and, to an extent is, taken as a relatively good reflection of the amount of environmental damage a new building will do; the US system LEED (Leadership in Environmental & Energy Design) is held by some, an increasing number, to be a better yardstick. LEED was developed by former BRE staff; BREEAM is generally thought to be somewhat arbitrary in its definitions and not good in how it relates to actual building performance: LEED is better in that regard.

A report carried out by UCL for the government led to the establishment of the on-line analysis tool Carbonbuzz for, *inter alia*, comparing actual to designed energy use in $kWh/m^2/year$; for some hundreds of education buildings.[43] The ratio was 1.48: that is, nearly half as much again as designed and predicted, and budgeted, a significant problem for University managers, clients nationally, and for the environment.

Increasingly, the matter of sustainability was becoming part of the design brief. The 2002 Estate Plan set higher priority for sustainability, and therefore energy reduction, and so a policy of managing design and procurement on a 'whole-life' basis was adopted to optimise the total cost of a new building over its life, rather than just to minimise the initial capital cost and carry higher running (e.g. energy) costs: optimisation of total cost in discounted cash-terms over a period of 10 years was introduced and no major project went ahead without that being documented to the Planning and Resources Committee. Research noted above[44] showed that for a building such as a university building, per pound of capital cost, of which about 10p is the cost of design, the cost of running the building (mainly maintenance and utilities) over a 15–20 year life is £2–£3; so the main lifetime cost of a building to a university lies not in its construction (which in any case, is generally paid by public funding or benefaction) but in the subsequent cost, which is paid for from the University's current account, very rarely by external sources. Devolution of departmental budgeting took another step forward from 2008 when the amount of electricity used was compared against a rolling average of previous years and the value of under-use credited to the department.

The move from procurement by initial cost to procurement based on a ten-year life had at least some effect on energy/carbon reduction, though given the rapid increase in the size of the estate that effect is difficult to quantify. In 2005/2006, as part of a Hefce-funded programme, the Carbon Trust carried out a review of carbon emission management in universities, including the report *Cambridge – Higher Education Carbon Management Programme, University of Cambridge Implementation Plan*. That report set out the uses of the various types of fuel and of water on the main University estate. The report noted that the unit consumptions of water and of oil were both declining quite notably, while gas and electricity use were both increasing, reflecting the following factors:

- Increasing size of the estate
- Recent new buildings having been highly serviced accommodation for chemical and biological sciences – and for engineering
- Conversion of boilers
- The success of the water conservation programme.

The total tons of CO_2 rose from 50,064 in 2002/2003; 52,661 in 2003/2004; and 55,728 in 2004/2005. That added to concerns both about the effect of that level of emissions on the environment, and also costs of buying the energy and of paying for Carbon Reduction Commitment. The section of the report concerning projections, started as follows:

> The Estate Plan 2005 addresses the question of the financially sustainable size of the estate, allowing for both the initial capital costs and the ongoing operating costs of accommodation. It concluded that the operational estate was broadly the correct size in total extent, but was already at that limit of sustainable size, allowing for the active capital programme already in hand. Future additions to the estate should therefore normally be accompanied by compensating disposals, and the emissions forecasts reflect this anticipated slowdown in the growth of the estate. It is also worth noting that although buildings may be expected to be more energy efficient, they are often more highly serviced and the buildings of more space, meaning that their total use may actually be higher.[45]

The report noted progress in upgrading building management systems, saving £150k and 1,130 tonnes of CO_2 per year. That programme was steadily ramped up.

Table 5.4 shows the changes in production to CO_2 as reported by the SQW report on sustainability in HE commissioned by Hefce.[47]

The increases in total energy- and water-use were resulting in 20–25 per cent increases in cost by the end of the decade (20 per cent up in 2005–2006 and 24 per cent predicted for 2006–2007).

Table 5.4   Changes in production of CO2

University	1990 total ton CO2	2005 total ton CO2	2005 ton CO2 per staff or student in fte	2005 kg CO2 per £ income
Bristol	25,210	43,993	2.124	0.154
Cambridge	41,250	69,858	2.696	0.078
Oxford	39,677	63,956	2.422	0.105
UCL	32,957	35,537	1.431	0.063
UWE	8,146	14,301	0.578	0.095

One problem faced by the University, and to an even larger extent by the Colleges, was the clash between sustainability and conservation. In England, there is a popular concept of conserving old and well-loved buildings exactly as they are ('preserving in aspic') except insofar as they need essential work as they get old; and there is a feeling, often backed up by legislation, that after such work they should look exactly as they did before. In some other countries, the USA for example, there is a clearer differentiation between four levels of conservation: conserving to the extent that no work is done, the building just quietly degenerates; repair as required keeping closely to original appearance; repair to keep the function of the building as it was but allowing changes to keep the building to current building standards; finally, allowing such changes so as to allow its viability to its owner as a going concern. Often, there is a real requirement for refurbishment of a building for sustainability, accountability or for maintenance reasons which is against the rules set for conservation and/ or for disabled accessibility by the Local Authority. Indeed, sometimes regulations for sustainability and regulations for conservation were and are mutually exclusive: University staff would sometimes tell respective City officers to talk to each other and come up with a solution, but that was never easy and sometimes not in fact formally possible. By 1996, of the University's 300 or so buildings, 52 of them were 'listed' (that is, nationally protected), 4 of those at Grade 1 (in addition to the very many Listed Buildings in College estates).

Procurement staff issues

Training

A training fund was established in the Estates Department, and Continuous Professional Development (CPD) registers were set up for all staff. Each member of estates staff had training objectives each year, and funding was provided for staff seeking approved external training/qualification: 40 per cent of course fees at the start and 40 per cent on successful completion of the training. The biggest training issues included how to think through the teamwork issues, particularly how to ensure that constructors make a real contribution to developing design and contract procurement, how to understand the relatively new and more managerial contract forms, notably the NEC contract, and how to assess risk in a business-like way. Twice-yearly liaison visits were set up with Oxford, which had an estate and a project schedule comparable to, if smaller than, Cambridge, so as to exchange experiences and ideas, and in practice to offer mutual comfort as similar experiences were shared. Liaison was established with several Ivy League universities in the USA; the Director of Estates visited some of them most years to exchange benchmarking and procurement techniques (in Ivy League universities, by 2001, building costs were slightly higher than in Cambridge at an exchange rate of $1.57/£) and he lectured students of both universities under the CMI (Cambridge Massachusetts Institute) programme. Senior staff of Cambridge University and MIT made exchange visits each way in July 2002, funded by CMI.

The University promoted staff training in companies and practices with which it worked. In May 2000, as new procurement practices were working through, in particular with early, collaborative contractor involvement and close definitions of respective roles within a procurement team, the University led and jointly won a £36,000 grant from Hefce to develop 'progressive hiring (procurement) practices'. The grant was to five universities led by Cambridge, to incorporate training of contractor/sub-contractor employees working on University projects: the selected area of research was the difficult contractual issue of how to encourage and reward good practice within contracts, particularly training of staff at all levels. After a great deal of discussion with contractors, construction lawyers and the Construction Industry Training Board (CITB) a guide was produced.[47] One example was that there was an element of money in 'Preliminaries' to pay for agreed training; in one contract that included part-payment of a Masters' course for the contractor's site manager. One contractor was helped to do a PhD on net profitability relating to different procurement routes.

Training brought more confidence and improved morale, and those made for more robust clientship; all estates departments take a lot of criticism from academics, much of which in fact relates to the performance of the construction industry, and better training helped estates staff to understand and to communicate problems of the industry and of themselves to academic colleagues.

When the Cabinet Office in 2011 set up the Major Practice Authority, its objectives were to capture best practice, learn from it, and feed it on to others.

Managing staff changes

To achieve such fundamental changes, it was necessary within the Estate Management department to assess who were in which of the four categories relating to change: those who would lead change; those who would support it; those who would change because they were told to; and those who would seek to subvert or block change. In October 1998 all members of the senior management team set out how they saw their objectives and challenges. In his opening talk to all estates staff, the new director set out the need for change under the headings of: the culture of the department (in particular a need for 'openness'), internal and community communication, input from external 'lay members', roles of those in the University being served by the department. The low time-ratio between thinking/acting on one hand, and writing/delivering reports on the other, had to change. That ratio was guessed to be about three times worse than that in most other universities' estates departments, with staff having to write frequent, long and procedurally-based reports as 'committee spake unto committee'. Of the eight members of the senior management team in the Estates Department, two who were nearing retirement took the University's generous early retirement package, and one moved to a post in another department, albeit some 15 months later.

In recruitment as in all matters, there was a strict policy of avoiding any form of positive or negative discrimination, despite some pressure otherwise.

Staff numbers

In December 1999, £280M of projects were in planning/construction; by September 2003 that had grown to £675M: at the peak, the University was spending over £0.5M per working day on buildings, capital, recurrent, rentals and staff costs. While staff became more productive and more accountable, the sharp rise in the number of projects and the requirements for accountability and health and safety required staff increases in planning, construction, maintenance and utilities sections. In 1998 there were 156 staff in the Estates Department; by 2004 numbers had grown to 232. Annual salary costs for 2002/3 totalled about £5M. Average numbers from the industry working for the University at one time grew to a peak of about 1,200, with the associated benefit to the local economy.

Community relations

Another aspect of building procurement was community relations, the importance of which related not only to the general relationship between the University and local people and their elected Councils, clearly an aim in itself, but also to how easily building projects could proceed. Local opposition, which became very rare, caused delay, and sometimes costly re-design. Of about 500 planning applications made during the decade being studied, only one (the project to get a new building for research by a world-leading team into how to prevent or cure Alzheimer's) was not granted by the relevant Local Planning Authority, and one was 'deferred' at its first hearing (because of a disagreement with a councillor about railing design). The lesson learned quickly from a planning application for lighting of the University athletics grounds (see chapter 3) was that the estate department had to think through much better the effects of planned building projects on local people, and how representative and effective would be any potential opposition. It was decided to recruit a member of staff to develop community relations strategies, and to ensure that estates staff understood the need for local support and that local people understood University needs. The cost was trivial compared with the saving of many hundreds of thousands of pounds previously spent on planning consultants and planning barristers.

One important channel of councillor briefing and local community consultation about proposed new buildings was via Development Control Forums, initiated by the Council, with meetings arranged by the planning officers at which applicants could, relatively briefly, set out the main aspects of and reasons for the proposed planning application, and interested members of the public could respond. The presiding officer would then chair a controlled discussion. Councillors would listen throughout and would ask questions of any party. Most thought these meetings worthwhile.

Another mechanism for achieving local consensus in property matters was the setting up by the University of the Cambridge Landowners' Group (CLOG for short). This included most of those landowners who were working up development proposals. It met a few times, and did achieve collaboration, and lobbying power with the local councils. By 2004, it also began to make less difficult the long-attempted liaison with those seeking to develop the extensive land to the north-east of the Huntingdon Road at the National Institute of Agricultural Botany (NIAB). The University also set up a development group with the two local councils and NIAB to keep all informed about wider consultation and progress on moves to develop the NIAB and the University's sites in the north-west, and how they would inter-act, for example regarding public transport, and the provision of buses, schools and shops.

One useful means of spreading knowledge, within and beyond the University, of what was being done in capital procurement and what was being planned, was achieved through bi-annual 'Open Days' with exhibitions and guest speakers: those included Sir Michael Latham; the Director of Facilities in MIT with her team; the editor of the main magazine of the industry, *Building*; and many senior representatives of the construction industry. For example, the opening session in the 2004 Open Day was an hour's staged discussion between the VC and two current/former Vice Presidents of RIBA: Richard Saxon CBE, the senior partner of BDP, then Europe's largest architectural practice, and Richard Feilden, a King's College alumnus and senior partner of Feilden, Clegg, Bradley. (Within a few days of that brilliant debate, Richard Feilden was tragically killed while he was cutting a tree that split.)

Considerate Construction Scheme (CCS) as described above became a standard requirement of all University project contracts: in 2003, the construction company Amec won a CCS award for its work on refurbishing the Chemistry Department. Another initiative coming out of the Latham working groups was 'Architecture Day' when building sites were opened with talks and tours for school-children: that has, alas, since stopped nationally.

By way of overview: the methods of procurement in the University were recognised by RIBA Journal in a poll by a national magazine of leading industry firms in 2002 which rated the University as the sixth best client in a 'UK Top Fifty Client' listing. The following year the Building Construction Industry award panel judged the University the winner of the Major Project Award (for the development for the Maths departments). In 2002 the University of Cambridge was awarded the Contract Journal 'Client of the Year' award, the Highways Agency coming runner-up. The judges' comment was:

> Cambridge has made great strides in the past 18 months and its entry demonstrates evidence of good working relationships. It shows a good use of KPI's [Key Performance Indicators] and a very good demonstration of 'Rethinking Construction' principles ... Cambridge completed 17 projects on time and within budget last year.[49]

The development for the Centre of Mathematical Sciences won the 2003 Prime Minister's Award for Better Public Building. Many other projects won RIBA and other awards, several specifically for sustainability. Of the 101 projects noted as built in the decade 1996–2006 (or largely designed and built soon after), costing altogether £747M, none, except the two which were set up and largely built before the start of the decade (Law and the Judge School of Management), were subject to any legal proceedings. Annual summaries of the projects completed during the peak years 2000 to 2004 inclusive average out-turn costs, over-budget (+) and under-spends (-), were: +4.9 per cent, -1.35 per cent, -5.5 per cent, -0.3 per cent and +1.8 per cent; these variations between budget and out-turn cost compare well with procurement in the public sector, and indeed with much of the private sector.

The Report of the Council on the Financial Position of the Chest, Recommending Allocations for 2003–2004 noted at paragraphs 40/4145 that,

> The rate of physical expansion of the estate remains at a significant level. In the three-year period to August 2003 the estate will have grown by over 11 per cent ... The quality of new and refurbished built space has been generally very high as assessed by the users of the buildings, and by peer-review within the industry. Over the last three years £206M has been spent on large capital projects and in total the construction costs have been within budget ... As predicted in previous years, the significant growth in the estate has had consequential effect on the cost of maintaining it at an adequate level. Funding provision for maintenance since 2000–2001 has been barely sufficient. Maintenance backlog has grown but the condition of built space is still better than for the average for other research-intensive universities. To the increasing backlog of routine maintenance needs to be added the pressures of legislative compliance in a number of key areas: Access for people with disabilities; Fire safety; Trade effluent consent; Incoming water supply; Asbestos; Animal accommodation.

The 2005 National Audit Office report on public sector capital procurement in the UK, as noted above,[49] concluded that there was £2.6bn of avoidable waste annually in UK public sector capital construction procurement; the University was held up in the report as an example of 'best practice'. As directed by a Parliamentary Committee, this startling level of public waste led to much activity by HM Treasury's Office of Government Commerce, where a team was set up to establish procurement guidelines and have them implemented. The Director of Estates from the University was seconded as its director, as the first of a series of construction advisors to government, so the experience of the procurement system used at Cambridge helped shape public procurement more widely; for example, it is clearly reflected in the rules set out in a policy document, 'The 2012 Commitments' for the procurement for London 2012 Olympics.[50]

For various recruiting reasons the post of Director remained vacant for some time, and then the department was split in two until 2013.

Notes

1 Department for Business Enterprise and Regulatory Reform (BERR), *Construction Statistics Annual 2007,* Table 2.1 (BERR, August 2007).
2 HESA, *Finances of Higher Education Institutions 2012/2013* (HESA, May 2014).
3 KPMG, *UK Construction Barometer* (KPMG, 2013).
4 From that debacle sprang the excellent and influential Construction Industry Council (CIC) formally established in November 1988, representing all main elements of the industry: designers, suppliers, contractors, specialist contractors, and later, clients and managers. Whilst Ted Happold who set it up wanted it to be comprehensive, and for a while it truly was, it now is the body for the many professional Institutions and thus represents the consultants; CIC no longer includes constructors – they form a separate group, Construction Council, within CBI.
5 The Latham Report, *Constructing the Team* (HMSO, July 1994).
6 I was reminded of this by the journal *NCE* (*New Civil Engineer*) issue dated 3 September 2014.
7 After the 1997 change of government, however, the very able civil service was replaced, slowing the momentum and losing some support for reform.
8 The Egan Report, *Rethinking Construction: Report of the Construction Task Force* (HMSO, 1998).
9 It was rumoured at the time that John Egan, head of BAA, agreed to take on that industry review at the request of the incoming Labour government so as to 'upstage' the Latham report.
10 D. M. Adamson and T. Pollington, *Changes in the Construction Industry* (Routledge, 2006).
11 The Housing Grants, Construction and Regeneration Act 1996. Its title shows that it is a piece of omnibus legislation:

> An Act to make provision for grants and other assistance for housing purposes and about action in relation to unfit housing; *to amend the law relating to construction contracts and architects*; to provide grants and other assistance for regeneration and development and in connection with clearance areas; to amend the provisions relating to home energy efficiency schemes; to make provision in connection with the dissolution of urban development corporations (UDCs), housing action trusts and the Commission for New Towns; and for connected purpose.

12 *The Levene Scrutiny into Construction Procurement by Government* (HMSO, 1995).

13 Initiatives have continued: in July 2013 for example, a consultation called the 'Construction 2025' was launched by the Minister for Construction. Most of that reworked old issues. As one example, it was, after a great deal of work, agreed in 2006/2007 by all central government spending departments, that they would pay project bills within 28 days: one of the main issues in 'Construction 2025' was 'prompt payment in 30 days'. In 2014 it was reported by a *New Civil Engineer* survey that only 5 per cent of specialist contractors' invoices were paid within 30 days.

14 The immediacy with which M4I was set up showed that it had been planned before the election, the Shadow Minister for Construction having been blind-sighted. It was widely rumoured that the appointment of Sir John Egan by John Prescott to do a new review of the industry was agreed during an encounter between these two men in the toilet at a construction dinner, and that the deal was that if Sir John did the review, then planning permission for Heathrow Terminal 5 would be assured: he did the review, and T5 went ahead.

15 Stationery Office NAO Report, *Improving Public Services through Better Construction* (HMSO, 15 March 2005).

16 *New Civil Engineer* (2014).

17 A College Bursar planning a dinner being held many years ago for the architect on completion of a new building when asked if the equally qualified contractor manager was coming allegedly replied, 'What? He'd have to leave his muddy boots outside.' Such stereotyping is not new: a letter by Marshal Vauban to the Marquis de Louvois on 17 July 1685 made reference to contractors attracted by tenders as 'all the ne'er-do-well wretches, scoundrels and ignoramuses.' The focus of his attack however was on cost-cutting:

> the frequent cutting of costs in your works. I affirm that these acts cause delays and greatly increase the costs of the works … the more so because a contractor who loses money is as a drowning man who will clutch at any straw. A contractor reduced to these dire extremities does not pay the merchants who supply with materials, pays his workers badly, cheats those who can be cheated, is served only by the worst men because these men offer themselves more cheaply than the others, uses only the poorest materials, haggles over everything, and is forever pleading mercy from all and sundry.

18 RIBA has extensive statistics, and see http://resources.alljobopenings.com/architect-jobs.

19 NAO Report H&SE, *Improving Health and Safety in the Construction Industry* (The Stationery Office, 12 May 2004).

20 *Fatal Injury Statistics*, HSE 1982/1983, 2012/2013.

21 The history of EMBS dates back to 1923 when an Estate Management Branch of the School of Agriculture was set up to teach estate management and also to provide advice on the management of estates of the University (and the Colleges). In 1954 it was decided to divide off a Department of Estate Management to both teach estate management to students, and to manage the University estate: its first Director was Noel Dean. The two aspects veered apart, and after Mr Dean retired in 1961 a Department of Land Economy was established, and a separate Estate Management Advisory Service was set up in 1962 with John Mills as its Director; in 1973 the name changed to Estate Management and Building Service. In 2006 the department was split in two: the planning and provision at North West and West Cambridge, and the running of other projects and the existing estate – that error was corrected in 2013 and the remit of the Estates Department restored to what was and is standard in universities given the obvious advantages of continuity of management through the planning, design, construction, occupation of buildings. In 2012, the name of the department was shortened to Estate Management.

22 In November 1962 in his inaugural speech as RIBA President Sir Robert Matthew

noted the need to increase the 'business-like responsibility of ... architects. This is a situation we are determined to change'.

23 Construction Industry Board, *CIB Reports 1–12* (Thomas Telford Services Ltd, 1997).

24 When a TV series of programmes on UK procurement in 2012 asked the question 'who was the single "hero" of the Olympics procurement?', Sir John Armitt (who many would say was the real hero of procurement of the 2012 Games venue procurement) said 'the NEC3 Design and Build form of contract'. Later, in 2005 and through 2006, HM Treasury held that ECC was to be the preferred form of contract for all government construction contracts. Later again, JCT finally gave in under pressures from an industry sick of being so litigious, and produced a more managerial and less adversarial ('Lathamised') form of its contract.

25 In 2012, 24 per cent of all UK construction companies had at least one dispute concerning at least £250k; as the recession bit harder, that went up to 30 per cent in 2013. In the USA, serious disputes are estimated to arise in 10 to 30 percent of all construction projects. See Federal Facilities Council Technical Report No. 149 *Reducing Construction Costs: Uses of Best Dispute Resolution Practice by Project Owners* (University of Texas, 2007).

26 MPSC (98)(67). MPSC was the early name for the Buildings Committee, and reports of both were numbered sequentially within the year.

27 The Russell Group is an association of 24 leading research universities in the UK.

28 One particular and pivotal battle over space management was with the Physics Department: the head of the department insisted strongly that space management in his department was tight and that there should be no outside interference. The estates representative mentioned that he'd noticed that a professor had two rooms in the department (plus his rooms in College). The head of department's defence of him needing more than one room in the department was dented when it was then noted that the professor in question had been dead for some time.

29 RIBA, *A Review of Architect Design Competitions and other Competitive Processes* (RIBA, 2014).

30 Dr Sebastian Macmillan, 'Added value of good design', *Building Research & Information* (2006) 34(3), 257–271.

31 CABE, *Design for Distinction: The Value of Good Building Design in Higher Education* (Commission for Architecture and the Built Environment, 2005).

32 Building Cost Information Service, Royal Institute of Chartered Surveyors.

33 Some projects spent more than budgeted because further work was asked for and further funding agreed for it.

34 At the time of writing, the budget for High Speed 2 rail project was being increased by some £10bn to £42.6bn mainly to reflect the raising of the probability of coming within budget to 95 per cent; this caused uproar among those not understanding the probability aspects of budget compliance on a capital project.

35 Cabinet Office, 20 December 2012.

36 Mark Way and Bill Bordass, *BSRIA: Soft Landings Framework* (BSRIA, 29 June 2009).

37 *Trends – how to maintain your 'green' buildings*, 24 November 2010. Available online at www.abc.net.au/radionational/programs/bydesign/trends---how-to-maintain-your-green-buildings/2969866 (accessed 2 March 2015).

38 University of Cambridge Estate Plan 2005.

39 AUDE, *2004 Statistics* (Association of University Directors of Estates, 2004).

40 Richard Saxon, *Be Valuable* (Constructing Excellence, 2005).

41 OGC, *Achieving Excellence Procurement Guides* (OGC, 2003).

42 The incoming brief for the new VC in 2004 from the estates department urged that the rate of capital procurement be reduced for the reason that an ever-increasing proportion of the University's annual budget was being taken up in premises costs. Also because of the problem of holding and recruiting new staff during a construction boom, some in

the Estates Department worked under huge pressure for years concerned about both their reputation and their jobs.

43 The CIBSE commented on its website: 'The CIBSE /RIBA benchmarking exercise CarbonBuzz shows that most new and refurbished buildings consume between 1.5 and 2.5 times predicted values. At a time when buildings are better insulated and individual technologies more efficient than ever, this is a damning statistic.' The full article is available at http://modbs.co.uk/news/fullstory.php/aid/12777/Why_do_buildings_ often_disappoint_their_owners_.html (accessed 2 March 2015). See also the February 2014 issue of the RIBA Journal on the same subject, available at www.ribajournal.com/ pages/feb_2014__intelligence_report_228577.cfm (accessed 2 March 2015).

44 Richard Saxon, *Be Valuable* (Constructing Excellence, 2005) and The Royal Academy of Engineering, *The Long Term Cost of Owning and Using Buildings* (RAE, 1998).

45 The Carbon Trust, *Cambridge – Higher Education Carbon Management Programme, University of Cambridge Implementation Plan* (Carbon Trust, 2005/2006).

46 SQW report, *Carbon Reduction Target and Strategy for Higher Education in England* (SQW, January 2010).

47 D. M. Adamson, *Contractor Skill Levels for University Construction Contracts* (11 January 2001, available from author). The Director of Estates was a board member of CITB.

48 This is the last paragraph on Page 20 of the *Contract Journal* Awards 2002 special issue. (In 2003 the winner of Building Magazine Awards for a major project was Canary Wharf; the University came second.)

49 NAO, *Improving Public Services through Better Construction* (NAO, March 2005).

50 2012 Construction Commitments, Strategic Forum for Construction 2005. This was a report set up by Government Ministries including DCMS, DTI, OGC/HM Treasury and launched by Ministers on 3 July 2006.

6   The Sidgwick story

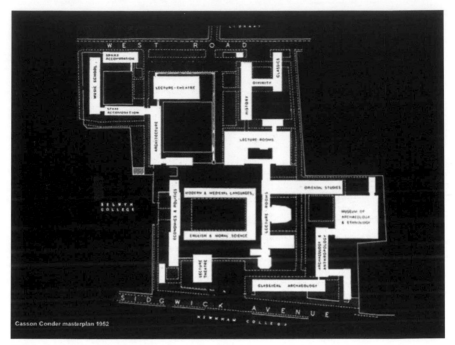

Figure 6.1   Sidgwick Site plan
(By permission of Allies and Morrison Architects)

It's not a long story by Cambridge standards. But over the years from 1948 when the site started to be bought progressively from Corpus Christi College, until the most recent building was opened in May 2012, the story of how the Sidgwick Site played out is more remarkable than any other site development. There was first a progressive site plan which although widely admired was ruined by the construction of a massive, ugly building; then there came a striking but flawed building dominating the centre of the site, on which it turned its back; then a quirky and more efficient departmental building; and finally around the end of the century, a new site plan with more comfortable buildings was implemented in an effort to calm the 'architectural zoo' of post-war architectural fashions plumped down 'hugger-mugger' on the constrained site that is Sidgwick.

To start at the beginning. Once the University came to terms with the need to find expansion space 'Beyond the River' by accepting that siting the University Library to the west of the river did not really represent the end of civilisation, then expansion by means of developing the Sidgwick Site for use by the Arts, Social Sciences and Humanities Faculties (Departments in other parlance) could be planned. The first and major issue was that some Faculties, notably staff in English and History, were adamant that they didn't want 'University' buildings because they wanted to continue doing their research and their teaching in their Colleges, hardly ever meeting together (and in some cases, rarely on a cordial basis). The outcome was that when the site was first developed, it provided for some Faculties but not others. So, the first aim was to provide better space for those Faculties which did see the benefits of consolidating in University buildings rather than only teaching across Colleges and dispersed sites; the second aim was to release space near the centre of the University estate for growing science/medical departments. All this was set out in the Report of the Syndicate on the Sidgwick Avenue Site.[1]

The University bought about two-thirds of the Sidgwick Site, the remainder to be later used for College student accommodation. The architects chosen to produce a master plan for the site were from the practice Hugh Casson, Neville Condor and Partners, and they went to the task with enthusiasm and skill. (Sir Hugh Casson had recently been co-ordinator for designing the exciting Festival of Britain site on the River Thames.) The strategy of the Casson/Condor plan, developed 1953–1955, was to have a central, raised building, with a clear view of and through a courtyard, and subtle pathways leading to buildings around the site. Courtyards are one of the two most distinctive architectural features of Cambridge Colleges through the centuries ('lanterns' providing light and natural ventilation being the other), and the courtyard of the 'Raised Faculty Building' at the heart of the site was the last to be produced for the University until near the end of the century when Ted Cullinan as architect for the Maths development brought the courtyard back: for about 40 years those two architectural traditions of the University got trampled by new Sidgwick buildings and by the West Cambridge master plan.

Dr Alistair Fair, Chancellor's Fellow in Architectural History at the University of Edinburgh, describes the Casson/Condor master plan as follows:

> It offered a loosely structured framework for future development that contrasted with the Beaux-Arts formality of Robert Atkinson's rival scheme. Underpinning the Casson/Condor master plan was the desire to create variety through contrasts in materials, building heights, ground levels, and surface treatments. This idea owed much to the neo-Picturesque 'townscape' philosophy favoured by the Architecture Review in the late 1940s and 1950s.[2]

The Casson/Condor plan was well received, and although Sir Hugh Casson and Neville Condor had not been given any promise of appointments to design any buildings on the site, they were soon appointed to design what became known as the 'Raised Faculty Building' (RFB) for the study of Philosophy (then known as 'Moral Science') and Modern and Medieval Languages, a building for the then Faculty of Economics and Politics (thought to be the first departmental building for the subject of Economics in any UK university) and two adjacent lecture theatres, one of which is the second largest lecture theatre in the University; the other, smaller, and in the words of a former Director of Planning, 'reminiscent of a gothic church in a modern German housing estate'.

The RFB is raised up on high-quality concrete columns, technically 'pilotis', with three storeys clad in good stone along three sides of a fine lawn. The building itself works well for its users. Two mezzanine floors were added and other enhancements effected in 1998–1989 at a cost of £6.543M. The Marshall Economics and Politics building, which forms the fourth side of the RFB courtyard, is built of rather tedious brick, as is the large, adjacent Lady Mitchell Hall lecture-theatre building, whose more exciting internal design was led by Michael Cain (the architect, not the actor of that name) who later designed the adjacent Classics Department building. In 1966–1968 there was added a building for the study of Oriental Studies, as the subject was then known; this also was a low-key, brick construction to complement the Raised Faculty building and the two adjacent lecture theatres. Buildings for the Faculty of Classics, and for examinations and lectures, followed suit so the south east side of the Sidgwick Site was coherent, and efficient for its users.

The Casson/Condor plan would almost certainly have continued to serve the University well: there would have been a comprehensive, coherent and probably popular and efficient campus for the Arts, Social Sciences and Humanities sector of the University. But the master plan was chucked out when Sir James Stirling, initially working with Gowan, was appointed in 1963 by a strong-minded professor of history and some of his colleagues to design a building for the History Faculty's library and offices. (Stirling had earlier entered design competitions for Churchill and Selwyn Colleges but had not been selected.)

The selection of designers then suffered from having the problems of a committee as a client (as they say, a horse designed by a committee often turns out to be a camel). The Sidgwick Site Committee considered some 15 architects for the History Building, visited six projects and commissioned sketch proposals from Architects Co-Partnership, David Roberts and Stirling & Gowan before they finally selected James Stirling's scheme. (Once a scheme is selected by way of a design competition the client is stuck with it: Chapter 5 suggested that it is much better to select an architect, by ideas and track-record, who then develops a design previously suggested in outline only, consulting with the client.) The resultant building was seen as technically innovative with two glazed wings over a base of tile and glass. The library layout was also innovative and, as a place to work, is liked by many, though not all. The building is certainly massive, overshadowing the Raised Faculty Building. At a late design stage the building was rotated through 90 degrees as a response to the site-acquisition problem over Mrs Lilley's garden; it would have been obvious that such a fundamental design change would exacerbate the high level of solar gain associated with such extensive glazing, and it was said that the engineering designers so warned. In the form of procurement then used, there was much less interaction between the client and designers other than the architect; the architect was seen as the spokesman for the entire design team when liaising with the client and with the building users. That now seems odd.

The building gave trouble from the start. It is said that just before the opening ceremony early in the summer of 1968, Stirling heard a hum and ordered that it be ceased: and so the high-level fans on which heat extraction was dependent were switched off, and then were left off over the summer; they were forgotten until a survey many years later found them rusted in place and inoperable. That was not the only cause of the intolerably high temperatures in the building: thermal gain from the huge areas of glazing was high so that staff did not like working in the building over the summer so few did. Had the glazed roof faced north there would have been little need for the extract system. Conversely, the building was cold in winter, and it was noisy in all seasons. And then the glazing leaked. Stirling, a great self-publicist, said that the design of the cascade of glass was analogous to a waterfall: wags soon noted that it was indeed a waterfall internally with the extensive water leaks being collected by arrays of buckets. Then much of the tiling fell off. The building was so unpopular that there was a strong move to have it demolished in the 1990s, but it was rapidly 'listed Grade II'[3] at the behest of the City Council. The University's Planning Statement for 1989/1990 to 1994/1995 (at para 11.1) notes that, 'Between 1985 and 1990 about £1.1M has had to be spent on major repairs to Stirling's building for the Faculty of History, and the maintenance of this particular building is likely to remain a long-term problem.' Remedial work continued over decades with much of its exterior and building services having to be replaced: for example, in 2005/2006 £1.25M had to be spent on yet more refurbishment.

But the damage caused by Stirling was even worse than inflicting a bad building: it killed the Casson/Condor plan for a coherent site. Further, subsequent buildings adjacent to it had to try to respond to it.

Revulsion among academics against the style of the 'Stirling building' was probably the first straw in the wind of a shift away from so-called 'signature architects' (or 'starchitects') and towards user-focused architects. Another straw in that wind came with the next building on the Sidgwick Site: the Faculty of Law building. Design began in 1990 with handover in 1995. The Faculty professoriate saw themselves essentially as 'the client' although they did not carry University responsibility for procurement. The design was led by Norman Foster and Partners; the engineers were YRM and the constructors were Taylor Woodrow Construction, with Drake and Scull as mechanical/electrical subcontractors, all contracted under the then-combative JCT contract form. Procurement was via a single-stage procurement whereby there was no contractor input until the designers (in practice the architect) said that design was 'complete' and ready to go to the market, which it did with a stack of drawings and specification about a metre high, tenders to be returned within a few weeks by a shortlist of up to six contractors so each had on average only a 1/6 probability of success. (See Chapter 5.)

The construction works for the new Law building were completed mid-1995, 26 weeks late. The building has some 8,000m² assignable space and the out-turn project cost was £22.1M.

The main reason why that building was so inappropriate for its site is that it is close to the centre of the site but turns its back on the central Raised Faculty Building. In response to the History building, design was based on a huge glass curved atrium with little acoustic treatment, and the noise levels generally, and between the circulation areas and the library in particular, were far too high. Some of the academic staff at first refused to move in because of the noisiness; the library was directly above the lecture theatres from which large numbers of students emerged and the noise throughout the main part of the building was intolerable, especially as there were no 'softening' features to absorb noise; subsequently, notice boards giving necessary information, and potted plants, were brought in to the irritation, it was said, of Foster. The entrance system allows cold winds to sweep into the building. Temperature control is a problem; there are eight big air handling units and regulation of them is difficult. And it is very expensive to run, probably the most expensive building on the site.

The legal disputes about design and construction into which the project descended lasted for a couple of years and were very expensive. Taylor Vinters were appointed as lawyers early in 1996; then a claims consultant, and legal Counsel. From a stormy meeting between the VC, Counsel and others in January 1997 it became clear that pursuing the University's dissatisfaction with the condition of the acoustics of the building, as well as addressing the claims by Taylor Woodrow and sub-contractor Drake and Scull for 'Prolongation and Disruption Costs', were going to be prolonged, and increasingly costly. A meeting of building users, University administrators and building designers in January 1998 agreed that the University would require the architects to design, at least partly pay for and have installed a vertical sound-reducing wall so as to 'insulate two-thirds of the building from noise contamination [although] … This option does sacrifice the western end of the building to existing noise conditions'. Fosters estimated that (at 1998 prices) this would cost £800k 'with some exclusions'. The glass wall from floor to roof was installed during the summer of 1999, four years after the official opening; the cost was confidential, over £1M.

The next issue for the University was to settle sub-contractor Drake and Sculls' claim at £156k, hoping that that would give an easier route to settle the very much bigger claim by the main contractor Taylor Woodrow for costs relating to over-run of time, and a further £50k for 'work in connection with the visit by the Chancellor'. Reports by quantity surveyors Davis Langdon and claims consultant Crowther on 11 August and 29 September 1998 recommended that the University should try to settle with Taylor Woodrow at £14.55M, and after negotiations, settlement was achieved within that figure, three and a half years after the building was handed over, and about five years after dispute started. Related legal and consultancy costs are not known.

The three main disputes (with Fosters, Taylor Woodrow and Drake and Scull) were among a larger number of construction disputes to be settled in 1998/1999, including the final legal dispute around the Judge project, and a claim over a research building on the Downing Site.

If the History Faculty building stopped in its tracks Hugh Casson's master plan for the Sidgwick Site, the Faculty of Law building next door killed it. By day, the building looks a bit like a beached whale. Indeed, the front of the building is attractive, even dramatic at night. However, the back of the building, which faces the centre of the Sidgwick Site, is dull; perhaps its most interesting features are the stickers on ground-floor windows telling people not to leave bikes there.

There was a view that a large number of young architects were employed at remarkably low salaries. It also appears that there was poor teamwork between the architect and the engineers.[4] The practice Norman Foster and Partners did not subsequently secure any other work in the University.

Raised Faculty Building

After the big issues around the creation of the Faculty of Law and the retro-fitting of its huge sound-limiting glass screen, the next project on the Sidgwick Site was the refurbishment of the Raised Faculty Building (RFB) in order to increase by 30 per cent the usable space, so as to provide social space and academic facilities for graduate students and visiting scholars, and to provide much better disabled access. A space-use review was carried out in 1996 by the architectural practice DEGW led by Dr Frank Duffy (who was President of RIBA during the Latham review of the construction industry; some say among the best RIBA presidents of recent decades); this practice had for decades been one of the leading architectural practices world-wide for optimising space-use efficiency. The practice was then bought out by a Dutch practice; but things did not go well, and the founding partners took over the practice and built it up once more.

The 1996 review of RFB space-use showed how the amount of working space in the building available could be increased by up to 67 per cent of the total internal area within the outer walls of the building by opening up areas which could not be used properly, and how two complete floors could be created from the storeys of peripheral galleries in part of the building. Work started on an affordable scheme in 1998 while the practice was in Dutch ownership. Soon there were problems in the flow of architectural information; when it was found that scarce design-time was being spent in designing a stand for a coffee machine while work on site by the 'pro-active' contractor Interior was being held up for want of construction drawings, there were discussions and then negotiations at high level, in an amicable and business-like way, following which DEGW handed over the role of architect to Fitzroy Robinson (which practice had earlier successfully taken over architectural work on the Judge School of Management from John Outram for similar reason). Work flow picked up, and was completed in time for the opening on 23 September 2000.

Up to that time, projects proceeded on the basis of 'cost-estimates' (CE) which rose as the project progressed, matching funding being found as required (said to be from 'goat-bags of money buried' around the place). The originally authorised CE for RFB was £4.4M, and the CE reached £9M at one stage. By dint of much value engineering (a good and early example of how escalating 'cost-estimates' could be brought back within a sensible properly-set budget), the project out-turn came to £6.58M. A good step towards a 'team' policy was achieved; for example, the Representative User's assistant was allowed by the contractor to discuss value-engineering with sub-contractors directly. The project delivered an excellent and much-needed enhancement of the building which had been so well designed by Casson and Condor some 40 years earlier, and without loss of any of its architectural qualities. Further upgrade-work for the Modern and Medieval Languages and the Philosophy Departments started on site on 1 June 2006, at a cost of £1.07M, the work designed by Woods Hardwick and Mott Macdonald.

Faculty of Divinity

The next chapter in the Sidgwick Story came in 1997 when design started in earnest on a new building for the Faculty of Divinity on the western edge of the site, it having been decided after sometimes acrimonious discussions about moving from the former home opposite St John's College (see Chapter 1). The architects appointed were London-based Edward Cullinan Architect, with Colin Price as the project architect; he came with a good reputation, in part because of his authorship of books on designing buildings for their use throughout their lives including 'Beyond Performance: How Great Organizations Build Ultimate Competitive Advantage'.[5] The Building Services engineers were Whitby Bird, a young and vigorous practice which had grown to be a national player, starting from an industrial estate near Bristol. The Quantity Surveyors were Edmond Shipway and Partners; the contract being by JCT 98 (the last major University project before the switch to the NEC/ECC – see Chapter 5), it was taken that in effect the architect would be the manager of the project; the client was represented by the internal client project manager from the Estates Management department. Sindall Construction company (later Bluestone, until taken over by Morgan Sindall) had the contract.

This was one of the last major University projects to be 'single-stage' procurement, whereby the builders were kept out of the process until the designers felt that they had completed the design and it was ready to go out tender (often with binding tenders required within only a few weeks); the tenders were then assessed almost totally just on capital cost rather than a balance of price and the quality of the tendering company. Partly because it was such a single-stage contract, there was an abnormally great amount of design done during the contract by the contractors (called 'Contractors' Design Portion', CDP). That usually works well enough if there is very careful control of the interfaces between that work and the design work done by the project designers; the contractors in this case have suggested that on this project that took a great deal of co-ordination effort by them.

The gross area in the building was 2,624m², and the stated assignable area (the area that can be used directly for the users of the building: viz the gross area less corridors, stairs, toilets, plant rooms, central store rooms etc.) was to be 2,000m². The starting budget was £7.754M at 1998 prices, so the unit budget costs were £2,934/m² gross, which then was high for that type of building (the period of very high construction cost inflation had hardly begun); this was partly because of the large area underground, and the high specification of the interior generally, the quality of the library and having a good underground lecture theatre. The external quality appearance is less valued by most observers than the internal. The Faculty of Divinity project came in nearly on time but 2.8 per cent over its final budget at £7.969M. The building was opened in early September 2000, with the official opening by HM the Queen on a very wet morning on 23 November 2000, followed by lunch of wild guinea fowl.

The Faculty of Divinity building was the first University project to be subject to a post-occupancy evaluation, 'poe' (see Chapter 5).[6] This poe methodology, like the reports themselves, was applied and made available to the industry more widely for all capital projects of over £1M in Cambridge University from 1999 until 2006. The poe reports went formally into the University committee system via the Buildings Committee; they were drafted by the Estate Management project manager in consultation with all attendees of the four discussions sessions (see Chapter 5). Crucially, they were finalised and co-signed by both the Director of Estates, as the officer personally held responsible to the University for all construction projects, and the Representative User (RU): that gave power and authority to the RU from the very start of a project as being one of two people who would sign and take responsibility for the final report on the project, a power that was respected by designers, managers and constructors alike.

In the journal *Architecture Today* the Faculty of Divinity building is described as follows:

> Against so much gravity [Law and History buildings] Divinity is a jaunty building. The horizontal banding of its aluminium sun-shades gives it a slightly art deco air, almost a sea-side levity: and the apparent incompleteness of its exteriors is a striking contrast to the absolute finality of the others. More importantly, though it is instantly recognisable as an 'ordinary' building in a way that its neighbours are not, giving it the sort of familiarity

that they would be embarrassed by. At no point does this building require us to believe in a particular doctrine of architecture, nor to accept any particular creed. After so much high-mindedness from its neighbours, this can be said to be a relief, and in this respect, the Divinity Faculty has more in common with the older Casson-Condor-designed faculty building on the site. The Divinity Faculty after two months' occupation is more at home here in a way that neither the History Faculty nor the Law Faculty have ever managed in their respective buildings. If the Divinity building is a building that makes its occupants feel 'at home', one might speculate as to why, at the end of the twentieth century, Cambridge should have decided to set aside a 40-year old tradition of 'special' university building? Perhaps we should see this change as connected to the University's need to make itself seem a less exclusive institution and to show that 'elite' need not mean inaccessible.[7]

The architect, Colin Rice of Edward Cullinan Architects, commented:

> The relocation of the Faculty of Divinity came at the start of the period covered by this analysis of procurement at Cambridge University. In many ways the success of the project illustrates that in the final analysis successful projects come from the strength of the team, regardless of procurement route.
>
> The Faculty set up a small 'new building committee' which in 1995 interviewed a small number of architects, choosing by chemistry rather than either a design or fee competition. The 'chair' was a former College Bursar, a wise and experienced client. The fundraising for the new building was championed by Lord Runcie with great success. A simple brief set a vision for the new building, with the aspiration that the new building should encourage the Faculty's sense of community: great emphasis was placed on the quality of the space for circulation; the gaps between the rooms that support conversations that lead who knows where and provides key 'innovation space'.
>
> In 1996 the Faculty was faced with a choice: to rush ahead for an early start on site or to take longer to develop the design and prepare a complete set of tender documents. The University's Estate Management's project gave a clear steer to use the time to prepare more fully in the interests of the project. With the advent of BIM and the power it gives to build the 'virtual prototype', the wisdom that 'thorough preparation is all' is now regaining currency. When the building is well described and where variations are kept to a minimum, the Contract can be 'left in the drawer' – for reference only if things go wrong.
>
> Some subtleties of communication helped. Before starting on site Sindall initiated a 'Project Charter', setting out their commitment to meet the 'customer's needs'. For example, the Faculty staff showed the site team round their existing premises, sharing their aspirations for the new building. Rather than just the usual procedural pre-contract meeting, a technical meeting was held where each consultant in the design team

explained to the site team the rationale behind the structural, environmental and architectural design. Good communication rather than forced 'bonding' events helped build a strong team.

Post-occupancy evaluation using the two day HEFCE method was carried out three years after completion. By that time the building had had a number of issues – a serious flood to the basement, failure of [the cedar wire-mounted] external blinds [which warped and caused distortion of the wires and jamming/rattling in changing weather], and front door operation. (A further ten years on and these have all been sorted). Soft Landings [see Chapter 5] should be used on all new buildings where modern systems force custodial staff up a steep learning curve.

The building had high sustainability ambitions which we would now call 'fabric first'. High ceilings, exposed concrete frame and natural ventilation (including night time purging) are all deployed. The architectural expression comes from the integrated aluminium louvres that give the sweeping horizontal lines that set the building in contextual contrast to Stirling's neighbouring History Faculty. Both shedding rain away from the façade and shading the interior they have helped create a strong identity that has stood the test of the first 13 years of the building's life.

For their part, Sindall afterwards commented on the architects: 'ECA was one of the best, especially Colin'. It was in part due to such teamwork that the project resulted in such a good building.

Sidgwick Site master planning

By this time, 1999, it was very clear to just about everyone that a review of the Sidgwick Site was badly needed: it had become as muddled as the Downing and the New Museum Sites, and for the same reason: individual building projects led by small groups of academic staff to meet, usually to meet well, their short/medium term requirements. The Sidgwick Site had become a jumbled mix of buildings of all styles and scales with few quiet academic spaces and no coherent theme. As an architectural 'zoo' with examples of each post-war species, it had attractions for visiting architects, indeed, visitors in general, but was not a good academic site for departments seeking harmony and synergy, to study and learn in peace, and have social space to meet in attractive places around the site. (The original Casson Condor plan had had all those.) Work on site planning had continued in a rather desultory manner by Casson Condor, and the then Dutch-owned DEGW[8] during the 1990s, but progress and outcome were slow and unimpressive, and based on a form of space planning analysis out of line with the analytical approach that the University had recently adopted.

The discussion about how to carry out such a review and produce a new master plan was carried through the University in an orderly and pro-active manner, the new committee system working at its best. A small group consisting of the Chairman of the Buildings Committee, the Chairs of the Councils of the relevant Schools ('Deans' in the parlance of other UK universities) representing all the academic Departments/Faculties/Institutes on the site, and the Director of Estates, met through 1999 and into 2000 and produced a client brief and an Invitation to Tender (ITT) which was sent to a selected long-list of architectural practices. The essentials of the ITT were to establish a framework for identifying on the Sidgwick Site the location and extent of practicable building sites, and how best to enhance the overall working environment. For the first time, all the short-listed architectural practices were invited to spend a half-day, all together, being briefed and being able to speak to senior academics and estates staff about the task and the ITT. There had been a long-established belief that it was somehow unethical to let practices know who else was on a short list, but of course jungle drums have always been so effective that anyone who wanted to know, knew within days if not hours of a short-list being decided who was and wasn't on it. Some practices were understandably shy about asking perceptive questions in front of rivals, so each practice also had a period of private questioning: following this open session, which was appreciated as real 'openness', there were inter-views of short-listed practices based on practice philosophy, management ethos and track-record, and initial thoughts. Marking by each member of the (rather large) interview panel was quantitated based on quality-price balance reflecting the report 'Latham Reform Implementation Panel No 4b' on balancing quality and price in the selection of consultants. Coming out of this process narrowly ahead was the practice Allies and Morrison, then a practice run closely by the two senior partners. They both had a good knowledge of Cambridge, one as alumnus, the other having taught there; when asked during the interview what would be their response if they were to be selected for the review of the Sidgwick Site review and master plan, the reply was, 'it would be rather like going back to one's prep-school and telling the headmaster how the school had gone wrong'. This appointment was notified to the Buildings Committee in October 2000 along with the appointment of Hanscomb as Quantity Surveyor, BuroHappold and W S Atkins as Structural/Civil Engineer.

The University's 'Needs' Committee set out approved assignable ('usable') space requirements as follows: English 3,136m^2, Criminology 2,800m^2, East Asia Studies 1,500m2, Land Economy 1,700m^2. (Of those confirmed needs, just the first two were to happen.) The Planning and Resources Committee on 18 October 2000 approved a financial commitment to initial development costs, and in principle to a matrix of budget estimates for those four develop-ments plus associated site infrastructure, £41.13M in total; it was noted that funding for a new building for the Faculty of Criminology was already largely in place while the other faculties were fundraising. (Some departments' fund-raising efforts were notably more determined and effective than others.)

The main conclusion of the Allies and Morrison Master-plan study, issued in November 2000, was that there were four areas on the site which could afford good space for new buildings. (See site plan at Figure 6.1.) The master plan report praised the Casson and Condor plan, and had an interesting comment 'Casson's watercolour sketches emphatically picture visionary spaces populated by crowds of spiky students and their bicycles, a vision appropriate to the forward-looking ethos of the post-war years.' It also commends Casson and Condor's clear view that 'the link between the Sidgwick Site and the University Library is emphasised' and the report adds that 'It is proposed that this significant City route be strengthened', and then calls for enhancement of the crossing over West Road, and of 'the arrival point' of the route to the University Library. (Unfortunately, the design of the Law Faculty building included a structurally unnecessary column smack on the axis between the heart of the Sidgwick Site and the route to the University Library; the roof could have been cantilevered instead.)

Noting the Casson and Condor plan for 'a new approach in which free-standing buildings were [to be] grouped to make a system of courts joined by walkways', Allies and Morrison said that, 'the new buildings should again be regarded as elements of an urban system which interact with each other and the site spaces, rather than as solo performers.' The report continued:

> The master plan of 1952 foresaw the continuing need for expansion of the site and realised the need to be able to include a variety of forms in the basic layout. To encourage this variety, the architects took the lecture halls out of the Faculty buildings and designed them as single buildings to exploit their demotic forms. Hence, the Lady Mitchell Hall and Little Hall were placed along the south edge of the podium ... Whilst the south half of the Site was realised largely to Casson and Condor designs, the north half starting with James Stirling's design for the Faculty of History Library in 1965 completely ignored it. The variety of forms was anticipated, the abandonment of proposed routed and spaces was not ... It can be argued that the architectural quality of the more recent Faculties for History, Law and Divinity have redeemed the bland sameness of the south half, and this is to be welcomed, but they have done this at the expense of the site as a whole. It would seem appropriate to complete the Site, albeit with contemporary architectural forms, in a way that also creates and defines spaces.

In 2000, the total built area space on the Sidgwick Site was 51,337m² for 6,045 staff/students. The Allies and Morrison report planned to increase the built area by buildings on the four plots by 13,570m² for the additional 82 staff and 397 students the brief anticipated. The Faculties of English and Criminology in their nineteenth century villas, which the plan recommended should be demolished (as had been long anticipated), had then had only 857m² and 1,277m² respectively: replacement by the new buildings were to allow considerable expansion in these Faculties. Development could be achieved without loss of any of the significant trees such as the Ginko and Holm Oak on the west and east sides of the site earmarked for the English Faculty, the Tulip tree in the Raised Faculty building courtyard, and the Mulberry and Medlar on the north-west corner of the Law Faculty building. The important lines of Plane trees down Sidgwick Avenue (gifted by the founder of Newnham College) would, of course stay, though disruptive and pavement-heaving.

The Allies and Morrison master plan set out to calm the site: to have more serene and less-self aggrandising buildings for the remaining areas available for construction, and for those new buildings to have as much in the way of courtyards as was possible, and to be de-conflicted from through-traffic, which was to be diverted into routes to the east and west around the site. The 2000 survey carried out by W S Atkins noted that the daily volume onto the Sidgwick Site was: 3,814 pedestrians and 2,019 bikes (one third of users of the site), 304 car-journeys and 68 journeys by small and large trucks. However, the number of car parking spaces, 200 in 2000 (with 308 badge-holders entitled to use them), was to be considerably reduced. This reduction strengthened the case for setting up a University bus service, long wanted by Estates staff on the grounds of reducing traffic and pollution. It was soon after this time that that battle was won, and the Uni4 bus service came into being, running from the West Cambridge Site through the main University area and eventually on to Addenbrooke's. Introduction was seriously delayed by difficult and frustrating negotiations over standards of service with the bus company Stagecoach (agreed to be a 10 minute, later reluctantly agreed to be a 15 minute service from 8am until 7pm along the West Cambridge to city centre route).

As well as requiring less-egotistical buildings, better spaces for calming the site, reduction in car parking and segregation of vehicle routes off to the sides of the site, the Allies and Morrison report anticipated some valuable details in design, such as water features: unlike US universities, UK universities in general, and Cambridge University in particular, were poor in provision of water features although those have a calming effect, their 'white noise' lowering the impact of noises around the site. The Estates department did a brief study of water features in leading US universities. The most impressive and successful example of calming by the white noise of water features, strangely, was found not in an Ivy League university, but in George Mason University, Virginia: on the upper floor was a library, on the lower floor were many catering outlets, all with yelling-out of orders and, at the time of the visit, plates being dropped.

Another issue in which US universities were way ahead of most UK universities was signage and path marking. For example, in Princeton routes are marked by paths, or parts of roads being built in various colours of brick, so one 'follows the yellow (or whatever) brick road' to get to this destination or that, a bit like navigating round a hospital by following lines. The Allies and Morrison plan had a good scheme for signage.

In 1999, £1.80M had been authorised to be spent on decentralising and upgrading the boiler system of the Sidgwick Site over the subsequent five years. The report incorporated implementation of that and more generally the necessary upgrade of utility provision so as to be capable of supporting the new buildings for the Faculties of English and Criminology, and for a 'research hotel' (which became a building to be a home for the Centre for Research in Arts, Social Sciences, and Humanities, CRASSH) and another building which at the time was to be for the Department of Land Economy. The A&M report, fairly, slated the quality of lighting of the site, and so provision for a new lighting scheme was also proposed.

A&M's analysis posited that Casson's master plan was based on the Colleges' external spaces and courts but not with the same level of enclosure, being much looser than a traditional court. A&M proposed the layout should now be more modern and informal with a strong connection through the Raised Faculty Building from Sidgwick Avenue to the University Library, while at the same time providing enclosure to external spaces to make them more related to and 'owned' by each Faculty. During consultation, A&M used a Mondrian painting to capture this idea.

The problem of the noise and irritation caused by skate-boarding was grasped: finding a solution proved to be unexpectedly and frustratingly difficult. The underlying issue was a social one: whose site was it anyway? Was the Sidgwick Site there solely for use by the academic users of the Site and their visitors, or was it also there for local people, including aggressive and noisy skate-board users, to enjoy as they wished? A sensitive aspect came up when the chosen solution of replacing certain smooth surfaces with studded anti-skate-board surfaces was challenged by those representing wheel-chair users, whose ease of travel would be impaired by rough surfaces. As ever, patience and a lot of compromise did in the end achieve elimination of skate-boarding and unimpeded wheel-chair surfaces.[9]

The Director of Estates and the Chief Planning Officer of Cambridge City Council, Peter Studdert, met regularly about every couple of months. At a meeting on 25 November 1999 agreement was reached on the proposed principles of development of the Sidgwick Site, generally as per the recommendations of Allies and Morrison; there was also agreement to create a more formal gateway to the site from Sidgwick Avenue which was haphazard and unimpressive, and a better site entrance from West Road; neither could be funded during the decade.

In the Cambridge University Reporter 309 of 13 December 2000 Council sought approval for the Allies and Morrison review and master plan, and that was duly agreed. When A&M won the competition to consider the master plan for the Sidgwick Site there was the presumption that they would design the first couple of buildings.

Although these were developed and built at roughly the same time, they have turned out to be very different buildings, reflecting the different briefs, funding and the process of design development. The first two of the buildings that were planned for the Sidgwick Site were the buildings for the Institute of Criminology and for the Faculty of English and the Department of Anglo-Saxon, Norse and Celtic (ASNC). The former had raised most of the capital funding required; the latter had not. It was possible to raise further funding for Criminology via the new Joint Infrastructure Fund (JIF) which brought considerable funding for new-build and refurbishment projects from government via Hefce, and Wellcome (see Chapter 2) and the University had capital money from Cambridge University Press and that helped funding of the new ASNC building, as it had supported the new Divinity building. The £10.3M Criminology bid to JIF in 1999 was successful; money was found for English.

Faculty of English

There was an English Faculty library in the Raised Faculty building and after decades of discussion (in the review of the Arts, Social Sciences and Humanities in 1945, there was the statement that 'It is the settled view of English that teaching takes place within the Colleges') it had been decided to put up a new building for the Faculty of English. Getting planning permission for this building was not easy. The head of planning in the City Council was still upset that the University had decided not to continue with Foster and Partners, even although the defects in the Law building were not disputed. It is understandable for a chief planning officer to like having it known that a famous, albeit controversial, architect was designing a building in his city, especially when the chief planning officer is one of a dwindling number of architect-planners (professionally trained planners or developers were taking that role in the UK). It is also fair to note that at that time, Allies and Morrison's name was associated in the minds of some planners with low-key buildings. Because of this, the professionally-close relationship between the chief planning officer and the director of estates in the University became more strained than at any other time; in one of their monthly meetings the former hurled down the table the A3 booklet of A&M's elevation drawings, roaring that they were the worst scheme he'd ever seen (the booklet being caught as it fell off the end of the table like an Australian pint of beer). The personal and professional relationships between the head of planning and the Director of Estates, however, remained cordial and strong; this incident was long remembered with humour. As time and design went on, the chief planning officer became increasingly happy with the buildings that emerged.

Overall, the progress of the Planning Permission application was quite good with an application for a 'building for the Faculty of English, the Department of Anglo-Saxon, Norse and Celtic (ASNC), and the Research Centre for English and Applied Linguistics' made on 3 March 2001. The start on site was on 1 October 2001 following site clearance and the demolition of the villa on West Road over the summer. Keeping to the demolition and site clearance schedule was made more difficult when in March 2001 a 'request for a ballot' in Regent House by seven professors and three other academics in other departments called for the villa not to be knocked down. The line of villas had been built around the mid-1870s, designed by the architect Richard Reynold Rowe, largely for purchase by academics (later Stephen Hawking lived there); Rowe also designed the Corn Exchange; the villas had attractive features inside (the fine red door of the Criminology villa, built in 1875, is stored in its new building) and some personal history locally, but no great external architectural merit, and they had never had been listed.

With a project budget of £15.16M at 2000 prices, unit costs were £3.58k/ m^2 and £4.66k/m^3 respectively; these are total costs including design and management fees (usually about 15–18 per cent of total cost), site development costs and taxes, but still quite expensive against relevant benchmarks. This was partly because of the expensive basement, partly because of high design standards and the emphasis on reducing in-use utilities costs (and CO_2 reduction). Working with architects Allies and Morrison were mechanical/ electrical design engineers BuroHappold and structural/civil engineers Whitby Bird, landscape by Joanna Gibbons.

Case study – English

The Faculty of English contains also the English Faculty Library, the Department of Anglo-Saxon Norse and Celtic (ASNC) and provides a home to the Research Centre for English and Applied Linguistics. The Faculty of English was the driving force and the administrator threw herself into the project. More academics were involved here than in Criminology, with Professor Dame Gillian Beer leading the campaign to raise funding and a consultative working group of about 12 people chaired by the Faculty Administrator, Claire Daunton. As far as developing the design goes, there were regular weekly and fortnightly meetings and also larger presentation meetings involving undergraduate and graduate students.

The main aim of the new building was to provide a heart and centre to the constituent parts of the Faculty of English which for many years were dispersed between the Library in the Raised Faculty Building and an administration base in one of the villas on the Sidgwick Site with lectures in various departmental lecture rooms and with supervisions in the Colleges. The new building was to bring these elements together while still expressing the identity of each constituent part.

One of the key elements of the brief was to think about the future flexibility of the teaching spaces needed by the Faculty and whether individual rooms or more open plan arrangements would be best in the long term. Therefore many of the partitions are demountable; allowing a transformation into larger rooms over a weekend while providing good acoustic separation between rooms. There are raised floors throughout providing totally adjustable cabling, which is very flexible compared to most University provision. There was also considerable thought given to integrating social spaces throughout the building, both the central social space and also smaller kitchenettes.

The main spaces are:

- seminar rooms
- offices, including offices for University Teaching Officers which are large enough for supervisions with a couple of students

- a social centre
- the Library
- a black box double height drama studio space for performance with a staircase link to the courtyard which provides the possibility of outdoor performances.

The built-area is 4,233m² gross including circulation space, toilets, store, plant rooms etc. and 3,256m² of 'assignable' space (i.e. space that can be actively used for research/teaching by active allocation to stated functions). For the first time in an architectural appointment, the University required designers to achieve not just the minimum net usable space, but also the 'balance ratio between that and the gross area'. The appointment letter (i.e. the contract) also specified the target unit cost per m² usable. Perhaps one could say this was conceived as a rational office building as opposed to the more idiosyncratic buildings the University had been procuring.

The building is five storeys including the basement, forming three sides of a courtyard. The south-east corner of the U has been cut back to mark the entrance and to protect an existing holm oak. The westward facing courtyard garden is an extension of the social space: when the doors are open this makes a good place for parties. A water feature and planting in the building's courtyard give a calm social space, and a line of beech trees and other external plantings reinforced a sense of peace greater than had occurred from other recent buildings on the Sidgwick Site.

One arm of the block contains the double-height library with top lighting to the reading room area. A series of rooms on one side of the reading room are used as teaching spaces and in the summer may be allocated to post-graduate researchers. Above the library is the ASNC (and Linguistics) wing with offices and a dedicated ASNC common room. The central section houses the drama studio, the social space opening onto the garden with seminar rooms and graduate study room and common room. The large seminar room is also used as a boardroom. The second arm contains more offices for the English Faculty.

The section through the library was designed to catch sunlight into the reading area. This works well in the library; however the ASNC offices above feel somewhat cut off from the garden though they do have their own terrace.

Relation to the outside is carefully considered. Many of the rooms open onto or overlook the garden or a terrace and are provided with full height French windows. South-facing windows have brises-soleil. The windows are extremely carefully designed and detailed with the A&M hall-mark of separating the glazed and light-transmitting elements from the opaque or louvred side panels which open for ventilation. All very elegantly proportioned and crisp and immaculate in appearance.

The cladding is in pink terracotta rain screen cladding (the alternative option would have been brick) whose colour has toned down over the years, somewhat to the relief of most users and passers-by. The colour choice was to mediate between the strong red of the History Library and the buff bricks of Music. A&M say: 'The composition of the elevations is dominated by the 3m module of the academic offices. This defines the dimension and the scale of all of the facades.'

The structure is a concrete frame with the concrete columns and slab soffits largely exposed, thus benefitting from thermal mass and night-time cooling and reducing the maintenance costs of repainting. However, this is one aspect that is unpopular with some, though not all, of the building users. Some dislike exposed concrete thinking it makes the building feel unfinished, 'rather like a multi-storey car-park', others feel the quality of the concrete finish is poor.

Internally the planning of the offices is based on a central corridor with offices either side. A&M worked hard on the corridors and the design of the doors to allow light and liveliness in the corridors while allowing occupants of each office to choose their preferred amount of privacy. Bob Allies describes this building being 'on the cusp' of thinking that academic buildings should not be very specific: what is the identity of the building and how flexible should it be? He commented that English and Criminology both responded to this approach in different ways, in response to their different briefs.

What do people think of the building?

The terracotta cladding was not immediately well received, but once its colour toned down, and the planting asserted itself, local opinion slowly swung behind the appearance of the building. What was never well regarded was the poor quality of internal exposed concrete finishes in the extensive exposed-concrete ceilings. Many academics like the concept of 'honest' architecture which shows the construction materials as they really are: but unfortunately, in the English Faculty building, nothing sensible could hide the imperfections of staining and inconsistency of colour.

Generally, the building is well regarded, heavily used and enjoyed by the users: they are happy with the air quality and the day-lighting too is considered good. The terraces and the courtyard garden are successful and well used in the summer, even though the best months are when the undergraduates are not around to enjoy them. The social spaces work well and are used in a nice way by students. But as far as staff go, the teaching in the Faculty still tends to be college-based, particularly in English, and some members hardly come into the building. So the tension between University and College still lives on in the Arts. The library is a great success and very well used.

Architect: Allies and Morison
Structural Engineer: Whitbybird Ltd
Services Engineer: Buro Happold
Landscape Architect: J&L Gibbons
Quantity Surveyor: Faithful & Gould
Contractor: Wates Construction Ltd

In the run-up to start on site there was a protracted series of staff changes, including senior staff, in the contractor, Wates, and that led to loss of teamwork and mutual trust, and delayed Wates getting a grip on the issues that always need to be resolved before the start of work on site. After work started, some unexpected underground services were discovered causing some re-design and delay. Delays cause tensions, and, unusually, the Representative User tried to take on some aspects of the role of client, without being able to take on the responsibilities of the client. The Wates site management on this project,

Figure 6.2 Faculty of English

however, was good despite those tensions, and despite a clear lack of sufficient support from higher levels in the contractor Wates. (Sadly, towards the end of the project the site manager's family was involved in a traffic accident. That caused a change of site management as he had to go to the country where his family was.)

Work was completed on 3 August 2004, 11 weeks late according to contract, but in time for the Department to get settled in before the beginning of term, starting with the move of the library. Negotiations continued after handover, and a negotiated settlement in 2006 made the out-turn cost £17.48M. At 15 per cent this project was by far the largest budget over-run of the 100 projects set out for this decade; most projects came in under or on budget, and this project and a few others with much smaller over-spends, brought the overall balance of all projects to almost exactly zero over/under total budgets.

Faculty of Criminology

The siting of the Criminology building was in itself controversial and at times the campaign by the University to get a building there had to be 'robust'. On one of the termly tours of building/development sites, the head of planning at the City Council stood on the steps of RFB brandishing his umbrella and saying that as long as he was in post, the University would never build on the site later occupied by Criminology. To be fair, and to his great credit, at the opening ceremony the chief planning officer publicly stated that, after all, the Criminology building was suitable for the site and had great merit. His pro-active advice and pressure undoubtedly led directly to design and planning improvements for many University buildings; the City and the University have much to be grateful to him for, more than they are aware of.

Design of the Institute of Criminology went more or less according to plan, and a project with a capital budget of £13.75M and annual running-cost estimate of £125k was set out in the Reporter (Reporter 1001) on 15 July 2001, but the problems of frequent changes of Wates staff took a heavy toll on the usual pre-construction discussions and agreements. The building has gross and assignable areas of 3,640 and 2,788m², and the project budget was £13.20M (£3,086 and £4,029/m²); the out-turn cost was £14.01M.

Case study – Criminology

The brief for Criminology is very different from English and ASNC and indeed is unusual for most University buildings. The Institute of Criminology has no undergraduate students, providing graduate research and in-house training only. Accommodation for police training is provided (in the basement) in spaces that can be divided up in different ways to suit the particular course. The Representative User was Professor Michael Tonry. The Institute, which had been set up in the 1930s, had a considerable investment fund to provide a new building.

The site is on the eastern boundary, rather off the beaten track and rather hidden behind the Raised Faculty Building, and

> is framed by unsympathetic neighbours, the offensively nondescript side of the Law building and the back of the Raised Faculty Building, as spoiled by an unsympathetic addition by DEGW Architects. In contrast to these buildings, the exterior of the Criminology Institute is reticently luxurious in dark purple-grey … Inside, there is … colour and a (great) feeling of richness.
>
> (Building Design 8 Oct 2004)

The main spaces are:

- the library
- space for in-service Police training courses in the basement
- individual offices with research offices at the top with controlled access
- Faculty boardroom
- internal bike store in the basement, with ramp access.

The new building cleverly creates a new courtyard by projecting forward the entrance as a pavilion with the main meeting room and café above. The library is on two levels (ground and first floor) with the main teaching rooms in the basement which extends the whole length of the building and also underneath the front garden court. The upper two floors contain offices, meeting rooms and research spaces including interview rooms for criminologists with criminals, sub-divisible in a variety of sizes. A ramp runs the length of the north side of the plan and the main staircase also runs on the long axis with a single line of roof-light above, throwing light down and opening up the levels to each other. The structure is concrete slabs and free-standing circular columns with the structural and planning grids separate.

When we look at the plan of Criminology we can see what Bob Allies meant as to the contrasting approaches in the two buildings. Criminology has large, generic, rectangular floor plates which can be used in a flexible way. The Director of Estates, stressing his role as client as different from the Institute's role as user, was interested in optimising the productive life of buildings with different, changing uses and this meant buildings had to be designed to be adaptable to accommodate changing research needs.

As with other buildings from this time onwards, there were good social spaces around the building, in this case also at basement level: day-lighting in the basement was from roof-lights up to the courtyard

and to the 'boules pitch' above. Ventilation is by natural cross-venti-
lation, and the doors are all wide enough for full access by wheel-chair
users.

The elevations of the main facades are of full-height windows and
vertical panels of louvres behind which shutters can be opened for
ventilation. The rhythm changes for the library where there is more
glass and therefore associated shading. The cladding is pre-cast concrete
and dark grey anodised aluminium, with metal reveals to windows on
the north but not to the south so the building looks subtly different
depending on where you are approaching from. The cladding of the
entrance pavilion uses the same materials but not to the same rhythm,
thereby expressing the different functions and announcing the entrance.
There is thus an amount of subtle variation within an overall harmonious calm. These materials relate well to the Raised Faculty Building
and makes a court with presence from what was once a left-over space.
Overall, Criminology, with its elegant corporate feel, presents itself as
very mature.

As in the English Faculty, soffits/ceilings are of exposed concrete,
but the quality of workmanship is slightly better, at least up to a
minimum-acceptable standard. In the new library there was installed a
then-innovative self-issue and security system based on Radio Frequency
Identifiers (RFID), the same system as is used in the Vatican. (Different
in use, however, as its books, as one of the librarians pointed out, could
only be taken out by the Pope himself.)

What do people think?

The general view, and that of many architects, is that the Criminology
building is more impressive than that of the English Faculty, with the
setting affording a solid but distinguished east boundary to the Sidgwick
Site by way of a quiet courtyard – a sandy boules court with trees and
an interesting water-feature – that responds well to the aspirations in the
original Casson and Condor plan. It also takes the eye away from the
featureless massing of the rear of the Law Faculty building.

Architect: Allies and Morison
Structural Engineer: Whitbybird Ltd
Services Engineer: Buro Happold
Landscape Architect: J&L Gibbons
Quantity Surveyor: Faithful & Gould
Contractor: Wates Construction Ltd

Why are the two buildings so different?

They both came out of the same architectural practice, at the same time. Bob Allies says both the practice directors worked on every aspect of both projects and certainly both show the preoccupation with window design and with proportional systems which characterise the practice's work. He feels it is not a matter of different design teams.

In that case it could be because:

- The clients (EMBS) were the same but the end users were very different, with the lead RUs having different interests, funding and ways of consulting their departments.
- The briefs for the two buildings were very different: English was to be used by many people including undergraduates; Criminology is not for teaching undergraduates but has specific training courses and research programmes.
- The Representative Users had very different attitudes to identity.
- Criminology had access to substantial funding which enabled them to achieve a more luxurious building.

Figure 6.3 Faculty of Criminology

Wates staff turnover continued, with reducing standards of site management. The German cladding company flew in, measured up the accuracy of the frame, showed that it was well below specified standards and then left site when they were due to start the work. (Once they did start, the employees flew into Stansted on Monday mornings, worked hard all week, sleeping somewhere, maybe on site, and flew back home on Friday afternoons.) Completion was on 20 September 2004, which was 26 weeks late per the contract, not quite in time to move in before the start of the new academic year. The cheerful opening by the Lord Chief Justice, Lord Woolf, was on 11 May 2005. (The publicity pictures, in the view of some, were impaired by a fishing gnome placed by a humorist on the bench above the water feature, noticed by no one until the end of the ceremony.)

The buildings for the Departments of English and Criminology were both late, and ran over budget, and were two of the three most problematic of 100 or so projects of this period of University expansion. Part of the problem was caused by the choice by the University of the contract-form PC Works with a Bill of Quantities (BOQ), less flexible and more formal than the ECC form of contract otherwise used for University contracts. It is not now clear why this contract form was selected, but it was almost certainly a mistake. When problems arose (for example, discovery of live utility services under the edge of the site) and there had to be changes to the extent of works, the chosen form of contract was unhelpful, and the contractor was later able to exploit that. The root cause of lateness and budget over-run though was the staffing problems in the main contractor, Wates, with a succession of different managers in the months before agreeing a contract and during work on site. Also, although staffing of the University's Estates Department was generally remarkably stable considering that there was a boom in the construction industry, and far better salaries (and perks) were on offer in the private sector, half way through construction of the buildings for English and Criminology, the Deputy Director of Estates in charge of these two projects left to take up a new job as Director of Estates of a large London university, and that caused a gap, with a significant loss of knowledge and expertise in the University's procurement team, and the difficult job of finding a successor prepared to come for the salary on offer. Well into construction, the Director of Estates asked the Chief Executive of Wates to visit; he came, clearly demoralised, and then resigned the following day, not of course just because of these projects.[10]

The dispute with the contractor was progressed as though it was going to be resolved through the nationally legislated adjudication process, and a six-month programme of adjudication was agreed, with one claim being assessed while the second was worked up. Lawyers were employed to advise, but settlement was agreed before it got that far, and the heavy legal costs common in legal disputes were avoided. The total project unit cost, of Criminology, was £3,086/m² gross, compared to £2,991 for English, so not much in it. It might be concluded that the tender prices were too low, and that the out-turn costs were reasonable.

Summing up the Sidgwick Site Master Plan and the two buildings: the national journal *Building Design* published a major article noting that the architectural press in 1952 had mounted an attack on recent Cambridge and Oxford buildings as having 'almost no contribution to make to the art of architecture and only reflect the barrenness and timidity of academic taste'. It then notes how

> By 1964 the university was well on the way to becoming a modern architectural zoo – one which contained some useful, domesticated creatures but also some wild and savage beasts. Today, thanks to the employment of Allies and Morrison as master planners of the Sidgwick Site, one important part of the zoo is being tamed and some interesting new animals added to the collection ... The English Faculty building can be seen as a return to sanity and to the civilised vision of Casson Condor as well as being a sensible, practical and humane design in itself. It also embodies a new approach by the University's Estate Management and Building Service, with an emphasis on better space for users and a whole-life costing while not forgetting design quality. It is also a deliberately flexible building, reflecting uncertainty about changes in teaching methods and about the balance between the faculties and colleges. The nearby new building for the Institute of Criminology has a similar character and structure but a more substantial feel as the budget was more generous.[11]

Faculty of Music

Meanwhile, a much lower profile but in many ways a more difficult project had been proceeding: the upgrading and re-modelling of the Faculty of Music's building on West Road at the edge of the site. (The original building for the Faculty of Music was designed by Sir Leslie Martin, he of 'The Martin Centre', in the late 1970s. From 1956 Martin was made head of the Faculty of Architecture, and Colin St. John Wilson was his assistant.) The first difficulty was the long and rocky road to getting enough funding, and then doing many re-designs to get a sufficiently good project within the very tight budget of £4.0M for the upgrading building works including £0.96M for revamping the noisy and unsatisfactory air-conditioning (cases of fainting were getting more common); the work was completed for a total of £3.22M. The architects for the project were Cowper Griffiths Associates, and the engineers Buro Happold and W S Atkins, and Needlemans as QS. The contractors, who also did well, were Marriotts. Work on site started on 30 July 2001 and finished on 1 January 2003, with the air-conditioning works (which included a lot of structural work in the roof, designed by Arup) done by contractors Knights Warner between 11 June 2001 and 1 October 2001. Despite the harsh pressures of getting a lot of work done on a small site for a necessarily limited budget, the team worked well together. There was a kerfuffle after changes in Faculty staff and then changes in ideas about the acoustics of the hall's seating, and some of what had been put in had to be taken out.[12] Some further work was done on the toilet area in October 2003 at construction cost of £190k.

Subsequent to the period covered by this book, the long-held aspiration to develop a building for the Centre for Research in the Arts, Social Sciences and Humanities (CRASSH), along with a new department called the Department for Politics and International Studies, came to fruition in 2011 with a 4,334m², £11M, Alison Richard building to the east of the Faculty of English building. There is a large atrium for informal meeting, and offices and seminar rooms. The building achieved a BREEAM Excellent rating. The style of the brick-clad building fits in very well; it was designed by architect Nicholas Hare (Allies and Morrison had grown very fast, changing somewhat its ethos, and some in the University worried about its response to the problems caused by Wates.) Construction was by R G Carter. The building was on the plot previously reserved for a new building for the Department of Land Economy, which was designed by Allies and Morrison but never got sufficiently funded. Some of the work designing that building was of later use on the building that did go up.

The Granta Backbone Network is estate-wide infrastructure; it comprises 32 km of ducts and 83 km of copper cables. It was installed prior to the development of the Sidgwick Site and all the developments described in this account at a cost of £3.6M.

Projects covered in Chapter 6

	£M
Faculty of Law – (pre-decade)	(22.10)
– acoustic screen	>1.00
Upgrade of RFB	6.58
	1.07
Divinity	7.97
Minor Works: re-boilering of site	1.80
Site master plan by Allies and Morrison.	Fees
English	17.48
Criminology	14.01
Upgrade of Music, incl subsequent minor works up-grade	3.41
Upgrade of History Faculty building	1.76
Sidgwick Site infrastructure	7.97
Total in decade	63
Granta Backbone Network (see Ch 3) (pre-decade)	3.6
Designed and planned in decade but aborted:	13.4
Centre for SE Asian studies/East Asia Institute	6.20
Dept of Land Economy building	7.20

Notes

1 Reporter 1949–1950 pp.831 et seq.
2 Alistair Fair, 'Arts Faculty Building, Sidgwick Site', in Marco Iuliano and Francois Penz (eds), *Cambridge in Concrete* (Department of Architecture, University of Cambridge, 2012, pp.56–58).
3 This refers to a list of notable buildings which by law may not be altered or demolished without special permission rarely granted.
4 There was a story that circulated about such relationships during the subsequent Foster-led building project at Stansted airport (the Law building is sometimes compared to the terminal at Stansted): BBC TV had recently re-run the film 'The Eagle has Landed', and as Norman Foster arrived late for a Stansted meeting, swooping down in his helicopter throwing mud and dust over everyone, the engineer muttered to the surveyor, 'the ego has landed'. Just a story.
5 Scott Keller and Colin Price, *Beyond Performance: How Great Organizations Build Ultimate Competitive Advantage* (Wiley, 2011).
6 Poe later became mandatory for all major public sector projects in England and Wales from 2012 as a crucial method of improving the generally dismal record of learning from experience in the construction industry.
7 Adrian Durant, 'Comfort and Joy: Cullinan in Cambridge' in *Architecture Today* (Issue 114, January 2001).
8 Previously, and subsequently, DEGW had a superb reputation for space planning, and for associated architecture, but not at that time.
9 The campaign against skate-boarders was not helped by the fact that one of the champions of the skate-boarding community was the son of a leading academic working on the site. The son discovered that the VC, Alec Broers, was a good ice-skater and tried to persuade him that he would get a really cool image if he would roller-blade down King's Parade to a Council meeting. He wasn't persuasive enough.
10 Later, when the Director of Estates was subsequently working as construction policy director in HM Treasury he went to visit the successor chief executive of Wates on an entirely separate issue, but before he could get a word out, the new CEO said that the first directive he had issued on taking over the company was that never again was it to take on a project unless there were enough staff of sufficient ability to cope; a revealing comment volunteered.
11 Gavin Stamp, in *Building Design* (Issue 1465, 8 October 2004).
12 Perhaps the happiest moment for the team was at the official opening concert when the (very musical) VC, Sir Alec Broers, after praising the orchestra's performance, commented on the new building works and added that the air-conditioning system had been changed, although 'from the quietness it must be switched off': in fact it was fully working. Then everyone toddled off to the new and rather expensive bar.

Table 6.1 Overview of capital projects on the Sidgwick Site

Chapter 6. Sidgwick Site.	Start	Finish	m2 gross	m2 ass	Architect	Svcs eng	Str engr	Contractor	£M
Law new building		17/06/95	8,000		Foster	YRM		Taylor W'row	22.10
Law--acoustic remedial work					Foster	YRM			1.00
RFB--refurb	01/07/99	01/07/00	5,277		DEGW (Dutch) then Fitzroy	Hannah Reed	Roger Preston	Interior	6.58
RFB--refurb	01/05/06	01/09/06	1,200		Woods Hardwick	Mott Macd'ld	Mott Macd'ld	Dean Boyes	1.10
Faculty of Divinity	01/12/98	01/10/00	2,889	2,000	ECA	Whitby Bird	Edmund Shipway QS	Sindall	7.70
Boiler decentralisation						EMBS	EMBS		1.80
Faculty of Music--air conditioning	01/06/01	01/11/01			Annand and Mustoe	EMBS	Arup	Knights Warner	0.96
Music Technology & Performance area	10/09/01	01/08/02		1,000	Cowper and Griffiths	Buro Happold	W S Atkins	Marriott	2.26
Sidgewick Infrastructure					Allies and Morrison	Buro Happold	Whitby Bird	Wates	7.97
English	01/07/02	03/08/04	4,335	3,057	Allies and Morrison	Buro Happold	Whitby Bird	Wates	17.48
Criminology	15/07/04	01/09/04	3,640	2,788	Allies and Morrison	Buro Happold	Whitby Bird	Wates	14.01
History --Seeley Library, glazing, temp con	19/06/06	22/09/06			John McAslan	Arup	Samuely	Jackson	1.76
Land Economy	Aborted		2,205	1,514	Allies and Morrison	Buro Happold	Whitby Bird		7.19
East Asia Studies	Aborted			1,500	Allies and Morrison	Buro Happold	Whitby Bird		6.19
CRASSH, Dept Politics & Int'l Studies.	15/12/09	15/05/10	4,344		Nicholas Hare			R G Carter	16.00
Cost of completed works, 1996 to 2006									62.62

7   The rise and rise of health research

After living for only 39 years, Dr John Addenbrooke MD died on 7 June 1719, having destroyed most of his documents in a great bonfire. Packed into that short life, he started his studies at St Catharine's College (then Catharine Hall) aged 17, planning to progress into the church, then directed his learning to medical matters so as to gain his MD in 1710; then he married and so resigned his Fellowship at St Catharine's and moved to London around 1711. He made no significant contribution to medicine himself but, inspired by the Quaker-led voluntary hospital movement, he left his fortune of £4,676 'to hire, fit-up, purchase or erect a building fit for a small physicall hospital for poor people' in Cambridge. Delays by trustees and prolonged legal rows so hindered the project set up to meet his wish that it wasn't until 1766 that a hospital of four wards, each with five beds, opened near the boundary of the town. Peter Richards, the editor of the magazine 'Cam', describes how

> money ran out before the hospital was finished, and with no endowment, the completion and running of the hospital relied on benefactions from landowners, farmers and clergy, and the income from the annual University Sermon in the University Church of Great St Mary's (£161 in the first year alone), and the profit from performances of Handel's Messiah the following night with seats at 5 shillings. Despite struggling finances, and the terrible nursing standards, the hospital grew apace: by 1820, to 46 patients, 100 by 1840.[1]

Growth in patient numbers was such that a new hospital, designed in brick, tiles and terracotta by Matthew Wyatt, opened on Trumpington Street in 1866. That hospital grew steadily, with major additions in 1824 and 1834, and a new top floor in 1915. However, by the late 1940s it was under such pressure that the decision was made to replace it with a new hospital on a site near Trumpington, to the south of the city, thereby creating an increasingly emotive question about what should be done with the original hospital once vacated: that story is picked up in Chapter 8. 'Old' Addenbrooke's was sold to the University in March 1985 for £5.75M.

Planning the new hospital was started by the recently established National Health Service in 1948 with a budget of £736k for an out-patients' department, Ward A, and a radio-therapy unit. It also had a major trauma unit to deal with big accidents. HM The Queen opened the first phase of the

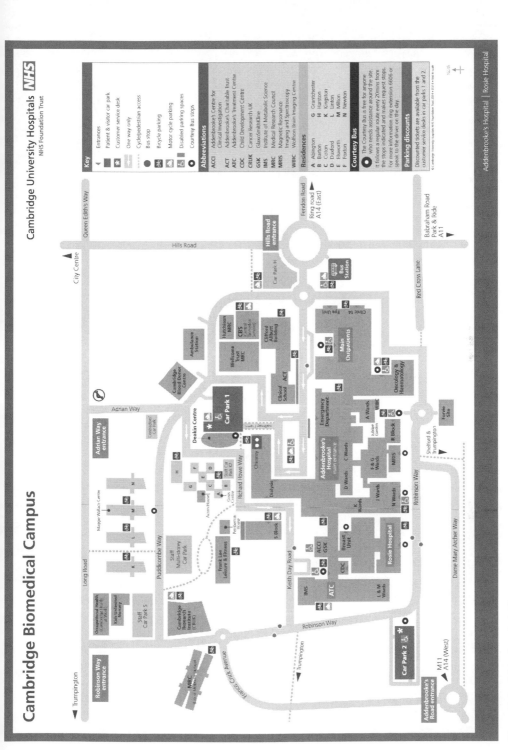

Figure 7.1 Addenbrooke's Site map

By courtesy of Cambridge University Hospital NHS Trust

hospital in 1962. Phase 2 was built by Mowlem with a budget of £13M, and the last patient transferred from 'Old' to 'New' Addenbrooke's in 1984 after its further expansions, first with the full ward block and then with the MRC Laboratory for Molecular Biology. (This was subsequently replaced by a new building about twice the size, for about 600 people including 440 scientists, designed by RMJM and built by BAM Construction, finishing in 2012 at a cost of £212M.) The newly founded University Clinical School opened in 1976.

Among the larger NHS construction projects were the Teaching Spur (1,502 m^2) in 1959, C and D Wards (26,826m^2) in 1973, Theatres 9–11 (3,233m^2) in 1983, F and G Wards (11,371 m^2) in 1984, Addenbrooke's Treatment Centre (32, 892 m^2) in 2006, the staff car-park described below, and the NCP car-park for visitors, huge at 27,170m^2, in 2008. The Medical Research Council opened its Laboratory for Molecular Biology in 1962, and it very soon became a worldwide research phenomenon. Further projects are noted below, taking the various sites in the Addenbrooke's hospital area in turn.

Forvie Site

In the early 1990s, the newly formed University Divisions of Neurology and Neurosurgery (later to form the University Department of Clinical Neurosciences in 2004) were located within the B spur of Addenbrooke's Hospital and were greatly enhanced by the creation of the Wolfson Brain Imaging Centre (WBIC) and the E. D. Adrian Brain Repair Centre (later, the John van Geest Centre for Brain Repair). WBIC is a research facility incorporated within the envelope of Addenbrooke's Hospital Neurocritical Care Unit and dedicated to imaging function in the injured human brain of critically ill patients. It is unique worldwide in bringing such technology to the bedside of critically ill patients: sick patients cannot go to the scanner; the scanner must come to the patient. The name reflected generous funding from the Wolfson Foundation (1994) supplemented by a £4M award from the MRC through the Technology Foresight Challenge, and support from The British Heart Foundation, the University of Cambridge and SmithKline Beecham (now subsumed into Glaxo SmithKline). The space rented for Smithkline Beecham was upgraded in a £2.79M project. A new 3T MR scanner was installed in 2006 (£1.8M including building work). Subsequently, the WBIC's success led to a need for more space for a second high field MR scanner (dedicated to metabolic studies and funded by the Wellcome Trust), for the expanding PET radiochemistry programme and for the growing number of scientists and the high performance computing facility. A design worked up by Frank Shaw Architects formed part of the later refurbishment and extended an extension of the WBIC. (Frank Shaw Architects subsequently did a lot of work for the University designing 'Cat 3 containment labs'; always challenging to satisfy the greatly increasing requirements of Health and Safety legislation as design and sometimes research requirement near completion.) In a four-phase project, the Institute of Public Health and the Molecular Resonance Imaging (MRI) Centre were developed; the final phase at £950k delivered 526m² gross space, designed by Cowper Griffiths Associates and Hannah Reed, and built by Rattee and Kett in 1997/1998.

In June 2004, the Parke Davis Building, then owned by the pharmaceutical company Pfizer which was consolidating its activities at Sandwich, was bought by the University for £3.5M. It was then refurbished at a cost of £12M, including some post-contract works, to house the WBIC Molecular Imaging Laboratory (a satellite of the WBIC), the Anne McLaren Laboratory for Regenerative Medicine and the BHF Cardiovascular Unit. The architect was Saunders Boston, the engineers Silcock Dawson and Hannah Reed; the contractor was Gleeson Building Ltd; main works were completed on 20 June 2007. There was also minor works refurbishment in the Brain Repair unit at a cost of £750k.

'Island Site'

Construction on the so-called 'Island Site' for the Wellcome Trust/Medical Research Council amongst the National Health buildings was extensive, and challenging. A University-led project called Phase I, of 10,246 m² gross (7,015m² assignable) with a budget of £24.8M was completed in October 1998. It was designed by Feilden and Mawson architects, engineering by Whitby Bird, built by Shepherd Construction (which was soon to build the first buildings at West Cambridge) and managed by the Estates Department of the University assisted by Edmund Shipway. The next building on the Island Site was set up jointly and then project-managed by the MRC, in collaboration with the University and Addenbrooke's. This building was completed in 1998 and the official opening by HRH The Princess Royal with the Director of the Wellcome Trust, and the Chairman and CEO of MRC, was held on 30 October 2000. Some 220 scientists worked in the building; one of the research units was the Juvenile Diabetes team, in part funded by the Juvenile Diabetes Foundation, a large charity in the USA/ UK. (Juvenile diabetes was then said to be the fastest growing disease in Europe, though in part that reflected rather the considerable increase in the proportion of children being screened.) The out-turn cost was £24.8M, on budget, and good value for a building that size; it was not perhaps a beautiful building. Its starkness and the shape led to it being referred to as 'the Titanic'. (A dark and rather ugly building put up at this time for a law firm in north Cambridge was sometimes referred to as the 'Belgrano' after the Argentinian ship controversially sunk during the Falklands war.) Design of Phase 2 was started by the MRC, in mid-1998, by Feilden and Mawson, WSP and Hannah Reed. Planned built areas were 3,800m² gross and 3,116m² assignable: Laing was the main, that is, the lead, contractor, starting work on site on 31 August 1999 and finishing in March 2001, three weeks late, with a project cost of £10.55M.

All building developments on the Addenbrooke's site involved both the NHS Healthcare Trust (whose title became 'Cambridge University Hospitals NHS Foundation Trust') and the University to varying degrees. Until 2004 there was a bi-lateral University/Trust development committee which worked extremely well. It was small: co-chaired (very effectively) by the University VC and the chairman of the NHS Trust. It was comprised of a few of the most senior clinicians, notably (in many ways) the Regius Professor of Physic, the chief executive of the Trust and a small number of administrators including the two responsible for development of the two respective estates. All major developments on the site were discussed pro-actively by this joint committee. There was clearly understood management of the respective objectives: teaching and research as applicable to patients for the University, and patient care and education of health professionals for the Trust.

Medicine/Virology

A project funded by the University following a successful bid to the Joint Infrastructure Fund (based on Wellcome Trust, Government (DTI) and Hefce funding, managed by Hefce see Chapter 2), was the reburbishment of the Medicine/Virology facility, the feasibility study being authorised late 1999 with a cost-estimate of £2.215M, firmed up by Autumn 2001 at £2.118M. Work by contractor Sindall Construction (later to be Morgan Sindall) started on site on 20 November 2000 to an architectural design by Feilden and Mawson, mechanical/electrical engineering design by Silcock Dawson and structural design by Hannah Reed. As with an increasing proportion of projects, there were significant technical challenges to meet the increasingly stringent regulations on 'health and safety': the top three items on the Risk Register all related to that:

• delay associated with decanting and construction of Category 2/3 facilities
• possibility of toxic/contaminated ductwork services
• asbestos within structure/services.

Nevertheless, work was completed on budget in February 2002, with one room shunted out into a further phase of work at the users' request.

Strangeways

A tricky and rather prolonged project was the development of the Strangeways Research Laboratory which had been 'born' in 1905 as a small research hospital founded by Dr Thomas Strangeways in Hartington Grove, Cambridge. There were two wards each of three beds, and the coal-shed was converted into a laboratory. Before long, the treatment programme and the research were going very well despite the scarcity of funding: Mrs Strangeways noted that 'we were spending one third of our income of £150 per annum upon it, and we had two children'.[2] (She noted later that 'The converted coal-shed was not suitable for some of the research, and part of the laboratory work was moved to the Royal Mineral Water Hospital at Bath where a room was fitted up with a pathological laboratory'.) A site for a new building on the current site in Worts Causeway, Cambridge, was bought for £300, a building costed at £1,600 was designed by a London architect who stayed with the Strangeways family each weekend during construction, and the unit moved in on 24 May 1912; it was renamed the Strangeways Research Laboratory in 1928. Over the years it expanded, as so many small and personally-led and funded scientific and medical research facilities did prior to the government taking on responsibilities for much of the development of research facilities. The Laboratory has had a distinguished history as a centre for research into rheumatoid arthritis and other connective tissue disorders. The Strangeways Trust was set up as an independent charity: one of five Trustees was Lord Adrian, and one of the Advisory Council was Sir John Cockroft, Master of Churchill College, and founder of the Cambridge Society for the Application of Research (CSAR). Now the sole trustee is the University of Cambridge.

In 1997, when the building became a centre for genetic epidemiology, a larger expansion project was conceived and then implemented. A lease had to be agreed from Strangeways to the University, and the site had significant access and use-restrictions (reasonably) imposed by the planning authority. Then a revision to the user requirements, discovery of an unknown mains service and late-delivery by the designers resulted in delay: the project was completed 13 weeks late. The new labs were handed over on 29 August and the offices on 19 October 2003 with 1,888m² of gross area (1,197m² of assignable space). The out-turn cost was £5.67M compared with the original budget of £5.59M and a revised budget of £6.10M. The architect was Feilden and Mawson, the engineers Roger Parker Associates and Hannah Reed, and the work was done by Shepherd Construction. The building housed the Department of Public Health and Primary Care, the Centre for Cancer Genetic Epidemiology, the Thyroid Carcinogenesis Research Group and the Foundation for Genomics and Population Health.

In 1998–1999, a £750k refurbishment project was carried out, and opened by the Duke of Gloucester on 28 October 1999.

Staff car-park

There was a lot of staff parking on the site, 3,200 spaces, but with rapidly increasing numbers of staff (over 8,000 car movements on the site each day) and increasing irritation from the local house-owners whose streets were being used by staff, and more particularly by patients and their relatives avoiding high car-park costs at the hospital, there was a strong demand for at least one large new car-park on the Addenbrooke's site. So the next University project to start on site at Addenbrooke's was a new car-park designed to provide 1,255 car-park spaces, with a very high level of security for staff and their cars, with 64 CCTV cameras, automatic entrance/exit control, high-security lighting and emergency buttons all over the place. Anshen/Dyer Architects, a London-based offshoot of the California-based company Anshen + Allen, and Mountfort Piggott Partnership were the architects, the engineering services and structural/ civil engineers were Faber and Maunsell Ltd and Zisman Bowyer and Partners; the external project managers Davis Langdon & Everest, universally known as 'DLE'. (At that time, DLE was on a buying spree hoovering up companies, so DLE was said to stand for 'Davis Langdon & Everyone'.) The constructors were Laing O'Rourke. Ray O'Rourke ran a small Irish concrete company which had recently taken over Laing, UK's oldest, and according to many, most revered family construction company, and whose heads, Martin and Chris Laing, were good friends to the University. Martin Laing had been a brilliant chairman of the Construction Industry Training Board, just about the only ITB to escape the Thatcher ITB axe.

Laing's had got into dire financial straits during disastrous construction contracts for the Cardiff Millennium rugby stadium, a Private Finance Initiative hospital in Norwich and the National Physical Research Labs; probably also a financial loss on the Defence Academy at RMCS Shrivenham. It was said that Laing went on sale for £18M, was bought by the Irish concrete company O'Rourke for a nominal £, with a dowry of £12M as cover for residual contractual liabilities.[3]

The design of the car-park was remarkable, necessarily functional but elegant and dignified by car-park standards. Planning approval was granted on 22 January 2003; work on site started on site on 14 April 2003.

Despite these issues, and a lot of technical problems causing the contractor to be awarded some extra time without costs, the project finished to time and budget of £9.2M; the first car rolled into the car-park on the day of the opening ceremony, 15 March 2004 with various of the local 'great and good,' shivering on a chilly day. The Chief Constable, whose tertiary education had been at the University of Oxford, awarded a Safer Parking Award as the car-park had been judged among the best car-parks that year by the national professional body, British Parking Association. The car-park worked well and at low cost, and staff did feel safe, a crucial aspect of car-parks.[4]

ICRF/CR-UK

Back in 1999, the 'Regius Professor of Physic' (in most universities, the 'Dean of the Medical School') was Sir Keith Peters, (certainly one of the 'powerful academics' mentioned in Chapter 2. He described himself as 'an old man in a hurry'; there is no record of anyone contradicting him) was engaged in, among many other fund-raising projects, discussions with the Imperial Cancer Research Fund (ICRF), Li Ka Shing of Hutchison Whampoa, a Hong Kong conglomerate, and others, to fund a huge new cancer research centre for ICRF, with some participation by University and NHS researchers, to be built on the playing field land owned by Downing College at the west side of the Addenbrooke's site.

A few days before jetting off to Hong Kong, the Regius asked the Director of Estates for some drawings of what such a totally unspecified building might look like to a potential benefactor. This task was achieved, to a degree that pleased the Regius and apparently satisfied the potential benefactor, by employing at two days' notice an architect well recommended by a colleague in Percy Thomas Partnership, one Ronald Weeks by name. The Regius guessed how much funding he might get from the potential benefactor and from other sources; the Director of Estates divided that total by the likely cost per square metre of gross area in the building, did some rounding to a figure of 10,000m² and then divided that between building functions on advice from the medics (figures from an earlier JIF bid were helpful in breaking down the total area into lab, office and other space) and gave that number to the architect as what was euphemistically called in the resultant document 'an unrefined brief'. One number quoting a built area split into sub-sets to give functional space require-ments, with a note on the use of a building, must be the shortest statement ever to be called any sort of brief: maybe the briefest brief (with potential for the Guinness Book of Records). Ronald Weeks visited the site on 4 May 1999, and then sat for three days at a coffee table in the Director of Estates' office with a water-colour set and a box of crayons. The water-colours and a few sheets of procurement plans were clutched by the Regius as he jetted off to meet with the potential benefactor.

Meanwhile, discussions had continued between the University, fronted by the Regius, and the ICRF Council, led by Sir Paul Nurse, later a Nobel Laureate.[5] On the morning of 8 December 1999 ICRF voted to agree to the proposal to re-locate ICRF lab teams and support staff to Cambridge, and into a new building at Addenbrooke's. Sir Keith Peters was away from Cambridge, whizzing and fund-raising, but emails being what they are, within a couple of hours all in the University who needed to know were responding to the news. Then on 19 September 2000, news broke of a donation of £16.5M from Hutchison Whampoa for this project.

This project was clearly already shaping up to be complex. As soon as funding was looking promising, negotiations started to buy land close to staff accom-modation on the north-west side of the site, and co-located with the site of the next NHS staff accommodation project. The owner of the land, Downing College, was using it for playing fields. The Master of Downing at that time was Prof Sir David King, who soon was to leave to take up post as the Chief Scientific Advisor to the Government and personal adviser to Tony Blair; a huge outbreak of Foot and Mouth was his first challenge. The Vice-Master of Downing College, Stephen Fleet, was the previous University Registrary, so he 'knew the system', and how much the University needed the site. In the last few days of David King's time as Master, a deal was finally struck after discus-sions, which Sir David called in his letter to the VC on 21 December 2000, 'lengthy', something of an understatement: there had been a huge number of interminable meetings over months, with the College and its land agents. Part of the complexity came from the strict requirement in government legislation about the need to replace playing fields when such land was developed. After a huge amount of deliberation by the University the decision was taken to do a post-purchase land-swap with Addenbrooke's in order to get a better shape of plot for the new ICRF building and for the adjacent multi-storey car-park described above. As the Addenbrooke's '2020 Vision' (September 1999) document put it, 'the University and the Trust, working together in an exceptionally complex land deal, acquired the former Downing College playing field, filling in a major lacuna within the site boundaries'.

The selected architects were Anshen Dyer. David Langdon & Everest were appointed as external project manager. Over Summer 2000 an outline planning application was developed, authorised by the Planning and Resources Committee. Managers at Addenbrooke's and many decision-makers in the University were convinced that it was important to look after the social aspects of development and so it was decided to package together, for planning appli-cation, the proposed cancer research building and 280 affordable ('key worker') housing units. That social commitment helped negotiations with the planners, but for the local community and its council representatives the main issue was traffic, and the addition of housing exacerbated that concern.

Pressure on the procurement staff of the Estates Department project team had been growing exponentially: by way of illustration, the agenda for the Buildings Committee meeting on 8 November 2000 had seven substantial 'decision-papers', and reports on another 31 current University projects of over £1M budget. Hundreds of millions of pounds were at risk through the unprecedented rush of projects in the University. As discussed in Chapter 5, recruiting was extremely difficult during this period of 'boom' in the construction industry, with salaries in practices and companies far greater than was allowed to be offered to University administrative staff whose salaries were very rigidly tied to academic staff salaries. Some senior estates staff were working daily from morning till well after midnight, thereby adding to the risk of error in the fine judgements that had to be made in negotiations with tough professionals from industry. Regularly but not frequently, the Director of Estates reminded the VC of the risks of the escalating capital programme and how those risks were increasing and had become high even by industry standards. If, as had a fairly high level of probability, one project went badly wrong and sucked in resources of time and available skills at the expense of other projects, they in turn would be more likely to go wrong in a downward and hugely expensive spiral. If a quarter of projects went 25 per cent over budget, as was not at all uncommon in public procurement, that could have cost the University about £45M in payments and costs: if even just half a dozen larger projects went 25 per cent over budget, and as a consequence of the diversion of management time to those projects, half of the other projects went just 10 per cent over budget, then the University would have been up for about £35M unplanned extra cost. The wider issues around overstretch, risk and the associate effects on quality of decision-making and on quality of life are discussed further in the final chapter.

Only once in the decade 1996–2006 did the Director of Estates make a strong case that the Estates Department should not take on a project, and that was this cancer research building project: he noted that most of the people using the building would not be University staff, and that it should be an ICRF or Addenbrooke's project, both by logic and by pragmatic management of University risk. The Regius, Sir Keith Peters, a close ally of the VC, was present at a meeting to discuss the issue. The discussion was hard-fought but the outcome, predictably, was that the Director was told to get on with it, which of course he fully accepted. No correspondence about that meeting was entered, nor should it have been, but it took the University beyond reasonable risk. As it happened, events took the project, and hence other projects needing the same staff resources, very near to, though not into, serious failure.

At the Buildings Committee meeting of 28 September 2000 a preliminary budget of £32M (for 30,000m² gross space in two phases) exclusive of site purchase, inflation, and various elements of equipment, was agreed as 'an initial outline budget'; this was prior to the planning application (and hence without knowledge of the level of cost needed, after negotiation, to meet local government requirements for infrastructure associated with the project, as per the procedure set out in Section 106 of the Town and Country Planning Act; see Chapter 3), and prior to the level of design that produces drawings and specifications for the planning application. By the Buildings Committee meeting of 8 May 2002, planning consent had been achieved although final negotiations of Section 106 commitments (particularly about the housing aspect of the application) were still to be settled. Purchase of the land from Downing College (and, as part of the deal, Downing's purchase of land at Laundry Farm, Barton Road), for the ICRF building and the adjacent multi-storey car-park was finalised for £8.0M in July 2002.

Laing O'Rourke had been earlier appointed on a first-stage contract to assist with design and tendering of works packages. It was now agreed to proceed to construction contract (the second-stage contract) once price-certainty, from receipt of works package tenders, reached 90 per cent of budget. (Details of how two-stage tendering works are set out in Chapter 5.)

The project was going well, but then a serious problem for its procurement arose after ICRF merged with Cancer Research Campaign to form Cancer Research UK; it soon became apparent that a very different approach to the project was being taken by the new CR-UK. The first straw in the wind came when during a late-Friday afternoon telephone call with the Representative User, Dr John Tooze, it became apparent that the call was 'on broadcast' and that there were others were in his room, commenting on the conversation. When asked about that, Dr Tooze said that a new set of lawyers had been brought in and had to be included in all significant discussions. At a meeting in London very soon after, the new lawyers sat next to Dr Tooze challenging most matters, and insisting that all agreements from previous meetings on acceptable savings ('value engineering') to meet budget were now 'at large', even though they had been minuted and the Minutes formally agreed: in construction law, these were not contractually enforceable even though clearly agreed. For example, a saving of about £100k from a marginal reduction in speed of a lift which had long been agreed was suddenly refused. This 'volte face' put huge stress onto all aspects of the project, and eroded mutual trust and confidence in the project right across the team. It was widely rumoured that members of CRC staff didn't want to relocate to Cambridge and that a senior official in CRC, now senior in CR-UK, had taken their side, and that she wanted to abort the project. It was not possible to default on the contract, so the tack, it appeared, was to make the project so difficult that the University would agree to abandon it. Whatever the truth of the matter was, from then on morale among project staff was low, notably lower than on any other University project. In complex procurement, mutual trust is all. Well, not all, but a great deal.

There were two other issues that complicated the project: archaeology and ground conditions. The site generally had been used by the Romans much more than had been anticipated by the archaeologists who had thought that it was far from the main Roman route through the site. It is of course delightful, and people are pleased and interested when archaeological finds are made, but it causes great delay and considerable extra cost in meeting the hugely increased requirements set by planning law. Secondly, site condition problems were compounded by extensive presence at a depth of about 17 metres of an odd soil which is classified as a chalk but often acts as a clay, and has very different properties, and needs different size and depth of piles to get enough bearing strength to support a building.

CR-UK added £1.4M (from interest already gained on their funding) into the project budget so as to get further facilities, and design and archaeological work moved on, the latter at a budget of £1.2M, including a fee of £200k to Cambridge University Archaeology Unit, which, although in the University, worked as a commercial organisation. It was difficult to get archaeological work done and accepted by the planners at that time by other than by one of a fairly small number of teams, including Oxford and Cambridge Universities.

Then further disaster struck. There had been subtle signs of tension between the architect team working on this project and their practice, and it was assumed that, if indeed there was such tension, it was because of the tension between the practice, Anshen Dyer, and its US holding company practice. Then, without any warning, word came that those architects working on this project had resigned from Anshen Dyer (for reasons totally unrelated to the CR-UK project to which they felt great loyalty) and were setting up their own small practice. The University thus had a hellish dilemma: if the design contract with Anshen Dyer were terminated, which would be legally expensive, and a new contract signed with the project architects' new practice, there would be very little professional indemnity cover available from such a small, new practice, and, further, this project would represent a very large proportion of their work, never a healthy position in an architectural business. On the other hand, if the existing design contract with Anshen Dyer was continued, none of the architects who had worked on the project would be available and new architects would have to get up to speed on a very complex project on which morale was already low, and without any help from the architects who had left and who had made it clear that they would not be co-operating with their former practice. Indeed, there was an absence of communication between them. Clearly, both options were serious, and bad. The Director of Estates advised a concerned Buildings Committee meeting that the lesser of the two evils was to continue the existing contract with Anshen Dyer, and that was agreed after much discussion. Matters were not helped when the new architects appeared to have much less enthusiasm than their predecessors, and when many of the key members of the project team left for other jobs; the Estates Department project manager emigrated to New Zealand to a good civil engineering job, the Deputy Director Estates, a crucial member of estates staff on this and many other projects, went to be Director of Estates of a London university, and replacing him in a buoyant construction market was very difficult. (Sadly, he was subsequently very seriously injured when a motorist drove straight out of a side road and knocked him off his motor bike, leaving him confined to a wheel-chair, but still active and appreciated.) The external project manager also left the area. A great deal of University estates staff time was spent on trying to get design back on track and restore morale and confidence in the project until that began to stabilise, albeit at a fairly low level.

Planning approval was granted on 22 January 2003 and the design team

Figure 7.2 CR-UK

of Anshen Dyer, mechanical/electrical engineers Zisman Bowyer and struc-
tural/civil engineers were, as planned, novated to work for the contractor, a
joint venture of Laing and Crown House. Enabling works started on site on
14 April, and progressed well. Visits by school children, in March 2004 for
example, were arranged and were popular as part of the National Architecture
Day (an annual open-day which came out of the Latham Review working
parties, as did the 'Considerate Constructors' scheme, now almost universal on
British building sites, making them less intrusive and annoying for neighbours).
Construction proceeded reasonably well for much but not all of the contract.

A cheerful 'Topping-out' ceremony took place on 19 April 2004. But then
the contractor hit a serious problem with its cladding sub-contractor, Pluswall,
and that persistent dispute set the programme back, as did late delivery of
design information and changes in CR-UK specification for the basement,
and new, demanding Home Office requirements which necessitated removal
of some work already done. The ground floor was financed, and was to be
occupied, by GSK. Completion of the works slipped from November 2005
to the end of March 2006 for the main building and 30 June 2006 for the
lecture theatre at the side of the main building. (As it turned out, many of
the staff who had been predicted to be ready to move into the new building
weren't; some building users privately expressed relief that the building was
late otherwise it would have to have been secured and maintained until it
could be occupied by the new staff.) Associated with the delay, the budget,
which perhaps had been pared down too much, was increased from £45.9M
to £47.0M, and the final cost including remedial Home Office required
work was £50.0M. Post-handover commissioning followed to plan, and the
building, despite all the catalogue of problems, came into use, and was found
to be well fit for its purpose; Her Majesty the Queen seemed impressed when
she conducted the official opening ceremony.

This was the most difficult and unhappy estates project of the decade, but a
good building was produced at a fair cost.

CIDEM/IMS

The final project discussed in this chapter was to provide accommodation
for a new Cambridge Institute of Diabetes, Endocrinology and Metabolism
(CIDEM), later to be known as the Institute for Metabolic Science (IMS). The
University was due to get the use of some 3,800m^2 gross out of the total space
planned to be 29,738 m^2, and was therefore the junior partner in the project.

This £75M project was for the University altogether different from others
described in this book. It was procured through the Private Finance Initiative
(PFI). That was because Addenbrooke's was leading the project and all NHS
projects since the genesis of the Private Finance Initiative (PFI) under the
Conservative Government in the mid-1990s had been required to go that
route; that is, the project to be funded and managed by a consortium which is
often led by a bank or insurance fund and includes a contractor and a facilities

manager (who in practice is often appointed at a late stage even though early appointment achieves much better results because whole-life performance can be made an important and informed part of the design procedure). The consortium contracts to put up the building and run it to a pre-arranged schedule of requirements and payments for 25–30 years. PFI is generally more expensive, and certainly much less flexible in design and in building operation, than non-PFI, but has lower initial risk to the client, and of course provides the capital funding. The Director of Estates had previously been invited by William Waldegrave, Cabinet Minister without Portfolio, to be a member of the then-Government's PFI advisory panel: having serious misgivings about PFI, he declined.

There had been a plan which was nearly brought into regulation that all Higher Education projects should also be procured by PFI. When rumour of an order forcing universities to use PFI got to the ears of the Association of University Directors of Estates (AUDE), and that in fact such an ordinance had been drafted and was due to be signed in the Ministry, a delegation went the next day to London and got it stopped; the grounds advanced were that there would be no collateral assurance of property value in the event of the client, in this case, a university, going bankrupt or unable to pay the high annual fees charged for running the building. If the building, for example, was a chemistry building in a city, who would want to buy a second-hand chemistry building if the local university had gone bust? So there would be insufficient financial collateral. In fact, PFI could have been applied to such functions as student residences, catering and conference centres, but PFI was seen by directors of estates as inflexible, cumbersome, bureaucratic and in the end significantly more expensive over the lifetime of a building,[6] and unnecessary when universities could usually find their own funding for highest priority projects; AUDE's intervention was notable and valuable in saving universities from compulsory PFI.[7]

The University paid for the ground floor, which was dedicated to out-patient care for children and adults with diabetes, endocrine and metabolic disorders; the novel thing about this was that it was the first time that the University had paid for the construction of 'clinical' (not clinical research) facilities. This emphasised the intimate connection between research and clinical practice, and facilitated the working practices of the many clinically active scientists based in the IMS. There was to be a dedicated diabetes out-patients' department, research facilities and genetics laboratories. The MRC also bought into this building process and paid for a floor for the MRC Epidemiology Unit; this has now come into the University as a University Unit.

To get the 3,800 m^2 gross (3,526 m^2 assignable; £4,254/m^2 capital cost) the University agreed, as the lesser of two evils, after rather protracted and unenthusiastic negotiations to pay one 'bullet payment' of £12.5M towards the capital cost and a further £2.5M to capitalise and pay-off the annual fees due during the concessionary period of 30 years and to cover its contribution towards the significant legal cost of setting up the PFI contract. The PFI contract went to a consortium known as 'Key Health'. The contractor was a consortium of Alfred McAlpine and Hayden Young. (Some years later, Alfred McAlpine got into a huge row with the contractor Sir Robert McAlpine over use of the name; serious financial problems hit the company and it went bust.) The consortium contracted to put up the building and run it to a pre-arranged schedule of requirements and payments for 25–30 years. The mismatch between the Trust's PFI form of contract and the University's one-off 'bullet payment' did give problems during both design and construction. For example, the University's Representative User (RU) was the brilliant Professor Sir Stephen O'Rahilly, and even he with full support of the University's estates staff had huge difficulty in achieving adequate design input; indeed, it was extremely difficult to get the University's requirements into the contract. Once work started on site on 1 November 2004 (some eight months later than hoped for) it was even more difficult to get permission from Key Health for anyone from the University to even visit the site to check that what was being built was what had been agreed and contracted. By mid-2005, what was allowed, under sufferance, was one weekly visit by the University Estates Department, and one visit per fortnight by the RU. Such a confrontational style ran counter to all that the University held dear in procurement. Work on IMS finished on 13 February 2007, and the Department of Clinical Biochemistry moved in from the end of October 2007. Right from the start the University was having to claim against the PFI consortium, through Addenbrooke's, for 'non-availability of facility'. The official opening was on 24 July 2008. The building is said to work well and has an efficient 'feel' to it.

Addenbrooke's 2020 vision

While these projects were progressing, by the late 1990s, the University, Addenbrooke's NHS trusts and the Medical Research Council were all facing increasing planning constraints on the Addenbrooke's Site. South Cambridgeshire District Council, which had had a reputation for opposing development in general, was particularly opposed to the proposed move of Papworth hospital on to the Addenbrooke's Site, and in general was not being at all sympathetic to Addenbrooke's' expansion plans; even relatively modest planning applications were repeatedly being opposed. The Medical Research Council was wanting to replace its Laboratory of Molecular Biology and there was no room on the existing site for that. The University was attracting ever more clinical research, and funding for such research, and was neither in a position to offer up land for others to develop, nor prepared to limit expansion to create buildings for research of such international acclaim. With Addenbrooke's NHS Trust, Cambridge University Hospitals NHS Foundation Trust as it became in July 2004, firmly in the lead, a crucially important planning document 'Addenbrooke's: the 2020 vision' was produced with support from the University, with the slightly tentative first version being launched in September 1999: updates which were clear and positive were launched in June 2001 and then in July 2004. Of the special significance for the University, the

2020 vision document called for 'explicit recognition that the site is suitable for the development of specialised medical facilities, and teaching and research, and that the area was suitable for the development of these activities into a major new biotechnology cluster of medical knowledge park'. Also, 'that the green belt should be relaxed about Cambridge to allow development particularly in the area to the west and south of the Addenbrooke's Site, with provision for additional housing, particularly affordable housing, appropriate to the size of the city and ideally within walking or cycling distance of Addenbrooke's'. The consequent 70-acre Cambridge Biomedical Campus includes the new $330M research centre for the pharmaceutical giant Astra Zeneca, and the huge development of social support, schools and housing for campus employees, and others.

Projects covered in Chapter 7

	£M
Brain Repair Minor Works Refurb	0.75
Cardiovascular Research Unit SMB	2.79
WBIC final phase	0.95
Island Site I	24.80
and II	10.55
Medicine/Virology	2.12
Strangeways development	5.70
minor works	0.75
Purchase £3.5M and refurb of Pfizer £12M	15.5
Staff car-park	9.20
CR-UK land	8.0
CR-UK building	50.0
CIDEM	15.0
3T MR scanner	1.80
Total	147.9

Notes

1 Peter Richards also notes that despite hospital rules, 'Swearing and 'rude or indecent behaviour' were common and nurses were regularly dismissed for being drunk or selling alcohol to patients. Even the patients sometimes tottered into town to return 'very much in liquor". *Cam Magazine*, 2004.

2 V. C. Norfield, *History of the Strangeways Research Laboratory 1912–1962* (W. Heffer & Sons, n.d.).

3 Ray O'Rourke and his brother spent half a day at Cambridge with the University Director of Estates discussing the state of public sector procurement in general and higher education procurement in particular during the period of negotiations with Laing; that would have been one of a number of discussions prior to Ray O'Rourke deciding to buy Laing.

4   All was not entirely sweetness and harmony; there was an additional chill following a terse pre-ceremony conversation in which the Chief Constable made much of a recent downward trend in crime figures in Cambridge, and the Director of Estates, who recently had had a third bike stolen and hadn't reported that on the grounds that that would achieve nothing, responded that the statistics about the downward trend related to reported crime rather than crime per se. But the show went on.

5   Sir Paul had a brilliant academic and public service career which started from walking a long way to school along rural paths and watching insects and other wildlife: his father was a mechanic in a local food-processing factory and his mother was a cook/cleaner.

6   On 4 June 2014 it was noted at the Public Accounts Committee that Balfour Beatty alone had already made £188.9M profit from PFI.

7   AUDE was set up in 1991 following a conference of a rather procedural and defensive organisation of estates directors known as the 'Building Officers' Conference' (BOC) with a mandate only to run an annual conference. Three new directors of estates, from Bristol, Belfast and Leicester, rebelled and forced a vote that BOC should be morphed into a professional organisation to share information and best practice; lobbying and argument went on through nearly all of the night, and the vote in favour was won amongst great acrimony, dominated by the tight group of Midlands universities which had controlled BOC. AUDE took on a secretariat and set up training, the establishment of best practice, and lobbying.

Table 7.1   Overview of capital projects on the Addenbrooke's Site

Chapter 7. Addenbrooke's	Start date	End date	m2 gross	m2 net	Architect	Bdg Svcs	Engineer	Contractor	£M
Island 1-- Wellcome/ MRC	17/06/96	20/06/98	10,246	7,015	Feilden Mawson	WSP	Hannah Reed	Shepherd	24.8
Island II--	31/08/99	02/03/01	3,800	3,116	Feilden Mawson	Sibley Robinson	Hannah Reed	Laing	10.55
Medicine/ Virology	19/03/01	22/03/02	936	750	Feilden Mawson	Silcock Dawson	Hannah Reid	Blueston/ Sindall	2.12
WBIC	4th phase	20/06/99			Frank Shaw Assoc				0.95
Strangeways Research Lab	05/08/02	10/10/03	1,732	1,100	Feilden Mawson	Roger Parker	Hannah Reed	Shepherd	5.67 0.75
IFRC/CRUK	21/07/03	30/06/06	14,000	9,844	Anshen Anshen	Zeisman Bowyer	Faber Maunsell	Laing O'Rourke/ Crown House	50.0
Staff Car Park	14/04/03	08/03/04	1,225 spaces	64 CCTV	Anshen Anshen	Zeisman Bowyer	Faber Maunsell	Laing O'Rourke/ Crown House	9.2
Pfizer bldg. refurbishment	early 2003	01/06/04	2,100		Saunders Boston	Silcock Dawson	Hannah Reed		15.5
CIDEM—Elective Care, Gen, Diabetic	01/07/04	31/08/06	4,426	3,526	Key Health	Hayden Young		Alfred MacAlpine	15
Brain Repair Minor Works refurbishment									0.75
3T Scanner	2006	2006							1.8
Cardio-vascular SKB, rented space									2.79
Cost of completed works, 1996 to 2006									139.88

8 Development of the University in the city

For centuries the University's estate was mostly concentrated around an axis along Trumpington Street to the Old Schools, with some chunky sites near that axis, notably the Downing, New Museums and Mill Lane sites developed during the nineteenth century. Lensfield and Tennis Court Roads at the south end of the axis were developed in the twentieth century. During the decade 1996 to 2006 there was much development along that Trumpington Street axis, and a lot of new buildings further afield in the city of Cambridge.

Scott Polar Research Institute

Starting in the south: in the mid 1990s, it was proposed to extend the Scott Polar Research Institute (SPRI) in Lensfield Road. This was originally built in 1933, formally opened by Stanley Baldwin, in the rain, on 16 November 1934,[1] and extended in distinctive yellow brick in 1966–1968. The Institute has the largest specialist polar library in the world and houses international-level research at staff, graduate and, on occasion, at undergraduate levels. An extension was designed by John Miller and Partners and Fulcrum Engineers and was built in the autumn of 1998 by constructors Haymills to a capital project budget (or 'cost-estimate' as the term then was) of £1.3M for the building works; the whole project was funded to £1.75M, largely made possible by the Heritage Lottery Fund. It was an exciting design internally, and it increased floor space by 30 per cent for exhibitions and library use; the museum has over 250,000 visitors every year and until this project it had no dedicated education space, museum shop, restaurant or adequate reception area. The development worked well and gave the museum a new life.

Case study – Scott Polar Research Institute (SPRI), Shackleton Memorial Library

The original Scott Polar building on Lensfield Road was designed by Herbert Baker in 1934 and can be found between the Faculty of Chemistry and the Catholic Church, both very large, imposing buildings. The Baker building is the public face with the Museum fronting onto Lensfield Road, and is much loved by the staff of SCRI.

This was extended in 1968 with a concrete and brick statue, discreetly tucked between the Baker Building and Chemistry, its lack of visibility belying its size. This consists of a two story link that led to a lecture theatre and laboratory/research areas, a basement with storage and freezers for low temperature research, and additional space for the library, archives, academics, other researchers and students. In 1987, additional space was made in the link for bibliographers and pamphlets.

In the late 1990s, the decision was made to increase the area available for the library and students. This resulted in the Shackleton Library, a three story rotunda to the north, and an addition to the second story for pamphlets and journals with research alcoves for the MPhil students. The pamphlets area was converted into the Friends' Room. The basement was extended to provide a new map room and compactor storage. Completed in 1998 and receiving an RIBA award in 1999, the handsome rounded extension of the Shackleton Library, with its bands of stone and brick, is the only visible part of the extension from Lensfield Road.

The Rotunda provides stair access to all levels in the institute. It also sets up an axis from the new entrance on the West side of the Link, through the reception area and past the Admin offices. However, this was never utilized and in 2009 further works moved the entrance to one side in order to enlarge the Museum, still accessed directly from Lensfield Road. To improve the vertical circulation a new hydraulic lift was installed from the Basement to the Second Floor, with exposed steel structure to the lift shaft which, from the Ground floor up is clad in glass, perhaps a visual pun on ice.

So the Rotunda is in practice a glorified staircase with shelving for books, a small table and chairs on each level. The librarian would love more shelving and points out the curved shelves are not convenient for arranging books, nor are the shelf heights sufficiently adjustable. The additional shelving amounted to some 443 linear metres and was to be filled over the next 25 years: in fact it was filled in 10! Moreover, the windows, which leaked very badly, are curved single glazing and there is concern about how the glass will be replaced, the light fittings are inaccessible for replacement of lamps and the large roof-light is never opened to create the chimney effect for passive ventilation and roof leaks were an early problem. Heating was to be by warm air but this was found to be too dry and dusty so a hot water radiator has been installed on the ground floor to heat the whole space. However the relative humidity cannot be controlled and is much too high at 75 per cent (recommended RH for library archives is 35–60 per cent).

An oddity of the steel staircase is that it works itself loose and the bolts have to be tightened up every few years. At the first floor the Rotunda links to the Friends' Room in the Link and thence to the old library on one side and to the main entrance to the library on the other. At the second floor, the Rotunda opens to the new reading and study room which, despite the interesting coffered ceiling and corrals looking out into the leaves of the huge Robinia Frisia tree guarding the entrance, feels somewhat utilitarian. This room is named the Wubbold Room in honour of Captain Joe Wubbold, a seafaring captain who came to Cambridge to take a Masters. It is thanks to him that the project came to fruition on site as it had 'ground to a halt' until 'Joe saved the project', as current staff put it.

This leads to a question about briefing and who is involved and who 'takes ownership' of any project. Speaking with members of the library staff who were in post during the project the following points arose:

- Nice shape (very pleasant from the outside) and light, but not suitable for a library.
- Heating columns through the rare book section were asking for trouble.
- Hoping for a good sound building which would enable changing uses.
- Too much light and humidity.
- The three storey space is too noisy.
- Funding limits resulted in ill-considered savings were made (such as replacing the library strip system for adjustable shelving with a kitchen cupboard type system which was not sufficiently flexible).
- Was the brief good enough? Did the architect test and question the brief enough? There was total involvement at the top but concern that too much input from too many people would muddy the waters.
- The Rotunda (a fashion statement at the time) looks like an organising element of vertical circulation and yet was never intended to be the main entrance to the library.
- Did the architects visit site during construction? There seemed to be very few drawn details.
- Project management: Joe Wubbold was a very good project manager on the Institute's behalf and managed to keep the library open during

the building works; the users weren't aware of EMBS doing any project management.

- The entrance doesn't work adequately being at the side of the building.

It is interesting how little the Feasibility Report sketch plans changed by the time the building was built, perhaps giving substance to the view that the project was 'done in haste' with very limited consultation of the users.

A contrast was drawn with the experience of the later project to extend the Museum in 2009. There were regular meetings with the architect (from EMBS) and, during construction, monthly meetings with the builder with reports to all staff members.

Architect: John Miller & Partners
Structural Engineer: Campbell Reith Hill
Services Engineers: Fulcrum
Quantity surveyor: David Langdon & Everest
Contractor: Haymills

Chemistry buildings

Along the Lensfield Road are the Chemistry Labs. Originally, briefly, the playing field which became the Sidgwick Site was proposed for the Chemistry building, which would have been totally unsuitable; the site that was instead selected, on Lensfield Road, was sufficiently central and convenient: landlocked, but with enough room for expansions. When opened by HRH Princess Margaret in 1958, Chemistry was the largest building on the University estate other than the University Library. But it was certainly not the prettiest, being so out of scale in its setting of residential Lensfield Road, even although the building line was drawn back from the road frontage as there was a plan that a Cambridge inner ring road would go down Lensfield Road. (The only section to get built before being cancelled on the first evening of a newly elected city council was Clerk Maxwell Road on the East side of the West Cambridge Site.)

Unilever

In the early 1990s there was a lot of discussion about the possibility of moving the Department to West Cambridge, but it was thought that the cost of a new building could not be found, so it was decided to stay in the Lensfield Road building and plan for refurbishment of the labs. Before that happened, however, a different project was proposed, funded by an investment of £13M from the international company Unilever for a new building, a research building for molecular informatics, plus an endowment for staff over five years; building procurement was to be managed by Unilever in conjunction with the Head of the Department of Chemistry. Notably, the controversial architectural practice of Danish architect Erik Christian Sorenson/Zibrandson, the architect for the recently constructed and adjacent Cambridge Crystallographic Centre for the

Figure 8.1   Unilever

Institute of Crystallography, was appointed. (That building is remarkable in its open and dramatic internal style, and for the very generous space per person working in the building, probably the highest on the University campus.)

The Unilever building is a large brick rectanguloid structure, and was to house the large Chemistry Department Library as well as its own research facilities. The capital budget was £6.67M. There were problems with the construction project. (It had been hoped, in the early stages in 1998/9 to forge a teamwork approach between Unilever and the professionals in the University's estates department who could help, but such input was, in practice, marginalised). Soon after it opened, the library area of the Unilever building had to be taken out of use as the floor had to be lifted and re-laid because the wooden floor blocks had shrunk, but the functional use of the building once completed by the contractor R. G. Carter was clearly of great benefit to the Department of Chemistry. It became a successful node for 'linking scientists of all strands of research worldwide' as a centre for informatics. It was opened on 15 March 2001 by Lord Sainsbury, the Minister for Science and Innovation (later to become the University Chancellor). From May 2001 until September 2002, the University estates department managed a £998k retrofit-out of the basement, and converted some of the basement-storage area into a laboratory.

Chemistry buildings refurbishment

In 1997, while the Unilever building project was still in progress, an eight-phase, estimated £30.5M refurbishment/extension of the Chemistry building was being planned, with a £28M bid to the Joint Infrastructure Fund (JIF) to improve the efficiency and safety of the Chemistry building and provide more space for research. In particular, the provision of fume-cupboards was woefully inadequate. In June 1998, the University's Planning and Resources Committee (PRC) noted that a £9M programme of refurbishment across the University had upgraded/replaced 228 fume cupboards of which 130 were in the Chemistry Department, but that left a further 247 to be upgraded in the Chemistry Department plus 59 elsewhere.

To develop a master plan for the site, the practices Erik Sorenson and Nicholas Ray Associates had been appointed in 1995; Nicholas Ray was on the staff of the Architecture Department of the University and also ran his own practice. Subsequently, following the success of the JIF bid, he was appointed for the proposed refurbishment, with engineers WSP, a safe set of hands, and Quantity Surveyor Northcroft (later taken over by Capita). This was about the last building project, along with the first phase of the Centre for Mathematical Sciences project, which was seen by the academic users of the building as a construction project to be managed by them. This approach, common in the University over the previous eight centuries, assumed that if, as was often the case, funding of the capital cost was stimulated and/or organised by the academics, then they should manage the projects: it was 'their building'. There was little awareness or experience of management, or consideration that the running/servicing costs of a building over the years, all paid by the University, much exceed (by 2–3 times over 15–20 years; see Chapter 5) the capital cost of constructing the building. Nor was there among those academics who expected to manage the project an understanding of the bizarre nature of the construction industry with its harshness, confrontationalism, waste, illogicalities, low efficiency and extensive and expensive use of legal settlement. (Otherwise it was perfect.) The Head of the Chemistry Department, a very high-profile and powerful scientist (soon to be appointed as the Government's Chief Scientific Advisor), had pushed through his own selection of key appointments, only in part following professional advice from the University's estates department. A workable *modus operandi* between him and the new Director of Estates was achieved after some frank discussions whereby certain appointments wanted by the Head of the Department went ahead but it was agreed that the University had clear responsibility for professional management of construction, thereby establishing that principle for this and future projects.

The Chemistry project started on site on 15 May 2000 (the first phase, oddly, being Phase 1b) with a total project budget of £34.45M. Further work was added during the eight phases of the project as more funding became available, and the out-turn cost of the augmented work was £37.08M.

The refurbishment, as was anticipated, was extremely complex and demanding as for any 'old' science building, especially as the building remained in operation throughout the works. A top imperative in the project was a substantial increase in the number and effectiveness of fume cupboards to meet increased research intensity and higher safety regulations. Surprisingly, there was virtually no local or planning opposition to the design, even to the ducts (as described in the case study), and planning permission was given by Cambridge City Council as Local Planning Authority at first bounce. Another requirement of the project was the provision of new space; a new two-storey link was built connecting the north and south wings of the building.

Early on, the contractor, Amec, had a dispute with one of its main sub-contractors about work quality in a basement; in the end the sub-contractor rightly accepted responsibility for some bad work and made that good at its expense, but the delay put a strain on the project programme and therefore budget. Amec was a large contractor, (at the time of writing even bigger with £5bn turnover and 40,000 employees in 50 countries), so the University needed to defend and promote its procurement standards.

The excellent quantity surveyor who had been in charge of cost control for the project had earlier gone off to work in another practice (as had been feared) and after a couple of phases the replacement quantity surveyor began reporting that the project was heading for a large over-spend, and so a group from the Buildings Committee reviewed the project. The first three conclusions of the BC review (January 2002) were that the completed work would meet the scientific requirements, that no waste was apparent and that the whole project could be completed within the original JIF programme. A further conclusion, however, was that late notification of additional expenditure from the design team limited the amount of value engineering (and hence cost-reduction) that could be done, and that there was a failure to alert those concerned of the budgetary issues associated with design changes and late information; long-running issues in the building industry. Although the basis for the higher estimates of project over-expenditure proved not to be sound and the project could be kept to budget, better change-control procedures were brought in. The project then proceeded relatively smoothly, although always difficult technically and managerially.

The constructors Amec did a good job coping with all the problems of such a complex and changing project, and won top prize for this project in March 2002 for the local 'Considerate Constructor Scheme'. Inevitably, however, on 18 April 2003, the chemists were very glad to see the back of the last contractors. The end-result of the project was a 'state-of–the-art' Chemistry Department to the extent of the planned project (some of the labs remained to be refurbished later). When the University Chancellor, HRH the Duke of Edinburgh, walked round the completed labs he was briefed on them and on University development generally and was, as ever, absorbing complex infor-mation very quickly and comprehensively, asking astute and sharp questions in a very pleasant manner, and enjoying and remembering the replies. A great man and a great Chancellor.

Case Study – Chemistry Department and the Centre for Molecular Informatics

Over the years, alterations had been made to enable the Labs to expand their work but this kind of make and mend could no longer solve the problems of modern Chemistry teaching and research. The Sorensen-NRA master plan proposed a bold strategy to deal with the serious demand for proper air extraction and to rationalise internal circulation. A new external zone was created to contain the rising flues. Air is drawn through ceiling plenums and exhausted at high level with flues contained behind fritted glass up to roof level where they emerge in high funnels clad in copper, rising to 31 metres above ground-level, and visible as green caps from miles away. Flues for supply and exhaust air are on the north and south sides of the South Block but are entirely on the south facade of the North Block to retain the Lensfield Road facade unchanged. (Surprisingly, there was virtually no local or planning opposition to the design, and planning permission was given by Cambridge City Council as Local Planning Authority at first bounce.) They create a visually striking composition which is a signif-icant signpost for the Chemistry Department. This approach utilised the advantages of the original building which proved very adaptable within this approach. The clear span was good, the floor-to-floor heights just allowed for the deep plenum and the Nuffield bay system of window spacings was utilised and fume cupboards and flues layout adapted to fit.

The fritted glass screen provided some solar shading which was beneficial as the south-facing labs had been over-lit and suffered solar gain.

Proposals to rationalise circulation were not implemented so the circulation remains somewhat awkward with continuous access through the whole building only on the second floor. Also proposed were photo-voltaic panels on the south-facing facade and a heat recovery system using exhaust air from the fume cupboards. The PVs were rejected on the grounds that a proposed extension (never built) would shade the panels and the MVHR (requiring special valves) was vetoed by EMBS's Minor Works as untried technology and therefore too risky.

Nicholas Ray Associates executed the nuts and bolts of the refurbishment of the main building in a number of phases, no easy task. Following commissioning, and a detailed measured survey, the detailed design of each lab was discussed with the individual research groups using computer generated models. The result is a pleasantly light and calm series of spaces.

Over a rolling programme of work, the Chemistry Department continues to upgrade itself. Later phases reduced the sq.m spend by omitting the suspended ceilings and exposing the air-handling ducts as can be seen in the large teaching labs which seem to provide a good environment. The problems seem to arise from running such a busy operation and moving large and often hazardous materials and equipment around a confined site, particularly should a lift be out of action. It is impressive how things are kept ship-shape and working.

The Centre for Molecular Informatics building is located on the south-west corner. The Centre was set up in 1999 to be a leader in Molecular Informatics working on blue sky science, commercially orientated projects and also to collaborate with other companies and funding bodies. The architects were the Danish practice Cornelia Zibrandtsen Architects together with Professor Erik Christian Sorensen, the architects for the award winning Cambridge Crystallographic Data Centre (1993) nearby in Union Lane. The executive architects were CMC and WSP North provided structural and services engineering, lighting and thermodynamic modelling.

The gross floor area is 2650m² on four floors of what seems a fairly simple brick box, accessed only through the main Department of Chemistry building through a two-storey link containing stairs and lifts. The three floors above the basement contain research accommodation for about 40 researchers, training areas and meeting rooms and a small jewel-like lecture theatre, the library and a cafe. The cyber-cafe is on the roof with an outside terrace and views across the neighbouring local terraced houses. The scale of the openings is carefully considered with the narrow widows looking west effective from the outside and the inside. The cafe is comfortable and relaxed, bright and acoustically comfortable: it is clearly popular and well used.

The library occupies the whole of the first floor and is a stunning space, 30m x 30m x 5m high. At its centre is a huge abstracted steel tree symbolising the Ash Tree of Scandinavian mythology with its roots in the underworld, reaching up to the light of the heavens and with the fruit of knowledge hanging down in the form of visually significant light fittings. But it is not just the iconography of the tree which impresses: the texture of the walls (brick frogs out, painted white) and the ceiling (myriads of red lacquered timber shaped slats) create an almost quilted effect and a surprising yet calm space. After completion, one evening (lucky it was 'out of hours'), one of the lacquered slats fell, distorting the top of a steel shelving unit. Every slat was unscrewed and re-fixed with longer screws. The original spec for the floor was for small fir blocks but it seems these dried out leaving unacceptably wide joints. They were then replaced with more conventional overlay timber flooring.

The librarians admire the building and feel it works well especially acoustically, though the air-circulation is a little inflexible. They open high-level windows in summer when noise from the outside allows. The Building Services Manager says the building is wonderful but 'impossible' to maintain with cooling circuits and wiring controls in particular being inaccessible, and replacement parts being difficult to source from Denmark.

So this project begs the old questions: how do we design poetic spaces that are appropriate and comfortable which are straightforward to build and to maintain?

Laboratories

Architect: Erik Christian Sorenson and NRA
Engineers: WSP
Quanity Surveyor: Northcroft
Contractor: Amec

Unilever/Centre for Molecular Informatics

Architect: Cornelia Zibrandtsen architects together with Professor Erik Christian Sorensen
Executive Architects: CMC (commissioned by WSP)
Structural and Services Engineers, including lighting and thermodynamic modelling: WSP North
Project Management: Unilever & EMBS
Contractor: R G Carter

Figure 8.2 Chemistry Departement: side and façade

Further refurbishments in the Chemistry Department were to follow: from 29 April 2002 to 31 October 2004 three lecture theatres were refurbished to conference-venue standard, at a cost of £3.1M. Then the Phase 1a lab upgrade project was carried through for £1.47M. These projects were constructed efficiently by the contractor Marriott; the architect was the Cambridge-based practice RH Partnership which produced sound designs.

The total cost of the refurbishments, £41.65M, might in the mid 1990s have bought at West Cambridge a good new building of the size and standard then needed. Was the decision not to go to West Cambridge correct? Probably.

Figure 8.3 Chemistry Department: fume and duct wall

By courtesy of University of Cambridge

Engineering buildings

Another department that decided not to move lock, stock and barrel to West Cambridge was Engineering:[2] one reason was that the large, adjacent land-plot at the Royal Cambridge Hotel was on lease from the University, and would revert to the University if the University decided not to renew the lease. (In 2014 most of the hotel car-park was taken back as the site for the new Dyson building for the Department.) Part of mechanical engineering did go to West Cambridge, including a geotechnical research centre which was refurbished in 1998–2000 at a cost of £1.29M. The refurbishment, designed by architect Mustoe, was successful but cost more than necessary as some components were custom-designed rather than standard products, creating future maintenance troubles. There was also a hiccup during the project as the engineers called for replacement of plant which when installed did not function properly; the original equipment was put back in and worked well. Partly because of irritation that the engineers wouldn't 'own up' and reduce their fees accordingly, and partly to gain experience of the adjudication system recently introduced under the Act of Parliament, the University went to adjudication, although the sum of money at stake was just a few thousand pounds.

Much of the 'light' side of the Engineering Department, notably electronic-based subjects, would also go to West Cambridge during the decade being discussed, but the rest of the Engineering Department stayed on its Trumpington Street site in its rather functional buildings around a concrete courtyard, with bits being added all over the place, and some areas being radically upgraded, necessarily at high unit-area cost. For examples, in 1999 R. G. Carter did refurbishment work at a contract sum of £730k; the top floor was refurbished when the electrical division moved out to the CAPE building in West Cambridge in 2006 at £690k for the Engineering Design Centre in the 1930s Inglis building. Refurbishment of the thermodynamics lab at £1.41M was done in 1999–2000. A difficult upgrade was carried out at some speed in 1999 in the Engineering building for Professor Hopper for his research in the field of computing engineering software: existing buildings on the roof were partly demolished and the remainder incorporated into new construction: to achieve this the steel columns holding up the building were extended through the roof and connected to a steel grillage above the existing roof. The capital cost was £841k, some £150k over budget, and the additional recurrent costs estimated at £2k per year: by 1999 whole-life cost procurement was gaining traction in the University, and those departments more aware of whole-life cost and CO_2 implications, notably Engineering, responded pro-actively. Refurbishment was also carried out at Prof Hopper's research group in the north wing at Old Addenbrooke's. Other works of note in the Engineering Department included extension to the south-west corner. This was designed by architects SMC and engineers Rolton Group and Andrew Firebrace Partnership and built by Marriot at a cost of £4.719M, the work being delayed until May 2007.

Faculty of Architecture and History of Art

Next to the Department of Engineering lies the Faculty of Architecture and History of Art fronting onto Scroope Terrace/Trumpington Street. Space was short in what was an old terraced house converted as well as was possible for teaching and studio work, with research offices and library. There had been some further refurbishment, but what was most needed was a newly built studio space. The area between the Architecture and Engineering buildings was chosen, and Freeland Rees Roberts, Scott Wilson and Max Fordham were appointed as designers. Work by constructor Dean and Bowes started on site on 17 July 2006, finishing on 1 October 2007, at a project cost of £2.964M, a low cost at £1,029/m² gross. It is not the most beautiful of buildings, but it functions well.

Extension of the Fitzwilliam Museum

Nearby lies the Fitzwilliam Museum, a grand construction built over a long period. The funding bequest was received in 1816, the site was acquired in 1821, the architect George Basevi was appointed in 1835 and initial construction was from 1837–1844; work was suspended because of funding problems and then resumed in 1870 and completed in 1875. Between the World Wars the area of the museum was doubled following substantial donations. Since then, the roof has needed frequent repairs as water-proofing was poor in several central areas (causes of leaks are really difficult to find as leaking water often runs far before becoming apparent).

The main need by the mid 1990s was for more space: the museum is one of the most vibrant in UK. The first attempt at expansion in the 1990s was to plan a large extension at the northern end, nearest to Peterhouse (College). That became a planning battle, as the College stoutly opposed the proposal on the grounds that some of its buildings would be over-shadowed: the College won. The museum left that field of battle and switched to planning an extension at the other end, south along Trumpington Street. A Heritage Lottery Fund grant of £5.62M helped.

Much of the running in this project was carried out by the Representative User, the Director of the Museum, often represented by his deputy. The designers appointed were John Miller & Partners as architect (a substantial practice before much declining in size and then closing) and engineers SVM Partnership and Campbell Reith Hill, neither of whom had previously done much work for the University. The contractors, Amec, which did a lot of University work, were brought in as part of the normal two-stage 'develop and construct' procurement and helped in the efforts to keep expected costs within budget, a continuous task as adherence to budget was not notably high on the architect's list of priorities.

The extra area of built space of the museum was 4,139m² gross, 3,037m² assignable. Work on site started in November 2002. It was completed on

19 December 2003, just three weeks late despite delays, the most worrying caused by the discovery that asbestos sheeting had been used for shuttering the pavement concrete in the old external courtyard, and then left in situ, so the concrete that had to be chipped out had a lot of old asbestos scattered about in it. Other delays were caused by some defects in the original building structure. The total project cost was £12.06M, which is high, but the extension has worked very well indeed, giving a new centre of activity which merges well into the existing areas. The design is rightly much admired by professionals and public alike. As well as the extensions noted above, there was refurbishment and upgrade of various galleries: Rothschild Gallery (£250k), 20C Gallery (£150k), Egyptian Gallery (£1M); later in 2009, the Greek and Roman Gallery was upgraded for £1M, a total of £2.40M. Such upgrades included installation of superb gallery lighting.

Case Study – the 1998 Courtyard Project by John Miller

This very successful museum founded in 1816 has continually expanded and evolved, a key requirement for more space being hampered by the restrictions of the site, surrounded by Peterhouse. Major additions included the extensions by David Roberts in 1975 to add the Library and the Adeane Gallery to the south, in his trademark modest style in yellow stock brick.

By the 1980s the Museum needed more space and John Miller & Partners were involved in looking at solutions. Original proposals were for extensions at both north and south ends but Peterhouse objected to the one and planning was refused for the other. The first plans would have advantaged the collections but not the 'visitor experience' or education, nor income generation which had now become essential.

A competition was run to select the architect and was won by John Miller & Partners, a very well-established, multi-award winning practice also then working on libraries at Scott Polar Institute and at Newnham College. John Miller had a fascination with and wide experience of working on museum buildings in England and Scotland. Also, together with Alan Colquhoun, he had made his mark as a theoretician.

By 1998 the chance of a very large award from the Heritage Lottery (HLF) was missed (Churchill Archives had recently been awarded £13M). Because of funding and planning issues, the Museum dropped the north end scheme and focused on the south end with a shop pulled forward to the street and the possibility of a gallery that could be opened at night. However, the city planners objected as they did not like the narrowed vista on the entry into town.

Therefore to achieve the requirements (in a 3000 sq.m scheme) the design had to incorporate the footprint of the courtyard, cutting back the Roberts building and remodelling some of the galleries. The recommendations of a very useful Conservation Plan by Cambridge Architectural Research were taken into account.

Although the Courtyard Project does not answer all space requirements it has provided:

- a new generous entrance space for meeting and orientation, with disabled access and level access to the new and larger shop and a new cafe
- orientation space for groups visiting
- a new public lift
- an education department with seminar room and studio
- curatorial offices
- conservation workrooms
- a temporary exhibitions gallery with exacting climate controls
- art storage with environmental controls
- additional WCs and cloakrooms.

The Courtyard Development opened in July 2004. John Miller described the biggest construction and design challenges as 'working within the existing floor to floor heights and incorporating services and new finishes within the limited ceiling voids and floor to ceiling dimensions'.

Besides these knotty dimensional issues, services presented a problem of another dimension: the tussle to deliver a controlled environment within the University's desire to control heating services centrally and the Museum's need to be able to effect quick changes in response to events on the ground.

Margaret Greeves (then Assistant Director), who was intimately involved as RU (Representative User) for the project, said the logistics of carrying out major works in a Grade I Listed building, full of precious objects, with the building continuing to function on a landlocked site (only 1m gap at rear) presented severe logistical problems, but all were harmoniously overcome thanks to very good teamwork.

The competition was organized by Kevin Bradbury for the EMBS assisted by Malcolm Reading Associates Architect-Project Manager (of the Wallace Collection and the wobbly Millennium Bridge) who was a useful 'Critical Friend'. The contract was the NEC form, favoured by both the Director of EMBS and the Heritage Lottery Fund. The whole team went on the course to learn about this form of contract.

The contract brought in the constructors AMEC at an early stage, enabling their construction expertise to be drawn on. The RU considered they were good managers who bought into the project. The EMBS input was also found to be useful.

This project is really a secret building. The modest entrance does not compete with the original imposing classical facade and grand entrance stairs of the main entrance to the Founder's building. The visitor segues past the reception desk and slips past the glass lift (which for the first time provides lift access to all gallery floors.) The old external courtyard has been transformed to contain the shop and cafe tucked under the mezzanine which is supported on large circular columns and a double height, lofty, light 'Orientation Hall' with space for cafe overspill, and for displaying sculptures or temporary exhibitions. The area under the mezzanine is low and potentially dark but the fluorescent tubes in coffered ceiling provide a pleasant light and the roof glazing over the double height space brings light to the edge of the courtyard as well as into the offices above.

The brickwork has been cleaned and retained, admirably demonstrating how special bricks were used originally to form the pilasters. This courtyard is the core of the project opening up circulation at this southern end of the Museum by cutting down the sills of the windows into the Korean Gallery and the Armoury. A French limestone floor is used throughout the space and unifies it. Above it the offices on the mezzanine look into the courtyard space and the conservation workshops above the offices look out above the glazed roof. A curved, glassy staircase leads down to the basement where the education department's seminar room and studio are located together with large banks of new toilets and an enormous (but very low) plant room. These are obviously extremely heavily used. (The toilets were subjected to a VE exercise, perhaps a misguided saving in one of the most heavily used parts of a public building.)

Most importantly the project has given an additional gallery, the Mellon, running parallel with the Adeane Gallery and with the potential of linking the two for larger temporary exhibitions. Other works re-configured the Reference Library on the ground floor and the Library Reserve in the basement. Despite best efforts the cafe kitchen had to be placed above the Library Reserve, and subsequent leaks call into question how risks can be avoided by not running services through archive areas. The old cafe on the first floor over the new entrance has been turned into 'The Friends Room'.

Overall, the Courtyard Project has clearly breathed new life into the thriving Fitzwilliam Museum by allowing schools and outreach work with larger exhibitions and providing a welcoming cafe and shop for visitors and disabled access to every level.

Funding

£12m contract
£5.62m Heritage Lottery

£1.00m Museum from a bequest
The Fitzwilliam Museum Development Committee led a successful
public appeal to raise the required 50 per cent partnership funding.
Contract period: 18 months plus 69 months fitting out and
recommissioning.
Architects: John Miller & Partners
Structural Engineers: Campbell Reith Hill
Mech and Elec Engineers: SVM Partnership
QS: Davis Langdon & Everest
Project Managers: Gardiner & Theobald Management Services
Construction: Amec

Changes since then

The introduction of facilities management and on-going projects
mainly to upgrade the lighting and ventilation of galleries. The
Facilities Manager cites the re-fit of The Courtauld Gallery (15)
(completed summer 2014) as a very successful collaboration between
the conservator, facilities manager, engineers including someone
with a theoretical background and English Heritage. She feels that
allowing time to get the brief right is imperative. In the Gallery 15 &
16 project there was one year of talk, briefing, design, tendering and
fundraising to lead to a nine months' contract period and a contract
sum of £1.25M. She considers that the innovative new air handling
project completed in the summer of 2014 will enable huge savings in
utility bills.

'Old' Addenbrooke's

Across the road there had been a very different building project: the
re-organising and refurbishment of the former Addenbrooke's hospital. What
to do with the building had been an issue since the University, led by John
Butterfield as VC, decided to buy the site for £5.75M, in March 1985, as its
health work started to shift to the new hospital site south of the city following
a relocation decision in 1948. The first buildings on the new site opened in
1961; the last move from the old site was in 1984.

For the University this 'Old Addenbrooke's' site was the last central site of
any size that could be acquired for University expansion, and it was bought
despite leaving the Land Fund with a £4M deficit. As noted in Chapter 1,
in the important policy document *Planning for the Late 1980s* written to the
University Grants Committee in response to its Circular Letter UGC 12/85
which set out further cuts, the Annex noted that,

Because it will be many years before the University can expect to have the necessary capital to develop the whole site, it is anticipated ... that the northern half of the site containing most of the existing hospital buildings will be let commercially. The southern part will ... be developed primarily for biological departments.

Strong and emotive arguments immediately broke out in the University, and in the city, as to what should be done with the old hospital's main building: demolition, keeping the facade and reconstructing the building behind, or leaving it with minimal change? Tempers ran high. (In the view of some academics, the controversy was an issue in the departure of the then Director of Estates.) In the end, the 'substantial reconstruction' option was selected.

The conversion of 'Old' Adenbrooke's was designed and carried out in 1992–1994; site possession was on 28 June 1993. The main facade was kept, and the Victorian era building massively refurbished; the late-Georgian extensions on the Tennis Court Road side, including the three-storey theatres, were demolished. The remedial works continued for a couple of years after completion and the last of the several legal disputes dragged on until 1999.

With influence from the benefactor, Paul Judge, the architect John Outram and engineer Max Fordham were appointed as designers, and on a single-stage basis i.e. 'full' design and then tendering. The design was remarkable: high atrium and bridges, mixtures of mostly highly coloured brick and concrete, large hollow columns containing services such as heating, narrow corridors and small offices, and two lecture theatres. The contract was set up using the then adversarial JCT form of contract. So conflict was near certain, and it happened soon, and extensively.[3] Work on site started on 28 June 1993 by the large and tough contractor Laing (then the largest family contractor in UK, about to celebrate its 175th anniversary), and it soon became clear that design information was getting far behind schedule, and anyway was not what the contractor needed: jokes circulated about working drawings appearing in the form of water-colour paintings. In a single-stage contract all the design work should be completed before work starts on site, but that rarely happens, and the contractor, having been excluded from the design process, can't easily 'fill in the design gaps' or, often, even understand the design strategy, and indeed he may not want to when there are juicy claims to be made.

Work on the project soon fell far behind schedule. So, part-way through the project, after a lot of soul-searching, and negotiation with the contractor, an architectural practice capable of practical design, Fitzroy Robinson, was appointed to take over those architectural duties, and review of some aspects of design, reducing, for example, the large number of colours in the ceiling of the main entrance hall. Following stabilisation and advancement of the design process, in September 1993 Laing was paid an interim settlement of £1.85M to agree to continue work on site, albeit far behind the original programme. The cost of the project was £15.7M. *Architecture Today* described the refurbishment as 'a collection of places, at once archaic and hypermodern'.[4] It certainly has enormous presence and excitement.

The top floor was not fitted out during the project: that was done later, starting on 5 January and finishing on 25 May 2001, a couple of months ahead of programme and within its budget of £1.02M. Being better working space than the rest of the building, it was soon occupied by those with the strongest claim.

One of the several large University legal construction disputes that ran on into the late 1990s related to the Judge project: Max Fordham (engineers) maintained that they had been underpaid. In those days, designers were (strange though it now seems) appointed on a fee-basis of an agreed percentage of the final cost of the project, so if a project ran over budget, they got paid more, rather than paid the same or less as one might expect in a performance-related culture, and they got paid more even if the basis of a budget over-run did not occasion any extra design work or cost. Max Fordham maintained that the final agreed out-turn relevant construction cost was £14.0M whereas it was £10,956,500. Why the real figures had not been made available to all parties is not clear. The inter-related dispute over responsibility for flooding (subsequent to completion of the main project, there was a flood in a basement on or about 1 January 1996) also ran on for years with three sets of lawyers (for MF, for the insurance company RSA and for the University). Repairs had been completed in March 1996 but with acrimony over who was at fault. The case was referred to the High Court of Justice, Queen's Bench Division, Technology and Construction Court by the University on 27 January 1997. Finally, four years after the building was handed over, the parties asked the Court to 'stay' action, and agreed late in 1999 to settle out of court. At least, the continued and expensive problems on this project showed clearly why building procurement had to change radically and soon. Which it did.

The next problem for the refurbished building started on 3 September 2002 when one of the 102 large concrete discs fell off the front of the building, a really serious danger in itself, and with the threat of further disc instability, Fitzroy Robinson architectural practice was appointed to investigate and to propose remedial work. The original architect John Outram was angry that the remedial action to what he called the 'blitzcrete' was taken by the successor architects and various specialists, and insisted that although the design was pretty and novel, that it was nonetheless strong and sound. The correspondence is fascinating with sentiments from a long bygone age.

The work stabilising the loose and suspect concrete was carried out with less trouble and cost than feared, and the exciting academic work in the Judge School of Management continued unperturbed. In 2013 further development and expansion of the Judge was being planned to a value of £30M.

Biochemistry buildings, Tennis Court Road

Meanwhile, a more measured project had been going on adjacent to the 'Old Addenbrooke's' Site: the Biochemistry Phase I building project on Tennis Court Road. (Many years earlier, the original plan for the new Biochemistry building on this site had extended to demolishing the villas on Bene't Place but that option was lost as the planning permission ran out of time and could not be renewed: that caused a big round of blame-attribution). The project started controversially. The University, unusually, decided to run a design competition to select an architect. The cheapest, proposed by the rising, Bristol-based JT Design Build,[5] was selected by the University. When the scheme came up for planning permission, it was passed by one vote, and that was when Councillor Liz Gard voted in favour after saying the project was 'an unfortunate chain of events which were bungled and mismanaged', a reference to the way that the University had made its decision. It took what the *Cambridge Evening News* called 'an eleventh hour appeal by the Vice-Chancellor who warned that the chance would be lost "for a considerable time" if councilors opposed it'.

After the planning approval, the University changed tack and architects R. H. Partnership and Engineers Whitby Bird and WSP were appointed as designers for this and the future phases of the Biochemistry project. That first phase, built in 1997–1998 by the constructors Marriott (later to be taken over by Kier), provided 5,750m² gross, 4,600m² assignable, at a project budget of £18.0M. It was named in honour of Dr Frederick Sanger, double Nobel prize-winner, and was formally opened by him in November 1997. Funding was mainly from Peter and Paula Beckwith, by the Wolfson Foundation and the Wellcome Trust. Later, a refurbishment of Professor Salmond's lab was designed by R. H. Partnership, WSP and F. J. Samuelly; the selected contractor was Amey and the budget £1.8M.

The second phase of this development, known as the Gurdon Building in honour of the popular and modest Nobel prize-winner Sir John Gurdon, was funded by Wellcome and the Cancer Research Campaign (CRC). The project was procured through the discontinued system of a single-stage procurement, and the then-confrontational contract form of JCT 98. The reason for accepting those two obstacles to sound, team-based procurement was that at the time the University believed that it was bound by the procurement system contracted for Phase I. Subsequently this was said not to have been the case. In any event, those two errors of judgement, if they were indeed errors rather than contractual requirements, made the project much more difficult, long, costly, and confrontational.

Another major problem for the project was the unfortunate circumstances around the site manager. The contractor, Sir Robert McAlpine, had done a superb job on Phase II of the CMS (Maths) project in significant part because of the considerable ability, personality and team-mindedness of the site manager; it was agreed verbally that he would move onto the Biochemistry project when it started on site on 6 August 2001. Alas, he took retirement. (US forms of contracting would disallow such a change without agreement and compensation). The site manager appointed, though dedicated and pleasant, was not as capable: he suffered from a painful back injury which prevented him from getting properly around the site. While the work progressed generally soundly, with some exceptions that had to be remedied, and stayed close to budget, progress fell further and further behind programme. As a consequence, almost uniquely among the 100 major construction projects carried through in the years 1996–2006, this project went to adjudication, the only project to do so during the decade, apart from the very small adjudication noted above (p. 168), and did so not just once but twice. The work delays that led to the first two adjudications on this project were in part caused by site errors (e.g. a huge 'setting-out' mistake which resulted in some of the foundations being dug and concreted in the wrong places), and in part precipitated by the sort of progress-delay problems that arose from single-stage contracting such as incomplete and/or late design information. The adjudication outcomes were that the contractor won a total of 19 weeks of the extra 58 weeks, the amount of time behind programme, he was claiming. Subsequent to the completion of works on 18 June 2004, the contractor tried again through formal adjudication proceedings with the extensive (56 lever-arch files of evidence) help of an expensive claims consultant: of the 60 weeks extension of time claimed, the contractor got just three days. The process of adjudication is safer, more predictable and less demoralising and hugely quicker and cheaper than the previously standard court or, often even worse, arbitration processes. Had the University lost, which never was at all likely but nevertheless always possible, it would have had a bill of probably well over a million pounds. The total project was completed with some additions at a cost of £23.62M including all fees, compared with the original budget of £21.65M.

The floor height of the first phase building was just 3.6m and that bound the second building on the site to that height although 4.1m would have allowed a better building. Nevertheless, the building when finally completed was, apart from some teething problems, such as the lifts not working properly, and the limited floor to ceiling heights, very good indeed for its purpose, and elegant in its style. The street scene of Tennis Court Road was much improved.

Downing Site

Across the road, work continued as it had for many years on the patchwork of rather higgledy-piggeldy buildings on the Downing Site. A complex project in 1997–1998, known as the 'Combined Facility', drew together some pharmacological labs near the site boundary with Downing College. It was carried out at a cost of £5.37M by constructor Willmott Dixon. The project didn't go at all smoothly, in part through a breakdown in trust, partly relating to sub-contractor appointments, partly through internal tensions (as in two leading protagonists' nose-to-nose yelling) within the University project management staff, and partly because the contractor moved his best people to another project (a swimming centre for the City Council). However, a fair deal was settled through bilateral discussion between the newly arrived Director of Estates and Sir Michael Latham, Chairman of Willmott Dixon (who was one of the Director's job-application referees). The refurbished labs worked well, and as it turned out years later (see Chapter 9), the University did well to keep such labs up to standard for many, though arguably not enough, years.

Around that period, near the end of the 1990s, numerous other refurbishment projects were carried through on the Downing Site, many of them designed as well as managed by the University Estates Department: Chemical Engineering, £1.1M; constructor SDC on the 1957 building, and then an upgrade for the Cambridge Unit for Bio Science Engineering from 11 February till 30 September 2002: RHP, Edmond Shipway and Rolph Partnership were the designers, and the constructors were French Kier at a budget £2.227M. (This project also came in on time and budget.) The fossil collection part of the Sedgewick Museum was upgraded for £320k in 2000/2001. More upgrade was carried out for the Department of Biochemistry on the earlier 1959 building on the Downing Site: firstly a project achieved by designers RHP, K. J. Tait and constructor Amey (which was about to switch to being a facilities management company, an industry with higher profits and lower risks than construction) at a cost of £1.80M; then the second floor of Genetics, design by RHP and WSP with a project budget of £1.45M, and the Cell Biology (drosophila-fruit fly) lab, constructed by Haymills to a design by RHP, Bedwells and WSP, at a project cost of £2.469M. When the Departments of Anatomy and Physiology were merged (with varying degrees of enthusiasm) in 2004, £3.70M was spent on adapting two floors including a lot of services upgrade. A larger amount, £4.33M worth, of refurbishment was done for the Department of Physiology, Development of Neuroscience (PDN) between 25 April and 5 December 2005; CMC was the architect, and the engineers were Andrew Firebrace Partnership and IBS. The contractor was Dean and Bowes, who did a good job keeping the project on programme without unduly disturbing all those working in the building throughout. Meanwhile, over in the Sir William Hardy building, work was done by contractor Bernard Ward (who twice had to be formally warned about low safety standards on site) refurbishing the Centre for Sustainable Landscapes (Geography Department) at £1.773M and space for Experimental Psychology at £680k. Another refurbishment project was carried out for the Language Centre in the Old Music School April to September 2003, at a cost of £629k. Research facilities for the study of plant sciences were upgraded in 1998–1999, complicated when the contractor, CEL Construction, went into liquidation in 1998; the architect was RH Partnership and the engineers Hannah Reed, and, despite the constructors' liquidation, the contract proceeded to a satisfactory conclusion at a cost of £1.15M. Phase 4 of the development was carried out in 2006 at a cost of £1.60M. Another project designed by RHP was the Wellcome Wing Biochemistry Skylab refurbishment, started on 5 November 2001 and finished on 31 May the following year, at a cost of £2.237M to relocate the research teams coming out of what became the Leverhulme Biological Anthropology Institute in Fitzwilliam Street.

In 2005–2006 there was a Downing Site project procured in a different way. The building used by the Systems (Stem Cell) Biology Institute was in need of substantial upgrade to afford the facilities needed for the new and expanding work in stem cell research. It essentially needed gutting, and there was concern about the level of knowledge of the services installed over the decades. Following careful study of risk profiles, it was decided, as one of only two design-build University contracts in this decade, to employ a contractor to establish through a 'design and build' contract what in detail needed to be done, and to price those works; the outcome of tendering was the appointment of V. M. J. Gleeson (whose chairman sat on CITB), although in 2005 the group made a £18M loss, primarily due to losses in the building division, which was then sold to the management. In March 2006, the civil engineering business was sold to Morgan Sindall. The project cost was within budget at £10.737M plus £520k for extra work required by and funded by the Representative User. The designers employed by the constructor company were Saunders Boston, Silcock Dawson and Hannah Reed. Work started on 9 May 2005 with satisfactory completion, within budget, a bit ahead of expected schedule, on 15 September 2006.

Centre for Human and Evolutionary Studies

Moving north along Trumpington Street and turning right down Fitzwilliam Street, past three college hostels at the far end, there is a building which at first sight seems not much more than the adjacent terrace-houses, but suddenly reveals a large courtyard and a considerable amount of building; something of a Tardis, bigger on the inside than on the outside. A bid was worked up in 1999 for Round 4 JIF to convert the Fitzwilliam Street Annex, then occupied by the Department of Biochemistry, to create a home for the proposed Centre for Human and Evolutionary Studies headed by Professor (then Dr) Rob Foley who would be the Representative User. The primary purpose of the proposed Leverhulme Centre was to establish a world-class research centre focusing on the integration of different approaches to recent human evolution. That bid did not succeed but following discussions between Dr Foley et al. with the Wellcome Trust and others in 2000/2001, the Trust agreed to fund up to 75 per cent of the building costs, with certain exclusions. The Centre would combine some of the University's human biological samples, notably the Duckworth Collection, and a number of research labs in a dedicated building. The plan was to remove all non-structural elements of the four-floor building, and some of the basement, and add an extension, with a new staircase and a new entrance. Sheppard Robson and Oscar Faber were appointed as designers (Oscar Faber was later replaced by Faber Maunsell by May 2003), and Marriott (which became Kier Marriott) as contractor. It had been planned that the building would be available for refurbishment at the start of 2002, but the programme to move out the Department of Biochemistry occupants was delayed by plans within that Department. In the surprising absence of any University-level mechanism to move departments around, that delay went on until June 2002. The design programme was therefore not on the critical path for a while; as it happened, that slack in the programme was useful as there was a need to redesign services provision following the decision to rotate the orientation of a seminar room by 90 degrees.

Top of the project Risk Register was 'the presence of mercury', mercury having been spilt over several decades, as it had been in many University buildings before the toxic effects of mercury were known. In 1994–1995, the University spent nearly £1M on removal of mercury from labs, but more remained. So a survey to assess the presence and implications of contaminants was commissioned and that reported that indeed there was mercury pollution, though not at a particularly high level. The relatively safe level of mercury was being carefully monitored, as it was in many of the University's science buildings at that time. The problem was that as the bio-chemists were still using the building and wouldn't move, a full survey (going below floor-boards, into plumbing and behind partition walls for example) could not be done adequately. A reasonable financial allowance was therefore estimated for mercury clearance as part of the project budget, which at that time stood at £4.21M.

Design proceeded and by April 2002 was at RIBA Stage E (now Stage 5). Work on site started on 1 July 2002, and it very soon became apparent that the mercury levels were very much higher than had been set out in the consultant's report: consequently work was suspended in October 2002. A series of meetings continued through that winter, and it was decided to recommend demolition of the building and construction of a new building; mercury removal cost outweighed the relative value of the building. The proposal was to appoint a design team immediately, to start work on site in September 2003, truly an ambitious programme, with a budget of £7.23M. That recommendation was approved by the University's Buildings Committee in January 2003 and by the Planning and Resources Committee the next month at an estimated budget of £6.02M (in addition to the 'sunk costs' of £1.21M already spent), later reduced to £5.43M. Throughout this period, and indeed at other times, the Higher Education Funding Council for England (Hefce) and the Leverhulme Trust were very reasonable and helpful when all was explained to them as it had been and was throughout all the tribulations.

A planning application for the new building of 1,300m² gross area (835m² assignable) was submitted in February 2004, and after some delays for discussion and a deferral about railings, permission was granted. The total project unit costs of the building were £4,171 and £6,493/m² respectively: high because of the high balance ratio, high specifications and complex fit-out. Work on site started on 26 August 2004, and despite problems with one of the Colleges which has a nearby hostel (whose Provost was difficult to contact; she was then reluctant to discuss how work might proceed), work on site was completed to a generally high standard on 5 December 2005. The building sits remarkably well in its setting, and works well for the users.

Case study – Leverhulme Centre for Human Evolutionary Studies in the Henry Wellcome Building, Fitzwilliam Street

The brief for this building was:

- to house the Duckworth Collection, the second or third most important collection in the world of human and non-human primate skeletal remains; casts of hominin fossils; and hair samples from around the world
- to provide research facilities for a growing and changing field, with increasing emphasis on genetics, ecology, archaeology and linguistics)
- to provide Laboratories for genetics, DNA and paleoanthropology, with writing up spaces and offices
- to encourage different disciplines to work together – interdisciplinary working rather than the more usual 'silo' approach.

The result was a discreet, carefully proportioned brick building which appears to speak to the fine early nineteenth century terraced houses lining Fitzwilliam Street with the vista at one end towards the Fitzwilliam Museum.

The public entrance from Fitzwilliam Street is through gates and railings with a design representing the evolutionary tree; this, together with the glass window to the ground floor seminar room and the glass doors which are etched with quotations from Darwin, are the building's public art contributions.

As the visitor crosses the small courtyard, the surprising 'Open-sesame' of the automatically opening front doors is welcoming. This was a response to the knowledge there would be no reception staff to manage access as well as trying to respond to Part M access requirements.

The entrance is directly into a generous light, airy staircase enclosure, but this is more than a staircase as it was conceived as a central part of the building: a dynamic space for serendipitous meetings. The large landings on each floor have displays of skeletons of primates and simple, comfortable seating.

Originally, the shell of an existing building was to be retained. In the event, when the Department of Biochemistry moved out, just before construction was to begin, it was discovered the existing structure was too contaminated with mercury. The courageous decision was made to pursue a new build on the existing footprint, with a very short programme to completion, retaining the original design team of Shepherd Robson (architects) and Oscar Faber (SE and M&E), and the same contractor, Marriott.

The building is roughly an offset T-shape in plan, four storeys on the street with a basement for most of the footprint.

The Duckworth Collection is housed in two separate rooms with moveable shelving, specimens in cardboard boxes, some with clear windows to see contents, not air-conditioned but with mechanical ventilation and humidity control. The basement also contains a large and well laid-out plant room.

On the ground floor with a street frontage is a seminar room on the street facade, which seats up to 100 with a moveable partition which allows the room to be divided. This room has shelving for displays and can be used by visiting scholars as well as for the regular interdisciplinary seminars, workshops and conferences. According to the building's administrator, this room functions well and is generally comfortable. At the rear, with its own separate, secure entrance, sticky mats in the lobby and positive air pressure, is a 'clean' lab for work with DNA. This is a specially designed and equipped lab, with positive pressure, isolated rooms and sterile tents, to reduce the risk of contamination to samples. Its design was important as this reverses the problems of contamination associated with most labs.

The first floor contains labs which have large windows and views out, very pleasant and quite often unusual for labs. One lab is for hazardous samples and therefore has negative air pressure and sophisticated extract system with filters. The other lab is for paleonanthropological microscopy, skeletal analysis and 3-D imaging. There are a number of offices of varying sizes. The aim was not to demarcate the territory of different teams; this has required positive management action to encourage the mixing up of groups.

On the second floor there are offices for senior academics, generously sized and light and pleasant. There is some overheating from solar gain in the south-west corner office. In the rear wing is an office for PhD students and post-docs in open-plan arrangement.

On the third floor there are three research rooms, a kitchen and toilets along the rear 'wing', and a large library which doubles as a meeting room and staff room. The building is cut back from the street facade and the top landing is a large glassy space opening onto a roof terrace with views across the rooftops. Planning considerations dictated that the terrace was kept back from the front facade. One of the main defects since occupation has been water penetration along the line of the balustrade, which took several attempts to get put right. The terrace is a very pleasant place, especially in good weather, and it is easy to envisage convivial gatherings here.

Construction

This is a steel framed building within brick and block cavity walls which obscure the frame. Deep beams in an even deeper ceiling void enable services and the air-ducts to be run around the building.

Environmental

Above the staircase are automatically opening vents which enable the stair to be used as a chimney to force through ventilation but there were problems with the operation and the automatic function has been discontinued. For cleaning and maintenance it is necessary to get in a team of abseilers.

Except for overheating in the corner offices and coldness in the PhD room, the building is generally considered to be comfortable. All heating is controlled remotely by the University maintenance unit from Laundry Farm and no information about the annual running costs is available. There is a perception that the compact fluorescent lamps to light fittings often need changing but no actual figures were available.

The overall effect is of a well-mannered building. Care has been taken over fenestration and there are nice ideas such as the staircase as a place for chance meetings and it is a pleasant surprise to see labs with windows looking into the street. In the architectural detail, however, the building on the inside feels slightly disappointing, an assembly of components which have not quite come together and been totally, satisfyingly resolved.

Security is an important issue for this building both because of the health issues with some of the scientific work and samples and also because of the sensitive nature of the Duckworth Collection. There was a clash between access under Part M and security: the process was described as prioritizing the legal requirements for disabled access.

This project was clearly loved and championed by the then Head of Department whose brain-child this was, as Representative User (RU), and one colleague; the RU reckons their input totalled some two-person years of time over the project design and construction period. This is a huge commitment but it was probably this detailed involvement which led to such satisfaction with the finished building.

There is also the question of technical innovation and new scientific methods and their effect on the requirements of a building. For instance, changes in scientific techniques in dealing with and analysing genetic samples, the increase in computer modelling and an increased emphasis on archaeology has changed the amount of use of certain types of lab space. Nonetheless the Centre has been pro-active in providing the facilities for use both new activities, such as testing skeletal material from more recent history, and the greatly expanded number of visitors to the Duckworth which has occurred.

Architect: Sheppard Robson
Engineers: Oscar Faber; later Faber Maunsell
Quantity Surveyor: Crump Newbury
Project Manager: EMBS
Contractor: Marriot – later Kier Marriott

Mill Lane

All the projects described in this chapter thus far were for academic purposes; there were also some projects for the administration. It is always hard for administrators to make a case for capital expenditure for the administration, but with such change and expansion in the University, more administrators and therefore more working area were needed.

The building seen on the left as one returns to Trumpington Street from Fitzwilliam Street is 'Kenmare', a Grade II listed former house used for the main office of the 'Estates Management and Building Services' department, renamed the 'Estate Management' department after the planning aspects of its function were for a few years largely and counter-productively transferred to another administrative department. Behind that dignified building, there was another fine building that once housed the Baily Grundy & Barrett Electric Company which, as described in Chapter 1, provided electricity locally to the town and to the University (generated in the basement; transmitted at low voltage, 110 V DC, along underground cables to Senate House and elsewhere).[6] The company was bought out by the Cambridge Electrical Supply Company but continued in operation as retailers till 1973. In 1999 the building, which had long been used as a store-house, was converted for use by the estates department so that the project managers and the accountants could co-locate there and work closely together, a key element in keeping projects to budget. (An effective and team-orientated accountant, Andy Clarke, was recruited in 1999, and he and his team took the University through its surge of projects.)

The University administration was on the move in other departments also. In 2001–2002, the Maths Department had moved from its old buildings in Mill Lane to its new home in the Centre for Mathematical Sciences building off Clarkson Road. One of its vacated buildings was converted between January and May 2001 at a cost of £1.7M for the use of the reconfigured Research Services Department, whose remit and importance was rapidly increasing as the research profile of the University grew enormously: this major part of the University's administration was moved into open-plan offices, the first of such a space-use change. Just across the road, Lecture Theatres 2 and 9 on Mill Lane were also upgraded in swift summer 2005 contracts: the architect was Alun Design and the cost £1.15M.

University Centre

Meanwhile, in March 2000 the construction company Interior plc (which had done an excellent job on the upgrade and mezzanine floors in the Raised Faculty Building on the Sidgwick Site, as described in Chapter 6) started work upgrading the kitchen area of the University Centre: an upgrade was thought necessary even although the Centre was not tapping the external market, and was running at a significant loss. It was not a happy project. The architect, RHP, had been set up as the project manager and that didn't work well, especially regarding the electrical/mechanical aspects after it transpired that the engineers hadn't adequately investigated what was in place at the start of the contract: there was a rather evil-looking duct that needed to be explored by crawling along, and hadn't been. It was sometimes difficult to get sign-off from the Representative User. There was, for example, ambiguity as to the balance between having a floor surface smooth enough not to harbour germs, but rough enough to prevent people slipping when the floor was wet. This project came at a time when a lot of large, high-risk projects were getting off the ground, and it was difficult to find enough managerial time to keep all projects on track; it was still not realised across the administration (far less among the academic leadership) that the University was in a large capital programme not just in a series of projects, and so there was little understanding of how problems on one project affected delivery of other projects. The University Centre upgrade project was soon running first a week late, and then longer, until finally the project juddered to a satisfactory completion, with a budget increased from £1.958M to £2.217M, not a great increase given the complexity of the project, but disappointing. The final report on the project ended with the words 'numerous defects being resolved'.

Figure 8.4 University Centre
By courtesy of info@Cambridge 2000.

New Museums Site

Along Free School Lane, there was a series of upgrades to the Whipple Museum of Science and its associated library, to convert a redundant lecture theatre into a library, and to enhance offices and substantially expand and improve exhibition space in the Department of the History and Philosophy of Science: the largest project there was designed and, with some £530k of minor works carried out in 2005–2006, funded at £1.154M by SRIF 3. (It is interesting to note that declaring the lecture theatre *per se* redundant followed a University-wide review of lecture theatres to attack the issue of every department wanting its own even if it could easily share.) The Department was very pro-active and the result excellent.

The stores of the Departments of Pathology and Biochemistry were merged and upgraded in 2006, at a cost of £1.4M. £1.91M was spent upgrading the Zoology Department, a complex job carried out while the building continued in operation. In 2005/2006 there was opportunity to be able to demolish the Arup building and transform much of the New Museums Site, with a link into the developing shopping centre and a revamped crossing into Mill Lane; two colleges were keen to get some of the range of new accommodation that would be possible. An outline plan was developed with a lot of analysis and discussion. The scheme once worked up was proposed by a working group chaired by Professor Malcolm Grant with the Director of Estates, and seemed to be gaining a lot of support. Then chairmanship and some membership changed; at a key meeting of the relevant working group on 22 June 2005 it became clear that progress to revitalise from the Arup Tower through to Mill Lane and the river was being lost.

There were minor works done to the Arup Tower after the Computer Lab moved out at the end of 2001. (When the Department of Materials and Metallurgy moved out to West Cambridge in 2013, the Tower had to be re-furbished; an initial estimate of the project was £55M.) The Mond building (once the office of the Russian physicist Piotr Kapitza, and with a stone carving of a crocodile, which was the name he used for Rutherford) was refurbished at a cost of £1.995M for the use of the Faculty of Human, Social and Political Science.

Not all projects came to fruition; some were designed and costed as a necessary step to applying for funding, and of course it was disappointing when projects had to be dropped after a lot of work. Such a project was a plan for the Department of Land Economy in Silver Street to move to a new building on the Sidgwick Site; this was worked up with designers Allies and Morrison, Buro Happold and Whitby Bird with a budget cost of £7.2M. However, after years of effort, sufficient funding was not forthcoming (despite the number of affluent alumni successfully managing development companies). Design proceeded to Stage 3 (formerly RIBA Stage C), and presentations were made to potential benefactors, but the project was abandoned. The riverside building of the Department on its site at the bottom of Mill Lane was refurbished under the minor works programme. The work was of good quality but just after it was completed there was a very high flood, and much of the enhancement, including all the carpeting, was soon under the waters of the passing River Cam; after the flood the dove returned with a twig and quality was restored.

Another binned project was to develop a University Visitor Centre in premises opposite King's College; this project was watched warily by colleges, in particular those suspicious that some of their interests might be prejudiced by 'the University' attracting visitors directly. From 1 May until the end of September 2004 planning and design did proceed quite far, with a project cost of £1.474M; but suddenly the word was passed that planning should stop.

Along the road, good quality was achieved in some substantial stonework repairs. There were three phases of upgrading the stone on the Senate-House; repair was needed especially after decades of petrol and diesel fumes from the nearby traffic going along what had long been a main traffic artery through Cambridge; the surface of the Portland stone of which it was built (like the limestone of nearby King's College chapel) went soft, and poor in appearance. Rattee and Kett, a Mowlem company, carried out the repairs effectively in 1998–2000 at a cost of £1.1M and to a quality that led to winning Natural Stone awards from the Stone Federation. £2.2M was spent repairing stonework along Free School Lane, Pembroke Street and Downing Street.

ADC Theatre

A series of projects was carried out at the ADC Theatre. Lack of stage and back-stage space, and the standard of general facilities, including disabled access infrastructure, was increasingly being found unacceptable as alterations since 1974 had been piecemeal as funding faltered. The original plan to carry through a huge project over a summer recess looked impracticable, so a four-stage programme was drawn up.

The first phase of the project took place over the University long vacation of 2003; foundations were reinforced and utilities were re-sited to enable the subsequent phases to take place. The foundations were found to be amazingly scanty, just a few courses of bricks below ground level. When asked why the building remained standing, the reply by the excellent project manager was, 'Habit, just habit'. The cost was £750k, less than it might have been.

The following year an extension was designed by Bland, Brown and Cole (who previously had been the architects for the less harmonious refurbishment project at the Arts Theatre. Barry Brown who had been in the ADC Footlights when an undergraduate, was Footlights President and was still making a good and appreciated input into the theatre). It was constructed, as all phases were, by Kier Marriott. This phase of the redevelopment, lasting from April to September 2004, required the closure of the theatre for one term. It saw the re-modelling of the existing ground floor areas, improving facilities for staff and customers alike, with long-wanted improvement to the foyer and ticketing area, and considerable upgrade of access for the disabled, in particular for those in wheel-chairs. That made a very material improvement to the appearance and conditions of the theatre. The facade of the building was remodeled; in particular, the foyer was enlarged and modernised with a new glass front, reducing the need for patrons to queue in the street and creating a lighter and more welcoming space. A successful web-based ticketing system was introduced. This phase was also completed on time and within the £550k budget.

The next phase, in 2005, focused on accessibility. It started in May 2005 and lasted until that September. It was undertaken at a cost of £500k. A lift was installed from the foyer to the first and second floors of the building. A new, airy corridor beyond the lift on the first floor created disabled-access to the bar and auditorium, as well as improving audience circulation in general, before and after shows. Previously, only one corridor had existed for entry to the auditorium. The bar was extended to meet the new corridor, reducing overcrowding, and a pleasant external terrace was created. This enabled the bar to move to fully non-smoking status, some time prior to such legislation against smoking in public buildings. By raising the upper flight of the main entry stair, it was possible to improve the raking of seats in the auditorium, giving better sightlines and better leg-room, and re-seating was carried out. The technical control rooms on the second floor were enlarged and slightly relocated to employ the new lift as the primary means of access. This enabled students with disabilities to be fully involved with many technical aspects of theatre productions for the first time.

In 2005/2006 designs were developed for a two-storey extension at the back to provide new changing rooms, a brand new multi-purpose studio space, named the Larkum Studio after Charles Larkum (1942–2006) who served as a popular and effective Chairman of the Theatre's Executive Committee from 1999 until 2006 and was instrumental in bringing about the modernisation of the Theatre through strategic planning and strong leadership. That work was carried out in 2008 at a cost of £1M.

Overall, the four-phase ADC project greatly improved enjoyment of productions, enhanced the standard of productions, increased ticket-sales and improved accessibility. Happily, the refurbishment programme was achieved in an era of brilliant productions, usually playing to full houses.

The ADC became one of the two fully equipped theatres with pit, flying tower and full facilities in Cambridge. The refurbishment won the Local Authority Award for the 'Best Technical Design and Construction in East Anglia' in 2009.

Kettle's Yard Museum

The cultural life of the University was also enhanced by developments at the Kettle's Yard gallery at Castle Hill. Kettle's Yard was originally conceived with students in mind: in 1954 the public-spirited Jim Ede envisaged creating 'a living place where works of art could be enjoyed ... where young people could be at home unhampered by the greater austerity of the museum or public art gallery'. In 1956 Jim and Helen Ede came to Cambridge in search of a 'stately home'. What they found instead were four tumbledown cottages nestling by the ancient church of St Peter. With the help of architect Roland Aldridge, Ede restored and substantially remodelled them. The gallery became ever more popular with successful exhibitions as well as display of the furniture and paintings in the house, and free concerts.

By 1998, the building adjacent was a University-owned shop being leased and used as a post office, with a tenant who was far behind in paying rent. When the tenant fled, allegedly under threat of violence from 'business associates', the University renewed its efforts to terminate the lease so that the space could be turned over to Kettle's Yard for expansion of its activities. Complex legal wrangles ensued, complicated further by a local campaign to retain the premises as a post office, even though the Royal Mail was refusing to re-issue the post office licence or issue a new one. Eventually, after years of legal and property struggle, the premises were recovered. During those years some refurbishment of the existing facilities was carried out. A major project to re-model the recovered premises and integrate them with the original Kettle's Yard buildings was finally approved in 2013/2014; at the time of writing it awaits funding.

Faculty of Education

Another exciting, and very different, project within the city was the development of a new building for the Faculty of Education in the grounds of (formerly specifically teacher-training) Homerton College on Hills Road, consolidating staff and teaching/research facilities from four other venues; this clearly gave an enormous boost to the capability of the Faculty, although some members of staff didn't want to leave the offices they'd grown to love during four years dispersed in different locations, including converted residential houses; no great enthusiasm was shown by some for being co-located with noisy, bustling students instead of enjoying personal offices with their plants and sometimes pet dogs.

The start of construction work was hugely delayed by the time it took to negotiate land transfer from Homerton College to the University; month after month negotiations moved with glacial speed (the phrase 'with the speed of a striking sloth on a Bank Holiday' was heard again) and it was difficult to understand why the delays kept recurring. Perhaps it was because such transfers from College to University rarely happened: much the same happened at Addenbrooke's with the transfer of land from Downing College to the University for the research building for ICRF/CR-UK. Anyway, work eventually started on site on 14 July 2003 to a design by BDP (Building Design Partnership, then the largest architectural practice in Europe), Mott MacDonald and Whitby Bird. The constructor was the national, now international, company Amec. The project, managed by a particularly effective internal project manager, was completed within its £13.75M budget, although

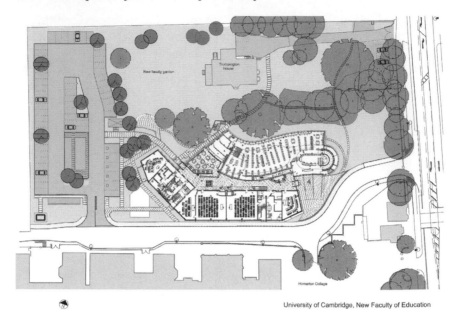

Figure 8.5   Education landscape and plan

By courtesy of BDP

there was trouble with the ground-source heat exchange units; their design was technically innovative, and despite much remedial work, they failed and had to be replaced at considerable cost by air handling unit chillers. That added to the work to be done, and with other factors caused completion to be three months late, with chillers still to be installed post-occupation.

Those problems were significant but temporary, and it was apparent from the start that the design of the building was striking, effective in achieving the aims of the project, and at an economical cost of £2,224/m² gross. The building was officially opened by the Chancellor, HRH The Duke of Edinburgh, on 21 June 2005 after many fine words.

Case study – Education

Education Faculty

This building, adjacent to Homerton College on busy Hills Road, was created as the home of the new Faculty of Education formed from the School of Education and the teacher training part of Homerton College. The reorganisation was effected in 2001 but it was not till December 2004 that the new building was ready for occupation. The site, on the north edge of Homerton's land, is in pleasant parkland stretching towards the listed Georgian Trumpington House and the curve and massing of the library responds to an existing tree belt.

The striking building, with an economical cost of £2,224/m², has three elements:

- a three-storey masonry block with teaching rooms and academic and research offices along the south facade
- the central 'street' with the library and a cafe on one side at ground level
- the sinuous timber-framed library.

The external elevations respect and enhance the two very different environments: the solid if unexciting brick buildings of Homerton College on one side, and the fine trees and lawns leading up to an old listed house on the other. The glass facade on the latter side curves its way through the trees, a design made possible by the glu-lam wooden beam structure chosen for that reason, and for reasons of sustainability over steel or concrete.

The Teaching Block

The more structured, rectilinear brick box 'for learning' has internally detached concrete columns supporting exposed, coffered concrete floors and ceilings with lighting and air-handling units suspended from them; an elegant solution though surprisingly the setting out of the vaulting and the columns is not coordinated. These teaching rooms are well liked despite getting too hot and having rather poor air quality. There were comments about lack of sound-proofing from outside with noise from the access road. In contrast, the sound-proofing through the east glass wall of the library is very effective and no disturbance from Hills Road's traffic intrudes.

The three floors of teaching rooms open directly from the street or balcony overlooking the street, but on the top floor, individual offices are double-banked on a central corridor with the inside offices having sealed windows and roof lights into the street space: mechanical ventilation seems to be provided by grilles through the wall into the street. On the second and third floor, bridges link across the street to a staff room, an art room, an open plan research space and offices, all within the footprint of the library.

Externally, the timber-boarded spandrels within and between the windows speak to the timber of the library but it is not clear this was a necessary device and it introduced complications in detailing. The brickwork has been peeled away to form two angled bays on the top floor of the two staircases, a nice attempt to signal the vertical circulation and get variety into the staircases which could perhaps have been celebrated further.

The central street atrium links the two main linear elements and was intended to foster as much interaction as possible, giving a heart to the Faculty previously housed in multiple buildings on separate sites. The street contains the reception desk at one end and the cafe and its kitchen at the other. Hung from the roof and charging straight up against the teaching block is the 'Stairway to heaven', an open-tread, very public stair giving access to each floor level. There is also a glass lift in the atrium and, in a more pedestrian way, two service and fire escape stairs are grouped together with lifts and toilets within the teaching accommodation.

The street starts at the main entrance canopy facing Hills Road and simply disgorges into the bike and car park, which is heavily used due to the peripatetic work and the need to connect to schools and other institutions. The result is that the building as a whole feels strangely turned in on itself, despite the setting.

The library, an impressive, curvaceous glass and timber building along the north side of the building, faces the gardens, a lofty space with exposed glu-lam frames with ramps and 'lily pads' inserted within the space, linked by bridges. The lower level of the library is such that the continuous work-surface is at about the level of the outside lawn, so that students can glance up from their books and see squirrels and other wildlife such as emerging worms.

The library admin and reception is under one of the lily pads, exposed chunky timber posts and joists which create a cosy, warm feel leading to an enclosed office with a curved end looking out through a glazed screen across some study spaces towards Hills Road. Unfortunately these admin areas have no dedicated heating or cooling and therefore electric heaters and fans have to be used to provide comfort.

The architectural promenade through the library was clearly an important concept in the designers' minds and makes for a dynamic and interesting experience for the visitor. However for the regular user this may not be such an advantage: a long route to reach the upper levels which were envisaged as providing some shelf-space but which cannot be used as such because of difficulties with the width of the ramp and the sloping surface; this means it would be hard work and dangerous to move trolleys up and down and to stock the upper shelves. These shelves are being adapted into display areas for exhibitions and outsize books.

The entrance to the library feels slightly cramped, and requires backtracking from the reception desk; strange as the library is one of the main elements of the building. Security for the library has obviously been an issue; the design did not allow for the library to be controlled from one point: fire doors and openable windows meant that books and equipment could easily 'walk'. To deal with this, various areas and routes have been marked 'Staff Only' and a window on a landing has been fixed shut; a pity because in sunny hot weather this would have helped provide ventilation. It is surprising that the strategy for such fundamental issues as access, circulation and security were not agreed at an early stage.

The librarian had requested a variety of study spaces, and these have been provided round the perimeter at different levels and on the lily pads. The study spaces on the sunken lower level where the worktop is level with the lawn stretching towards the listed Trumpington House are much enjoyed. The library is very popular with students consistently scoring the highest (over 90 per cent) rating for very good and excellent: this of course reflects the satisfaction with the service provision as well as with the physical spaces.

The interlinking of the three main elements creates a lively, dynamic series of spaces but does also create a problem of acoustics – not just intrusiveness of noise from the cafe being funnelled around the street and apparently into the library, but also the impossibility of having confidential discussions without going into an individual private office. After completion, areas of acoustic board were fixed to the brick walls at high level near the cafe. This has helped somewhat but the curious amplification of intimate conversation levels on the ground floor still continues.

Perhaps today switching and control would be considered in a different way: lights at the study spaces cannot be individually switched as apparently they count towards the overall illumination of the space. All the controls for the lights in the library are in the inner admin office, so for a period, the librarians had to use a torch to get out of the building safely. It took years for a timer switch to be installed!

The environmental strategy was considered holistically: the thermally weighty teaching block to the south shields the street and library from the sun, the lofty street's stack effect provides natural ventilation and where mechanical ventilation is required energy recovery is used. The library had to be sealed for security reasons and therefore is air-conditioned. The teaching block is kept shallow to allow for natural ventilation and daylight and one can imagine this all being in balance if the solar shading on the south facade had not been omitted as a short-sighted cost saving. The feasibility of using ground water to enhance free cooling was explored but the trial boreholes failed and this was abandoned.

This building was joint winner of the David Urwin Heritage Award for the most sustainable and efficient building in 2006. In the figures submitted for the RIBA awards, the energy for the first year in use for space and water heating was logged as $150kWhr/m^2$ and energy for electrical requirements as $111.4 \text{ kW hr}/m^2$

Client consultation

A user group of five to six representatives was set up which met the architects and design team regularly, at fortnightly intervals at the height of activity, and a larger workshop was held to discuss preferred options. However, changes in personnel led to some loss of continuity. Nonetheless, the EMBS's John Woods expressed himself satisfied with the consultation process, both for EMBS and for the Representative User.

The variety of shapes and forms lead to difficult junctions (e.g. at the edges of sloping elements (soffits to ramps, ceilings at roof level) and are in places not happily resolved. It could be that the develop and construct form of procurement with novation of the architects to the contractor, who also instigated a late Value Engineering exercise, militated against the happy resolution of issues on site. Indeed to the users it appeared the architects vanished and defects were never properly dealt with.

However, these niggles aside, the users are happy with and proud of their building.

Architect: Building Design Partnership
Structural & Civil Engineers: Whitby Bird & Partners
Buildings Services Engineers: Connell Mott MacDonald
Landscape Architect: Building Design Partnership
Acoustics: Building Design Partnership
Quantity Surveyor: Turner & Townsend
Main Contractor: AMEC Capital Projects Ltd

Figure 8.6 Faculty of Education
By courtesy of BDP

Figure 8.7 Faculty of Education
By courtesy of BDP

Fenners Cricket Facility

Next to be considered is the development of sports facilities at 'Fenners', a University cricket centre near Parker's Piece.[7] It was decided in 2002 to proceed with design and construction of a new cricket school, a large cricket practice area (some 2,360 m²) where wicket conditions could be varied, plus offices and support areas. The work, done by contractors Marriott, took from 14 July to 15 December 2003; the building was designed by S&P Architects, with W. S. Atkins as engineer. The cost was £1.1M. Getting planning permission was a long and weary battle: green space is highly valued especially around the middle of Cambridge. Once work started on site it went very smoothly. On the day of the opening ceremony, rain was stopping play in big matches, so the value of the indoor centre was well illustrated.

Plant Growth Facility

The final projects in this chapter on city area projects are those in the University's Botanic Garden. The Botanic Garden moved from what now is the New Museums Site to its current site between Trumpington Street, Brooklands Avenue and Hills Road/Bateman Street in 1846. By the end of the twentieth century many of its buildings and greenhouses were not good for their purposes: elderly and inflexible by modern standards. There was a plan to build a new Education and Interpretation Centre, and a very exciting and highly sustainable design was worked up by ECA (Edward Cullinan Architects), the designer of the Centre of Mathematical Sciences and other Cambridge buildings, with Max Fordham and Buro Happold; this design was taken to RIBA Stage 3 (formerly Stage C), but, sadly, funding was never achieved so that exciting project was put on hold, and then cancelled. The budget for fees to get that far was £255k; some of the work done was useful as user-requirements were worked up for the building that was erected later.

Several major greenhouses were refurbished or replaced, but the main problem/opportunity came in relation to the facilities for Plant Growth research: how plants grow, and how they get and respond to diseases. A new and very sustainable building with wide open spaces in two 'halls' was designed by R. H. Partnership with engineers Faber Maunsell and F. J. Samuely & Partners. The plan dimensions of the plant growth element were 48m by 19.5m to provide space to house the specialist growth rooms and associated facilities (soil storage/preparation, potting/harvesting, autoclave and wash-up). The walls were made from replaceable harvested timber (albeit transported from Germany), the interior finish of the curved roof was made of insulated stress-skin panels with a stainless-steel coping to protect the timber frame and channel rain-water for sub-soil distribution. The walls were made from prefabricated panels using blast furnace slag to replace 40 per cent of the cement. The ends of the building have trellis work for climbing plants.

The insulation was 15 per cent above minimum standards, and there is a very sophisticated heat recovery system. A retaining wall close to the building protecting the chillers and condensers was made from old tyres packed in earth. This was one of the first University buildings to be BREEAM (Building Research Establishment Environmental Assessment Methodology) rated 'Very Good'.

For a while the building was strongly opposed by neighbours in the adjacent flats, and one can understand their concern. However, over the months, with a great deal of discussion, opposition gave way to guarded acceptance. Indeed, the building when completed had very low profile, and it was designed to be controlled by just one technician plus at times a very small team, so little extra traffic was generated. The work was carried out well by the constructors Willmott Dixon, whose Chairman Sir Michael Latham took a great deal of interest in this innovative project. It was about this time that the fruits of the very extensive training in teamwork and collaborative working through all levels of his company were reaping benefits: certainly this building went up in good time, September 2004 to August 2005, within its budget of £6.0M, and in a context of very professional and collaborative relationships, still not very common in the construction industry. A second phase to this plant genetic science project was planned but shelved through lack of funding. But then late in 2005 a headline appeared in the *Financial Times*, on the morning of a Buildings Committee meeting, that the Gatsby (a Sainsbury) Fund would donate £26M for the purpose. A project was set up accordingly, with the budget growing steadily through design and construction to a stated level of £82M, out-turn cost rumored to be even more. The building has attracted comment concerning clarity of the entrances and external identity of the purpose of the building, and an uncomfortable main stair, but in many respects the building provides excellent facilities, is handsome, indeed opulent, and is so well-regarded by the architectural hierarchy that in 2012 it won the Stirling prize, as had the nearby residential development in Brooklands Avenue.

Disability Resource Centre

Throughout all the rush of new and refurbishment building works over the decade there ran a programme of upgrading buildings to afford better access for the disabled. In 2001, legislation removed the previous exemption of educational institutions from the requirement to provide disabled access to buildings, and in each of the succeeding years statutory requirements increased substantially. There was close co-operation between the Disability Resource Centre and the estates department in carrying out a detailed survey of 160 University buildings, 470,000m^2 of assignable space, comprising 361 access audits. In 2001, Hefce made provision for funding £56M for the HE sector; £761k was for Cambridge, in a first tranche, then £1.662M, to which the University added £2.0M. A further £4.08M was warranted between August 2004 and June 2006. Work done under this programme included:

- External enclosed platform lift to the core area of the Department of Zoology.
- Access to the main lifts of the University Library.
- Access and facilities for disabled people in the Lady Mitchell lecture theatre.
- New lift in the Mill Lane lecture rooms.
- Ramp access to Senate-House.

In February 2005, the Buildings Committee approved a further ten-year programme for works to provide disabled access. That started in the year 2005/2006 with £2.786M for eight projects, the largest of which were for the Department of Chemistry (£400k), the Department of Anatomy (£352K) and the Fitzwilliam Museum and Grove Lodge (£393k). For the following year the total allocation was £3.384M. The biggest projects were: the Department of Physiology (£613k), University Library (£681k) and further work on the Chemistry Department (£475k).

Property transactions

As well as the construction of many new buildings, there was some rationalisation of the University estate properties along Trumpington Street, and round the corner into Bene't Place. Several of these properties were long used as University offices although they were largely unsuitable for that, having restricted and old-fashioned domestic room layouts with steep, old wooden stairs connecting them. Further, their use as offices meant that there was negligible contribution to the life and liveliness of that part of the city in the evenings: few people came or went, or went anywhere near there, to enjoy themselves, so that 'out of hours' the area was rather 'dead'. Selling these properties would raise capital funding, and it was decided to sell them for use as restaurants or dwelling houses. The most notable was the sale of Number 15–19 Trumpington Street to a hotel/restaurant chain. Negotiations dragged on for many months. The main issue was parking: there was no on-street parking and so there had to be valet parking to a car-park elsewhere, and there was endless wrangling with the city planning authorities as to exactly many seconds cars could stay on the street before being whisked away to be parked. When finally a deal was just about to be signed, that hotel/restaurant chain was taken over by Hotel du Vin: negotiations had to start all over again. Finally in 2006 the sale of the leasehold at a price of £2.8M was agreed. Round the corner, after a new building for the Faculty of Education was provided, 17–19 Brookside and Braeside were sold in 2004 for £2.8M, 2-5 Bene't Place sold in 2009 for £550k to a sixth-form College and 1–2 Fitzwilliam St was sold . Further south, down Trumpington Street, a house was purchased for the Department of Psychology, the last property to be bought with authority of the University administration only sought when it was too late to do anything but honour the deal.[8] From then on the University, whose central budgets paid for all running costs, made sure that properties were neither bought nor built without due authority.

Other property leaseholds sold included 6 Chaucer Rd (2010), the Martin Centre, at £3.3M, The Three Horseshoes pub at Madingley for £503k in 2001. A very long-running and huge development scheme near Wellingborough which contained development of some University-owned land continued right throughout the decade of this book, and long beyond.

Projects included in Chapter 8

	£M
Scott Polar	1.75
Unilever	6.67
Basement fit-out	1.00
Chemistry phases 1–8 and 2 L/T upgrades	41.67
Eng Geo-tech	1.29
Thermo Dynamics	1.41
Roof works	0.72
Prof Hopper's lab	0.84
South-west corner	4.72
Arch new bdg	2.96
Fitz Museum extension	12.06
Gallery refurbishment	2.41
'Old Addenbrooke's'	15.71
Top floor fit out	1.02
CRC/Wellcome Bio chem. Phase I	18.00
Phase II	23.62
Centre for Human and Evolutionary Studies	1.21
	5.43

Downing Site

	£M
Combined Facility	6.21
Plant Sciences	1.15
	1.60
Chem Eng	1.10
Bio Chemistry	1.78
Sky Lab	2.24
Genetics 2nd floor	1.45
Language Centre	0.68
PDN Centre	4.43
Cell Biology	2.47
Sir Wm Hardy – two projects	4.43
Physiology	3.70
Neuroscience (PDN)	4.33

New Museums Site

	£M
Path/Biochem	1.40
Zoology	1.91
Mond	2.00
Plant Science – 4	1.60
Systems Biology Institute	11.35
Bio Science Eng	2.27
Upgrade of UC kitchen area	2.22
Conversion of Maths DPMMS for RSD	1.70
Mill Lane lecture theatre	1.15
Whipple, History and Philosophy of Science	1.68
Stone refurbishment, S House, Downing St	3.30
ADC Phases 1–4	2.80
Faculty of Education	13.74
Fenners cricket	1.10
Plant Growth	6.00
Upgrades re Disabled Access 2002–2006	10.25
Total	241.1

Designed, planned, aborted

	£M
University Visitors' Centre	1.47
Education and Interpretation Centre	11.16
Total	12.63

Other minor works

Upgrade of Bailey Grundy Building, Kenmare
Land economy re-fit
Arup refurbishment

Property sales

15-18 Trumpington St
28–29 Trumpington St
17–19 Brookside
Braeside
2–5 Bene't Place.
1–2 Fitzwilliam St
Education Building, Shaftesbury Road

Notes

1 The Institute, originally financed from public contributions, was enhanced by Sir Edgar Speyer, Honorary Treasurer of the British Antarctic Expedition in the aftermath of Scott's last expedition.

2 In 2014 the opposite decision was made, predominantly to move to West Cambridge.

3 Joint Contracts Tribunal; a body then made up of designers and contractors without any client membership at all: post-Latham that gap was remedied under pressure from the national Construction Clients Forum, CCF of which the Director of Estates was vice-Chairman.

4 *Architecture Today* (November 1995).

5 The first significant design-build construction company, inspired and run by John Pontin who through his evening-class studies came to realise the disadvantages of design-led procurement; he developed the first model for Early Contractor Involvement team-led procurement. This design, and that for the hotel in Pembroke Street which did go ahead, were certainly not the company's best efforts. John Pontin was the first lay member on the University of Bristol Buildings Committee, from around 1990.

6 The first use of electricity in Cambridge was in 1884, funded by Lord Kelvin, to afford lighting in Peterhouse to mark its 600th anniversary. (Ironically, that College has no electric light in its Hall today). About 1896, Senate-House got electric light, a decision, then unusual, to keep up with emerging technology.

7 Parker's Piece is called after Edward Parker, a College cook who leased the area of pasture in 1587; in 1851 Cambridge students levelled enough of it to form a cricket pitch. In 1848 University students formerly of Eton, Rugby and Harrow schools met to decide on a set of rules for football as these were always changing and argued about. The 11 rules agreed were written out and copies were posted on trees around the area, and when in 1863 the Football Association was formed nationally, it was those rules that were adopted as the basis for the national rules of the game of football.

8 The opening was graced by the brilliant Prof Richard Gregory from the University of Bristol. Long retired, he opened his remarks by saying that 'it is very nice to be here: in fact at my age, its very nice to be at all' Cogito ergo sum!

Table 8.1 Overview of capital projects on the Central University Sites

Chapter 8--City Projects	Start date	End date	m2-gross	m2-ass	Architect	Bldg Services Engr	C/S Engr	Construct-or	Cost £M
Scott Polar	1998	1999	631	511	John Miller	Fulcrum	Fulcrum	Haymaills	1.75
Biochemistry - Phase 1	1997	1998			RHP	WSP	Whitby Bird	Marriott	18.00
Biochemistry - Phase 2	01/08/01		6,745	4,135	RHP	WSP	Whitby Bird	Sir R MacAlpine	23.62
Biochemistry - Refurb	01/04/05	01/12/05			RHP	R Tait	Samuely	Bluestone	1.78
Unilever		15/03/01	2,650	2,000	E Soren/C Zilb				6.67
Unilever basement fit-out	21/05/01	21/09/02	425	380	CMC	WSP	WSP	R G Carter	0.98
Chemy--Phase 1b	01/05/00	01/11/00	All Rfurb	All Rfurb	Nicholas Ray	Oscar Faber	F J Samuely	Amec	3.65
Chemy--Phase 2	01/03/00	01/11/00			Nicholas Ray	Oscar Faber	F J Samuely	Amec	3.72
Chemy--Phase 3	01/12/00	01/11/01			Nicholas Ray	Oscar Faber	F J Samuely	Amec	5.49
Chemy--Phase 4	01/02/01	01/10/01			Nicholas Ray	Oscar Faber	F J Samuely	Amec	7.69
Chemy--Phase 5	01/03/01	01/12/00			Nicholas Ray	Oscar Faber	F J Samuely	Amec	4.73
Chemy--Phase 6	01/01/02	01/10/02			Nicholas Ray	Oscar Faber	F J Samuely	Amec	7.13
Chemy--Phase 7	01/05/02	01/12/02			Nicholas Ray	Oscar Faber	F J Samuely	Amec	3.29
Chemy--Phase 8	01/05/02	01/11/02			Nicholas Ray	Oscar Faber	F J Samuely	Amec	1.38
Chemy Pt1a labs G51	01/03/03	01/09/03	All Rfurb	All Rfurb	Nicholas Ray	K J Tait	F J Samuely	Marriott	1.45
Chemy LT 2 and 3	01/04/04	01/09/04	418	340	RHP	Roger Parker	F J Samuely	Marriott	3.10
Eng Dept			Rfurb	Rfurb	Adrian Pettit				0.69
Eng Dept-research accom			Rfurb	Rfurb	Annand and Mustoe				0.72
Eng Dept-Thermo-Dynamics	01/06/00	01/02/01	650	525	Annand and Mustoe	Bedwells	Hannah Reed	Sindall	1.41
Eng Dept--Earthquake	01/12/00	01/11/01	537	508	Annand and Mustoe		Hannah Reed	Mansell	1.29
Eng Dept-Prof Hopper roof extension	01/09/99	01/04/00			Bidwells		Hannah Reed	Sindall	0.84
Eng Dept – SW corner	2006	2007							4.72
Arch & History of Art-new extension	01/07/06	Oct 06	2,535	2,028	Freel'd R Roberts	Max Fordham	Cameron T Bedf'	Dean & Bowes	2.96
Old Addenbrooke's--Judge Institute	01/06/93	Legals 99	Rfurb	Rfurb	J Outram/ Fitzroy	Max Fordham		Laing	15.71
Judge Institute--6th Floor fit-out	01/01/01	01/06/01	810	525	Fitzroy Robinson	Roger Parker	AFP	Deane Bowes	1.02
Fitzwilliam Museum-courtyard dev	01/11/02	01/12/03	3,000	3,000	John Miller	SVM	SVM	Amec	12.06
Fitzwilliam Museum -galleries refurb	Continued				User consultants				2.41

Table 8.1 Continued

Chapter 8--City Projects (Continued)	Start date	End date	m2-gross	m2-ass	Architect	Bldg Services Engr	C/S Engr	Construct-or	Cost £M
Leverhulme Bio-anthro	26/07/04	19/07/05	1,300	835	Shepard Robson	Faber Maunsell	Faber Maunsell	Kier-Marriott	6.63
Univ Centre kitchen refurb	01/03/00	12/01/01	611	545	RHP	Roger Parker		Interior Svcs	2.22
Downing Site--Combined Facility	1998	April 1999	2,176	1391	Sibley Robinson	Sibley Robinson		Willmott Dixon	6.21
Plant Science --2	01/08/99	01/03/00							1.15
DS-Chem Eng--Cam Unit for Bio-science	01/02/02	30/09/02	605	576	RHP	Rolph Partn'rship	French Kier		2.23
DS-Chem Eng--3b			840		Andrew Firebrace				1.10
DS-Language (2)	April 2003	Sept 2003							0.68
DS--Genetics 2 Floor	01/06/99	01/02/00			RHP	Haymills	WSP		1.45
DS--Systems Bio Institute refurb	09/05/05	31/09/06			Saunders Boston			M J Gleeson	11.35
DS--Genetics--Cell Bio Drosophila	01/02/01	01/12/02	830	780	RHP	WSP			2.47
DS--Sir Wm Hardy--Centre Sust Landsc and Experimental Pschology	01/05/03	01/12/03	2,422		Nicholas Ray	EMBS	Hannah Reed	Bernard Ward	2.45
DS--Sir Wm Hardy--			1,183		Bidwells			SDC	1.98
DS--Plant Science--4	01/06/06	01/10/06	485		JGB	Tait	Hannah Reed	Bluestone	1.60
DS--Biochem refurb				1214	RHP	WSP	Samuely		1.45
DS – Physiology Anatomy		2004							3.70
DS - PDN	April 2005	5/12/05			CMC	Andrew Firebrace	IBS		4.33
Dept of Biochem Cardiovascular Unit	01/04/05	Oct 05	577	421	RHP	Tait	Samuely	Amey	1.78
Wellcome Wing Skylabs	01/11/01	01/07/02	788	735	RHP	Bedwells	Samuely		2.24
Whipple extension/refurb	01/12/06	01/02/07	180		JGP/EMBS	Silcock Dawson	Cameron Taylor	Kier Marriott	1.68
New Museum Site--Arup									
NM Site--Path/BioChem stores	01/05/06	31/12/07							1.40
NM Site--Zoology	01/04/06	01/06/06	670		RHP	EMBS	AFP	SDC	1.91
NM Site-School Humanities & SS; Mond	01/10/06	01/05/07	855		Bland Brown Cole	Tait	Cameron Taylor	Haymills	2.00
Mill Lane refurb for RSD	01/01/01	01/05/01	2,800		EMBS	EMBS	EMBS		1.69
Mill Lane L/Theatres	01/05/05	01/09/05	200		Alun Design	K Tait	AFP	Dean & Bowes	1.15

Table 8.1　Continued

Chapter 8--City Projects (Continued)	Start date	End date	m2-gross	m2-ass	Architect	Bldg Services Engr	C/S Engr	Construct-or	Cost £M
Stone Facades-- Free School Lane/Pembrk St/Senate House	Prog of work	1998	2000					Rattee&Kett	3.30
ADC	01/06/03	01/10/08			Bland Brown Cole	Roger Parker	AFP	Kier Marriott	2.80
Kettle Yard-- Phase 1					Jamie Forbet				
Fenner Cricket training area	01/07/03	01/12/03	1,160		S&P/Thurlow C C	W S Atkins	W S Atkins	Marriott	1.10
Education Faculty Hills Rd	01/07/03	01/11/04	2,005		BDP	Mott Macdonald	Whitby Bird	Amec	13.74
Plant Science Botanical Garden	01/09/04	01/08/05			RHP	Faber Maunsell		Willmott Dixon	6.01
Disability Works	01/01/01	01/12/06			EMBS				10.25
The Sainsbury Laboratory	15/09/08	04/10/10			Stanton Williams	Arup	Hannah Reed	Kier Marriott	82.00
Sainsbury associated re-ordering									4.40
University Visitor Centre.	Aborted				EMBS	RPA	Cameron T B'd		1.47
Plant Science Education Centre	Aborted		2,528	1,896	ECA				11.21
Plant Science Herbarium	After	decade							2.01
EMBS Bailey-Grundy refurb			1,999						
New Museum Site--Arup			2,000						Minor works
									Minor works
Cost of completed works, 1996 to 2006									224.6

9 The road out of town

By the mid 1990s Cambridge was getting full. Growing demands for academic space in the University, along with housing and commercial pressures (1,000–2,000 new jobs in Cambridge per year), accelerated the need for new buildings right across the city. Down the generations the University had 'planned long' in acquiring land beyond the built-up city, and so, unlike other universities in the top ten world research universities, it had over the decades acquired, at good prices, contiguous space on which it could build. And so it did.

University Library

The closest expansion area at the edge of the University in the mid 1990s was on land around the University Library, the 'UL'. As described in Chapter 1, the University had bought the land being used as a large war-emergency hospital as the site for its new library; that was designed by Giles Gilbert Scott, the designer also of Battersea power station, Liverpool Anglican cathedral and the British telephone-box. It was partly funded by the Rockefeller Foundation. (Whether it was Rockefeller or Scott who decided to change the draft design so as to include the 48m high tower that could be seen from the main University so as to draw people over the river to what was then a distant site, no one really knows. In any case, the University wanted more space for books than envisaged earlier and some say that that was the real reason for the addition of the tower.)

The UL is one of six copyright/legal deposit libraries in UK and Ireland[1] with a legal right to claim a free copy of every book published in UK and Ireland. Although many books are not needed or taken, and an ever-larger proportion of books are published in digital format,[2] the pressure for expansion of the library was enormous as it received about 100,000 volumes per year. Whereas in 1748 in the previous library in the Old Schools there were two librarians, and 700 people were allowed access to the 45,000 books, by 2008, there were 327 members of staff, 35,000 registered readers and 100 miles of shelves for 2 million of the 6 million books held by the library.

Following a donation of £3M by Mr Tadao Aoi (made possible by his success in setting up a chain of stores in Japan), an extension to the library, the Aoi Pavilion, was built by the contractor R G Carter to house the UL's collection of Japanese, Korean and Chinese works, formally opened in March 1998. The Aoi Pavilion and west book-stacks basement project cost £7.4M.[3]

Next, as part of Phase 2 of the development of the UL, a new exhibition and re-vamped reception area was developed, and built in 1998 at a cost of £900k. Design for a five-storey (plus basement) extension to the west side was developed. This library expansion project continued by pushing out the corners of the then existing library. There was also demolition of a further part of the 1970 extension for future development. This project was carried out concurrently with the forming of the basement area to a planned, new west book-stack. The first corner, the north-west corner, was extended in 2000/2001 to give more space for the Departments of Photography (Imaging Services), Rare Books and Manuscripts, more shelf-space and two new large reading rooms, one for rare books and one for manuscripts, including the University archives; the cost was £5.444M, a tad under budget (which was reduced as the VAT liability was less than expected). The work was designed by Harry Faulkner-Brown Partnership, a 'guru' of library design, Roger Parker Associates and F. J. Samuely and Partners, and with the constructor R. G. Carter formed a team that stayed together, throughout the UL expansion programme. The south-west corner was developed with the same team at a cost of £6.00M in 2000/2003 to house administration, inter-library loans, the Official Publications and Microform Reading room, and a new Digital Resources area.

Both corner-expansion projects ran to time, or nearly, and to budget, despite the challenges of working to such high technical standards of finish and temperature and humidity control: for example, walls had to be very thick, solid brick, to maintain the stable climate within the building. Worse, for most of the project only one source of acceptable clay for the bricks could be found, so matching bricks were expensive, at least until another suitable clay-pit was located.

Once the south-west and north-west corners were developed outwards, next came the projects to fill in between them so as to complete the development of the west side of the Library. When the design came to be reviewed it was decided that it should be five-storeys, the same as the 1960s extensions, and two storeys less than the original building. The first half was constructed from 2 February 2004 until 5 May 2005; the second, mostly designed and prepared in 2003/2004, was built from 15 October 2008 until 15 April 2010. The costs of those two projects were £5.7M and £7.00M, with the same designers and constructors as for previous UL extensions. Meanwhile, a large-scale, four-stage replacement of 1960s mobile bookcases used to store part of the printed research collection started in November 2005, being completed ahead of schedule in August 2008; the work was designed by Howe Partnership, RPA and Samuely, and constructed by the same contractor, R G Carter, on budget at a cost of £2.7M.

Centre for Mathematical Sciences

The biggest and the most pivotal University capital project of the 1990s was the Centre for Mathematical Sciences (CMS). Cambridge had for centuries been famed for Mathematics: Newton was the first Lucasian Professor from 1663; other luminaries include Clerk Maxwell, Babbage, Eddington and Dirac, and the most recent was Stephen Hawking, the famous cosmologist and author of *A Brief History of Time* and much else.

Discussions started in earnest within the two Maths Departments, the Department for Applied Mathematics and Theoretical Physics (which was established as a department in 1959) and the Department for Pure Mathematics and Mathematical Statistics (1964). On 27 November 1992 there was a meeting of four Maths professors and Sir Michael Atiyah (the first Director of the Newton Institute, which started in 1990, and Master of Trinity College, Cambridge). It was agreed to form a working group which became the Clarkson Road Working Group, comprising the head of the Newton Institute and the heads of the two Maths Departments; the second paragraph of the Minutes of that initial meeting noted that,

> There was unanimous agreement that the first priority was to secure the future of the Institute. It was also agreed that this may best be achieved as part of a larger campaign for funds, encompassing also the move of the departments. Arguments to support this view will be assembled, to present in due course to the Secretary General.

(That was the Secretary General not of the United Nations but of the University Faculties.) Two of the key professors were in the Newton Institute (inaugurated in 1992), and were also Fellows of St John's College: an example of the power of an ancient Cambridge College with its traditions and huge resources, synergising with a modern, world-famous research institute up there with the Institute for Advanced Study at Princeton (to where the driver of the project, and Master of St John's College, Professor Goddard, moved during the project).

The two Maths Departments had previously moved into the former printing shop and warehouse of the Cambridge University Press in Mill Lane starting in 1964/1965. In turn those premises soon became overcrowded and unsuitable, with long, dark corridors, low ceilings and cellular offices. The Department of Computer Science grew out of those departments; it moved out to the New Museums Site, but overcrowding in Mill Lane still got worse. A move to the Clarkson Road Site would not only allow expansion but would co-locate the Maths Departments with the Newton Institute.

The letter from the Working Party chairman to the Secretary General, dated 28 September 1992, set out effectively the case for doing a feasibility study on a move of Mathematics to Clarkson Road:

> It appears to me that the main reason for the increasingly cramped conditions suffered by the mathematicians is that their research funding is

growing rapidly at present. Part of the increase relates to new permanent staff including professors, but I guess that it is mainly caused by success in the competition for specific research funding, mainly from the research councils. Research Councils support neither 'bricks and mortar' nor their consequent upkeep.

Interestingly, the letter goes on to note that the cost of running a building needs an endowment as big as the capital cost of the building to provide enough income to cover running costs, and suggests that therefore fundraisers should raise that as part of funding a new building. This understanding of operational costs was way ahead of its time, and at the high rate of investment income then current, the estimate of endowment needed for running costs was about right. (Persuading the crucial University Planning and Resources Committee to switch to whole-life cost procurement took two years, starting six years later.)

Various sites were considered in the study, but a paddock for a single horse off Clarkson Road was clearly the best, and when that was supported in principle by the General Board, the Cambridge-based architect Duncan Annand was appointed to carry out such study and design as was needed for fund-raising for a new Centre of Mathematical Sciences, CMS. As usual, this architect did an excellent job. A leasehold on the paddock was bought, at high price, from St John's College. The deal was for a 173-year ground lease with an option to renew for a further 175 years as the University wanted the freehold and the College didn't want to part with it, so that was the compromise: such is Cambridge.

Much of the early discussion, and tensions, related to how big the building might be. In August 1993 that was set at 9,200 to 9,500m^2 working space. Also in 1993 it was proposed that the site might be best used as a mixed academic and commercial centre, but that proposal was not pursued; whether it should have been is discussed in Chapter 12. Soon afterwards came the news, not welcomed in the Departments, that even if the new buildings were to be funded by benefactions raised by the Departments, the extent of space in the buildings still had to be agreed by the General Board of the University.

After site selection, the launching of fund-raising and some provisional agreement on the extent of the new buildings, design could start in earnest. On 21 July 1995, the London-based practice Edward Cullinan Architects was appointed as the outcome of interview after assessment of track-record; that was to prove to be a wise choice. Ted Cullinan and the project architect Johnny Winter came up with imaginative and widely appreciated designs.

The mathematics project leader (the 'representative user', RU; see Chapter 5) followed a conviction that as the Departments had conceived the project and were raising much of the money, then they, the departments, would manage and direct the project as client, even although none of the staff there had any experience of building design or procurement, or of the fragmented and inefficient British construction industry as it was; it would however be the professional staff in the estates department (whose jobs were to set up and manage contracts), who would be the legal 'client'. That tension persisted and was frequently a source of irritation and some inefficiency until 1999. At an early stage, the RU referred, in a letter to the external project manager, to 'our architect' and got a clear response by return of post from that consultant that it was the project's architect, and directions to him had to be authorised: once users of a projected building start giving instructions directly to the designers, then budget and managerial control are soon lost. (That had been an issue also at the start of the refurbishment of the Chemistry Department building, and it was to be an issue on the procurement of the new building for the English Faculty.)

But despite such niggles, the client brief was developed, and then design of CMS proceeded apace. At first the design was based on, as Ted Cullinan later described it, 'a second side to the street with a series of three-storey houses surrounding catering and library facilities in a two-storey central core'[4] to provide space for future expansion of the Departments on the site, the houses had been 'extended to the perimeter of the site'. The design included space for the Department of Computer Science ('the Computer Lab') as it was planned until a late stage that that should be included in this project so that it could leave the very inefficient working space of the Arup Tower on the New Museums Site (that has small floor areas and lots of floors: academics, like most people, communicate better sideways than vertically).

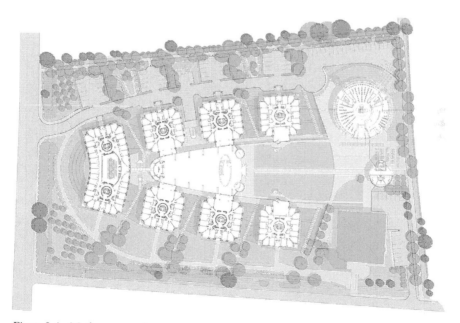

Figure 9.1 Maths master plan

By courtesy of ECP

The first designs were based on a large, rather monolithic building. But in a letter of 25 November 1995, Prof Stephen Hawking asked the architect Ted Cullinan to design 'not a single unit but containing various sections of different sizes ... which could each appeal to potential donors as a distinct unit ...' For that reason, and for planning issues of 'massing', design concept switched to individual, relatively small pavilions more in keeping with nearby houses than was a single monolithic structure; six departmental pavilions (named C–H) around a central building, with a larger office building (Pavilion 1/2, subsequently B) at the western end: Ted Cullinan used to refer to the central building as 'the mother pig with her piglets snuggled around her' (see fig 9.1).

It was decided in March/April 1997 that the Computer Science Department would not in fact go to the Clarkson Road site but to a large new building at West Cambridge, and that took some pressure off site capacity. However, design issues were well tested during the planning application process and local consultations and negotiations with the planning officers of the city of Cambridge. Local residents, many of them current or previous academics who had bought their houses before the massive increases in the prices of Cambridge houses, were highly effective both individually and collectively in assessing and, when deemed appropriate, opposing planning proposals. In this case the proposal was for transmutation of the field with its horse and lots of nice green grass, to a large-scale development with lots of traffic, lights at night and noise and mud during construction. Being part of the University, they knew well which levers to pull and into which wheels to inject sand, within and beyond the University. In the leader of the city's planning department they had an ally on the matter of the scale of this development, and also on control of irritants such as light spillage from the site, and the invariably sensitive issue of increased traffic. The VC got involved, perhaps without deep enough briefing, and got caught in the cleverly aimed cross-fire of the neighbours. At a pivotal meeting with local councillors and neighbours on 4 June 1997, the head of planning, Peter Studdert, pushed hard for removal of the built-space that had initially been in the plan for the Computer Science Department, and was then being retained in the design as space for future expansions. He said that 'the scheme [was] heading for refusal on the basis of site density.' There had already been considerable compromise on the issue of massing and the University had agreed that the building would be sunk deep into the ground to reduce its impact: the quantity surveyor estimated that the very approximate extra cost of thus reducing the above-ground height and having a green roof was some £4M. However, the neighbours were on the front foot and the University on the back foot in the argument regarding massing. The decision was made to reduce the scale of the project by eliminating the expansion space. It was, however, agreed that a separate and substantial library building would be included. The total built space now proposed was 22,667m^2 assignable net space, 30,550m^2 gross; although reduced, this was a far cry from the 9,200 to 9,500m^2 assignable net area internally authorised in August 1993.

Funding of the project was mostly by benefaction: included Märit and Hans Rausing Foundation, Charles N. Corfield, The Dill Faulkes Educational Trust, Garfield Weston Foundation, Gordon and Betty Moore. The balance came from internal University money, and the fruits of University bids to the Hefce-led Joint Infrastructure Fund (JIF). (See Chapters 2 and 4.)

On 7 January 1998 Planning Permission was granted by the City of Cambridge: the end of the first, long chapter, and the start of the next – construction.

Ground breaking for the first of three phases was on Saturday 4 April 1989 by the contractor Laing Construction, who had won the contract for the contract sum of £17,609,638 to build the central core building, the office pavilion at the end of the central core and the first of the six pavilions alongside the central core. Soon after the contract was signed, funding came through for a second pavilion and that had to be added into the contract: it is always risky adding work after a contract is signed, but in this case with a lot of discussion and negotiation, it went quite smoothly.

Procurement being the single-stage process whereby constructors were kept out of the process of design and brought in at the last possible moment, (rather than the team-work approach with early contractor involvement in the detailed design stage, as was to be used on Phase II of this project; see Chapter 5), Laing was on a steep learning curve trying to figure out the rationale and details of the design while on site. That was even more difficult than normal as the company was feeling the strain of a series of heavily loss-making contracts around the country. The group finance director said that construction of the new Millennium Stadium in Cardiff lost the company £26M.[5] The private finance initiative hospital project in Norwich was making a large loss, as did the National Physical Laboratory contract. There clearly were resource shortages within the company, and those got worse when in October/November 1999 there was a major company reorganisation with a lot of redundancies.

The company claimed that design information was coming to them too late and thence causing delays to the programme: seven weeks for one of the pavilions (the other two were handed over on time) and ten weeks for the central core building. The designers accepted responsibility for two weeks of those delays but no more. Delay nearly always causes over-run of cost as more people are working on the site longer than budgeted. In this contract the University's construction budget was £18.24M, and the contractor claimed £19.52M: after negotiations balancing accepted delays, and unjustified delays for which the contractor was responsible, the 'interim final' account cheque to Laing for the construction works was agreed on 27 March 2000 at £18.62M. That was serious, although not a great over-run compared with the large cost over-runs of projects completed by the University in the 1990s. Such over-spends were much reduced by using team-based procurement: this was the last major project done by the University on the basis of designers working right up to tendering the construction contract without input from or information to the builders. The total project out-turn cost was £25.55M against a project budget of £23.75M, an overspend of 7 per cent.

Completion and hand-over of a new building always has some challenges, but here those were 'in spades'. When the first pavilions were nearly complete, the mathematicians decided that they wouldn't wait for formal completion and handover of the building. Handover is an important contractual stage when the client assumes responsibility and liability from the constructor, and insurances and warranties start accordingly. When word spread that the mathematicians had unilaterally hired removal trucks and simply moved in (during a holiday), there was much alarm and some despondency about the contractual mess that could result. In the end most matters were reasonably sorted out, but the handover was not good, and those moving in never got much benefit of instructions about how to operate the building; nor for that matter the making-good of defects that the contractor would have had to repair but could now say had happened after the occupants moved in and had been caused by them. Nationally, many, indeed most, buildings operate well below the efficiency they are capable of (typically about 30 per cent less well) but a good hand over and set of operating manuals do help a lot. It's an ill wind that blows nobody any good: that experience was the background for the invention of the 'Soft Landings' handover procedure which was described in Chapter 5.

An existing procedure new to Cambridge was the Post-occupancy Review of Buildings and their Engineering study. PROBE studies were carried out on completed buildings by analysts Bill Bordass and Adrian Leaman (building scientists) with an industrial psychologist. They reviewed in depth how well/badly a building was performing, and how it was assessed by its users compared with similar buildings in a large national database. The deal was that the assessment was free, but the assessors could, and did, publish as they saw fit, subject to correction of factual error. To commission a PROBE study was risky; a lot could be learned and the way the building was operated could be improved, but a damning report would be unwelcome. The Estates Department got agreement from the designers and constructors that a PROBE study should be done, after the users had settled into the building. As noted above, the mathematicians had moved in before the building was ready; the site was still like the Mons battlefield in the third week, mud everywhere, and there was another contractor on site working on Phase II. And in the first place quite a lot of the mathematicians hadn't wanted to leave their old, central building (near to a much-loved bun shop). Then the PROBE team turned up, weeks before the Estates Department had wanted or expected. The PROBE people first worked (with only a few puzzled mathematicians aware of their presence) over a weekend when the building could be securely closed, pumped up and then tested for, among many other performance matters, air-tightness (at that time not a regulated matter). After a lot of tests, the team interviewed many of the staff. As it turned out, a bit surprisingly, the results were good, and the buildings compared well with comparable educational buildings in other universities in both operational and performance aspects.

Long before that first contract for the Centre for Mathematical Sciences was finished, design-refining and planning were in progress for the second phase; the construction of the Betty and Gordon Moore Library for sciences, and a small gatehouse building. This new physical sciences, technology and mathematics library was funded through a £8.075M donation made by Dr Gordon Moore, chairman emeritus of the Intel Corporation and owner of various companies, and said to be then the fourth richest man in the world. He said: 'The University of Cambridge has a long history of doing leading research into some of the most fundamental questions that mankind asks. I find this exciting and uplifting.'

There were two hiccups in the procurement process of this second phase. Firstly, it became apparent during the Stage I (design and tendering of works packages) that the selected Stage I contractor, Laing Construction, was neither acting as a pro-active member of the team nor was forming the professional relationships that are needed for modern procurement: to be fair, that was understandable given the problems of the company. It was therefore decided, uniquely in the history of Cambridge University procurement, not to award the Stage II contract, for construction, to the contractor that had worked on Stage I, the design and sub-contract tendering stage. Rather, it was decided to go out to limited competitive tender, to be based on quality/price balance, to Laing plus two other constructor companies. The tender was won by Sir Robert McAlpine. That turned out to be a fortunate decision: the contract proceeded in a very collaborative way, and the work was well done, on time and on budget.

The second hiccup was that Mr Corfield, a benefactor who had given £2.5M towards the cost of the Phase I construction, generously decided after the Phase II contract was well through its preliminary stages to fund Pavilion 7, at a provisionally-estimated cost of £4.4M. There was a dilemma: whether to continue with the contract for Phase II and add into the contract at this late stage the further large chunk of work to build Pavilion 7, or whether to wait and build it later as a separate contract. As already described for the Phase I contract, it is risky to add work mid-contract, but that choice was made. The risk was exacerbated because Gordon Moore made his benefaction contingent on all of the cost of the library coming from his money, that not a penny of anyone else's was to be put into the contract and, therefore, it was not to go at all over budget. So those making the procurement decision were between a rock and a hard place. After a lot of discussion, and wise advice from the external project managers, Davis Langdon (since bought out by the American giant Aecom), it was decided to negotiate Pavilion 7 into the contract.

Mainly because of the quality of all members of the team, designers, consultants, constructors, client and representative user, and the collaborative way the procurement was set up, progress was sound despite the contractor McAlpine having to share the site with Laing Construction who were still completing the main core building and the last pavilion in the first phase contract. The budget for the Betty and Gordon Moore Library and the

Gatehouse was £9,894,800, and the out-turn cost was £9,842,800, a small underspend despite the hiccups; the budget and out-turn cost for Pavilion 7 (then re-titled as Pavilion E) were £5,206,000 and £5,283,645, a small overspend so just on balance overall. Starts on site were 1 February 2000 and 1 October 2000 respectively. Both aspects of the contract were finished on time, 29 June 2001 and 15 January 2000. The circular Betty and Gordon Moore Library was fitted out internally with lurid colours (fuschia and pistachio), bright and very cheerful, if not to everyone's taste (indeed, patently, not to the VC's). The quality of construction work was better than for Phase I. Of course, that was not just down to the better team and better procurement route: the designs were similar to those in Phase I and lessons were learned from that.

The pavilion was similar to the previous pavilions. The library floor area is 3,339m² gross, 2,787m² assignable.

The last phase of CMS, the last three pavilions and the underground works for a fluids lab, was procured through the by-now normal University two-stage early contractor involvement (ECI) route, using the Engineering Construction Contract (ECC, part of the New EC suite of contracts). The design carried out by the design team again included input from a contractor selected on the basis of quality balanced against cost of overheads and profits, and design assistance fee. The Stage I contract for design and tendering work was won by Laing O'Rourke: Ray O'Rourke's small Irish concrete business had in 2001 bought out debt-ridden Laing Construction, Britain's oldest and most famous family builder, headed by Cambridge alumnus Martin Laing; it was said that the asking price was £12M, that the sale was for £1 and that there was a dowry payment by Laing of £18M to cover risks from current contracts taken over as part of the purchase. Ray O'Rourke and his brother spent half a day with the Director of Estates on site prior to the purchase of Laing, discussing the British procurement situation. Laing came out of construction then, and a year later came out of house-building and into investment and management, and public sector infrastructure, mostly by PFI, with profits coming from concessions to run facilities rather than from the construction.[6]

This last phase of the CMS project had a budget of £17.943M, of which £12M was funded by JIF, the rest by benefactions. Work started on site on 2 April 2001 and was completed on 23 January 2003, ahead of the contracted date and £355,233, 1.9 per cent, under budget. Very unusually in the world of construction, the Final Account was agreed verbally before the end of works and formally signed off just five weeks after handover of those last three pavilions; the contractor's site huts left site well before the work was finished. (Final account agreement usually takes many months, sometimes years.) With the changes in procurement methods, the three phases had been 7.6 per cent over budget, on budget and 1.9 per cent under budget; over time, on time and ahead of time. The reflected the sequence of procurement systems.

The CMS project had been predicted to take some 15 years to fund, design and build, but when potential benefactors saw what was coming out of the ground, funding sped up hugely. The total cost of the project at 2003 prices was £61M, and for that the University got 18,741m^2 gross internal built space, accommodation for the 1,012 staff and students who moved in, and such high quality that top-flight academics from around the world were attracted to come to work there. A national study, funded following a successful Cambridge bid to the Hefce fund for innovation in university administration, was carried out at that time seeking to find out if correlation and causality could be found between the level of money spent on the margin on new university buildings, and the benefits on the margin of building-users, productivity and success in attracting top students and staff to the new building. The results were inconclusive (unlike similar studies for hospitals which found both correlation and causality).[7] Chapter 5 sets out further research into correlation between design quality and building performance. There clearly was a correlation between marginal quality of the new building and successful operation and recruiting/ retention of staff/students, but that correlation could not be proved to be causal in any quantifiable way. However, even although not proved as generally causal, the local evidence for this CMS project regarding marginal correlation of quality and operational value is clear, and is quoted by best-practice agencies and by the government.

It had been a big, difficult and pivotal project. An accord between the academics who inspired and conceived the buildings, fund-raised for them and then moved in, and the professional project managers in the University administration, was achieved: not easily and not without confrontations, but achieved it was. That set a pattern, and from then the overall University handling of capital construction projects had a proven procurement model and it seemed natural to follow it. Secondly, this project shifted procurement from the previous confrontational single-stage procedure to a team-based system of all parties getting involved early in the project, each knowing what all the members of the team were and were not tasked to do. Those who couldn't, or wouldn't, work collaboratively were not employed. The run of over-budget and time, litigious projects through the 1990s was broken. The management of this project was noted more widely: it won (at a black-tie dinner in a posh London hotel with over a thousand from the industry present) the top national award for the Building of the Year 2003; it also won the 2003 Prime Minister's Award for Better Public Building, and locally the David Irwin Award. Overseas fact-finding construction visitors such as the US federal GSA (General Services Administration) were sent up to Cambridge by London civil servants, and the Director of Estates invited to visit and advise GSA. Articles appeared in the foreign press about the CMS project, and widely in UK. Senior architects have said that that it should, or at least might well have won the Stirling Award but for the nature of the selection for that award: the then-President of the RIBA rang up the Director of Estates at a horribly-early hour on the day of the short-list announcement to express his disappointment at the absence of the CMS project.

Running costs of about 200–300 per cent of capital cost over 15 years would be normal for such a building of that type and era. The running costs of CMS buildings were very low: about 6–8 per cent of capital cost per year. The annual cost in 2011/2012 was about £600k. That is partly because of sustainability features not then common. In summer, cool night-time air passes along the underside of the building slab, and by day, when the automatic building management system deems it good, convection and radiation takes the coolness into the building. Windows are deeply recessed and shaded by overhanging roofs; natural ventilation sweeps hot air up around the staircases for four storeys: lanterns work well as they always have. To meet neighbours' demands, all the windows have blinds that automatically close at dusk so that there is no light spillage from the buildings: in fact, the height and density of the much-praised peripheral landscaping means that blinds control became largely irrelevant within a few years of completion. The extensive tree/bush landscaping looked rather bleak for a year or two, then started to look really impressive, softening the squareness of the building structures. The grass-covered roof adds to insulation (and it has hardly ever leaked) and affords an elevated grass courtyard. Interesting to note that the two architectural features that survived in the University from medieval times to the 1960s, lanterns in the roof for light and ventilation, and courtyards, are both featured in the CMS project: the traditions of nearly eight centuries, lost in the brutalism of 1960s designs, had made a come-back.

Case study – Centre for Mathematical Sciences, CMS

The Department's brief called for a series of small offices in clusters corresponding to the existing size of their existing Maths groups, e.g. Number Theory, Cosmology, Particle Physics and also laboratory space for the Fluid Dynamics group, library, teaching spaces, computer rooms and a refectory. The requirement for private study in quiet spaces together with the need for creative exchange of ideas across multi-disciplinary boundaries was a principle requirement for the new Centre. Additionally, the brief called for the design to be both stunning, and low-energy in its operation.

The master plan developed with a central, semi-underground building to house reception, mail and admin, a cafe and auditorium (Pavilion A) which is conceived as the social hub. A double pavilion B formed the head with the main auditorium, and three pavilions on either side of A (forming the six piglets snuggled round the mother pig, in Ted Cullinan's word picture). Each peripheral pavilion provides 40 study spaces wrapped around a vertical core of lift and stairs: at each level the landing wraps round the stairs separated by a concrete and glass block curved wall and in four of the pavilions the common room links directly with the main social hub. At the eastern end are the circular library building, a new gatehouse and the existing Isaac Newton Institute.

The library is a free-standing circular construction, three floors above ground level, brick-clad reinforced concrete frame to first floor cill level, with cladding above, and zinc-clad roof. One floor below ground level extends into a rectangular plan beneath the footprint of the building above with a trapezoidal extension allowing natural light to enter the lower ground floor. Natural ventilation and cooling as part of an integrated building management scheme are provided in the upper two floors.

Cars are restricted tidily to the north side of the site but circulation to the main reception can be convoluted and confusing for the uninitiated arriving by car and even for those on bikes as there are several possible points of entry.

Fitting into a suburban site

The tension between placing such a large volume of building on what is essentially a suburban site has been carefully tackled and the buildings avoid a feeling of an edge-of-town industrial estate by manipulation of the levels, placing the large volume of the social hub semi-underground, and by the use of quality materials. Brick and repeating string courses of profiled stone to the two lower storeys are topped by dark grey window frames and cladding to the upper two floors with projecting zinc covered roof planes. These curve and swoop and morph from the square plan form of the pavilions to the circular lanterns, a hint of the circular staircases inside. These glazed lanterns are an important part of the ventilation strategy and also pour light into the central lift and stair cores. This strategy is used for each of the seven peripheral pavilions and serves to knit the whole complex together visually.

Pavilion A, under a curved turf roof, has a barrel vaulted form with muscular concrete ribs tethered with a huge hinge joint to concrete bulwarks rising from the floor. Panels of slatted timber fixed below the concrete vault, except under the spine of roof glazing, provide both visual warmth and also acoustic absorption sufficiently effective to enable discussions and supervisions to be given in this large space.

The library and gatehouse share the same materials as the pavilions. Natural ventilation and cooling as part of an integrated building management scheme are provided in the upper two floors. The floor area is 3,339 m² gross, 2,787 m² assignable. There are 7,000 m of shelving, and a very advanced 3M book-detection system. Reader places around the periphery have the advantage of natural light. Bookshelves radiate out from the centre except in the basement stack where a more compact parallel arrangement allows greater density of books.

Intelligent buildings/intelligent client – briefing

One of the key priorities in providing a new building for the departments of the various mathematical sciences was to encourage dialogue between the 26 separate groups. Professor Goddard wanted a building that would increase the chances of serendipitous meetings and the architects devised forms which would encourage mathematicians to use chance encounters to discuss and develop ideas. Hence the generous circulation spaces within each pavilion, visually linked and with ample space for two or three people to stop and debate and if necessary scribble on whiteboards and then perhaps move together to the common room or the social hub where tables had wipeable surfaces for scribbling on. This gadgetry has perhaps been superseded in these days of laptops and PDAs but the architectural intention holds true and provides a hierarchy of spaces for stopping and talking.

Ted Cullinan describes how the plan evolved:

> [we gathered] the offices on galleries round stairs and a lift, limiting the number on each gallery, often to people with a shared discipline. Three floors of galleries go to make a pavilion and on the ground floor, somewhat visible from all three floors, is a coffee room/seminar/meeting room for larger, pavilion-sized meetings, arranged and un-arranged. The meeting room of each pavilion overlooks a large shared space where you will come across people from the other pavilions and where you eat, relax, have meetings, work things out together, or organise big gatherings.
>
> On top of the large shared space is a fully fledged grassy garden which all the pavilions open onto as well, via their upper level terraces. So you can enter or leave your pavilion for the still wider sphere of Cambridge and the world via your pavilion's own front door, via the shared central space, via the basement lecture room or via the roof garden: an embodiment of the thought that in harmonious institutions (and cities and almost anywhere) there should be, at the least, two routes to everywhere so that you can choose whether you meet people or not and whom you meet without offending.

Strategy for environmental design

Environmental design aimed to reduce emissions through passive systems: exposed mass and earth sheltered volumes, the avoidance of air conditioning, a thermally efficient envelope and structure, deep overhanging eaves, automatic blinds and high levels of natural ventilation. The original concept for the environmental control of the Centre was for the building to provide its own passive means of ventilation using either stack effect or cross ventilation.

The design developed into its current form of a series of individual academic offices wrapped around the perimeter of the pavilions, benefitting from natural daylight and ventilation. The pavilions themselves are wrapped around the central core, which acts as the focus for the whole Centre and where the meeting of various scientific disciplines and exchanging ideas can occur. This is a fine example of the 'social programme' layout and the environmental design aligning through development.

The capacity of the perimeter rooms to take advantage of the cooling effect of the ventilating air is increased by exposing the concrete structure of the building and exploiting its thermal capacity. In the summer when the building is fighting a combination of internal heat gains and high external temperatures, the cooler night air flows along the underside of the concrete slab, thus reducing the material's temperature and storing this coolness for the following warmer day.

Although there is a sophisticated building management system centrally controlled by the facilities manager, it was also considered important to give building users individual ability to over-ride this by, for example, manually opening windows in offices. This has undoubtedly led to the high level of user satisfaction and perception of comfort particularly within the offices, although some lag is experienced in the heating and cooling of the larger spaces (The PROBE report in Building Services Journal July 2002 has full details).

A post-occupancy forum was held in March 2005 and its report makes for cheering reading in its record of an open discussion, a description of a complex yet successful project and people unafraid to discuss areas where things had been less than perfect, notably disabled access, AV provision and unrealistic expectations for natural ventilation under all weather conditions.

Architect: Edward Cullinan Architects
Structural Engineer: BuroHappold
Services Engineer: Roger Preston & Partners
Landscape Architect: Livingston Eyre Associates
Quantity Surveyor: Northcrofts
Project Manager: Davis Langdon & Everest
Contractor Phases 1 & 3: LaingO'Rourke
Contractor Phase 2: Sir Robert McAlpine

Figure 9.2   CMS aerial view
By courtesy of University of Cambridge

Figure 9.3   CMS side elevation
By courtesy of University of Cambridge

Figure 9.4   CMS front
By courtesy of University of Cambridge

Madingley

After that major building project of local and national importance came several smaller building projects: the animal behaviour unit of the zoological department, which had been founded and set up near the edge of the village of Madingley in 1950, had some upgrades and refurbishment, and then a new building known as the Avian building was constructed. This was in fact a replacement building for one that was derelict and which had to be demolished to comply with planning conditions concerning other parts of the site. The work of the Avian Laboratory had been increasing, particularly non-invasive investigation into conflict-resolution within family groups of birds, and the ability of birds to assess past experience and project that into remedial actions for the future. Also, the existing facilities had been declared unfit by Health and Safety officials, and the Avian Laboratory worked under Home Office licence. So a planning permission was achieved on 17 December 2004, not easy in a local area so sensitive to environmental change. (Something of a furore was caused when around that time the University felt that running pubs was rather far from its core business, and so sold 'The Three Horseshoes', now an expensive restaurant.) At a capital cost of £800k, recurrent cost £5.7k per year, a new 251m² building was put up, starting in March 2005; the contractor was R G Carter from Thetford and the design was carried out in-house by the Estate Management Department. The floor plan of the building is some 220 m², with various aviaries attached around it.

Meanwhile, at Madingley Hall, a beautiful eighteenth century Grade I Listed building country house with a fascinating history (see Chapter 1), there was a series of refurbishment and conservation projects. Some of those were to maintain the integrity of the external envelope, both roof and walls, and some were to develop the facilities for continuing and extramural education. Nearby, there was upgrade work to three of the houses rented out by the University.

Development of the University Farm, Huntingdon Road

Up at the University Farm along the Huntingdon Road (see North West Cambridge site plan, Figure 11.1 in Chapter 11) beyond the Travellers' Rest pub there was a series of small projects. The first in the decade being studied was the refurbishment of the farm's dairy unit at a cost of £1.4M in 1997/1998. The lead designer was architect Smith Wooley and the constructor R G Carter. It was quite a large project spacially: 3,000m^2 gross space, 2,550m^2 assignable. The second project at the farm was first mooted officially in 2000: a small building to house ten scientists plus support staff for the study of potatoes, to replace old and unsuitable accommodation spread over several buildings. The project was sponsored by the Institute for Agronomy (a research adjunct with some relationship to the Farm and to the University) in partnership with the potato industry. (The project soon became known in the Estate Management Department as 'Spudulike'.) The project got to site by 14 July 2003, finishing on 27 February 2004. The cost was £1.166M, the designers JGP Architects and W. S. Atkins and the constructor Crump Newberry, all of whom did a sound job.

The other project that went ahead on the farm at that time was the demolition and replacement of three substantial but outmoded buildings: bull pens, 'large yards and garaging' and the pony shed/feed store. The work, funded by Hefce under SRIF 2 capital grant, was carried out in 2005/2006, also by constructor R G Carter, at a cost of £1.525M, and then commissioned by the Home Office. Design was by Alun Design Consultancy and Andrew Firebrace.

307 Huntingdon Road

The last project in this chapter also relates to planned work in the same area – but there any similarity ends. This was a hugely emotive and resource-hungry project managed under local, national and international spotlights, virulently and violently opposed by 'animal rights' groups threatening violence and carrying out extensive intimidation.[8]

It was the project to provide new facilities for research into such degenerative conditions as Alzheimer's, Parkinson's and Huntingdon's diseases; the research might also have helped scientists understand more about the underlying causes of certain learning difficulties, and some aspects of brain injury. The research team at Cambridge was internationally eminent in this field. A bid to JIF (see Chapter 2) was worked up in 1998 and given high priority by the University, and in 1999 the success of the bid was notified: an award of £19.1M. The facilities were due to be built on University land among the University farm buildings off the Huntingdon Road to the north-west of Cambridge, just into the area of South Cambridgeshire Local Authority. It was in an area of green belt which the County Structure Plan soon was to recommend be removed from the green belt for development by the University.

The design was for a sunken building surrounded by a 'ha-ha' ditch and a raised bank to reduce impact on what was still green belt. The floor above ground level was scaled to suit the adjacent existing single storey building. The size of the building initially was to be 4,086m² gross, 2,239m² assignable, plus a service basement and plant room. It was to have a grass roof for environmental reasons, and the courtyards were to be covered to give extra working space and privacy from the elevated M11 off to the west of the site. The building would comprise offices, labs and storage for lower primates (small monkeys) to be used in the research work. As design progressed and needs were further defined, the size of the building grew to nearly 8,000m² gross including plant rooms, service areas and the courtyards. Including conversion of space in existing buildings, the total space to be available for this research would be 12,522m² gross.

By the end of 1999 the concept-design was approved by the University's Buildings Committee, and extensively discussed with the Local Planning Authority (LPA). The main issue was the need to get support from the LPA for building in what was still the Green Belt. Permission to build in the Green Belt is rarely given, but the LPA accepted the University's need for such facilities, and accepted that the building's outline design would be appropriate. The LPA wrote in a letter to the University that 'it is the Planning Officers' view that you have done sufficient to justify the necessary departure from [green belt] policy'. Given the sensitivity of the project, that important letter was sent on to the JIF managers, who continued to support the project. The programme was set for Stage C (now Stage 3) and Stage D (Stage 4) design reports to go to the Medical Research Council in June and August 2000 respectively, with start on site in December 2001. The budget was set and approved at £23.02M, with a need for value engineering to get the projected cost down to that. The planning application went in on schedule, on 7 September 2000. Thereafter, during the period leading up to the application being decided, there was a huge amount of lobbying against the use of animals in testing of drugs for humans, organised demonstrations and a very high level of mobilisation of police resources.

There had been an animal research facility on this site for over 50 years, and of over 20 applications for development of the site over the years, all had been approved (except two which were withdrawn as no longer required) and, as noted, officers had written that the University had made the case for allowing green belt development. Nonetheless, the planning application for this project was first deferred by South Cambridgeshire Council Planning Committee and then refused. The formally stated grounds for refusal were that:

- The proposed development was not appropriate in the green belt.
- The development would cause harm to the green belt through its visual impact, (increased traffic movements, the removal of some old research/ farm buildings in poor condition) and the encroachment into the open area to the rear of the existing buildings.
- There were no proven exceptional circumstances to justify departing from the green belt policies.

More of the discussion at the planning meeting was about the nature of the research than about green belt or building issues.

Discussion in the University continued to confirm the validity of the project, and a further planning application with a modified design for a building costed at £24M was made on 13 July 2001. That was refused on the entirely different grounds of security: that the police would be unable to keep sufficient order or keep the highways sufficiently clear of disruption, and that another site would be more suitable. Refusal of permission was notified on 18 March 2002. No reason was given for the total change in grounds for refusal, and that was later to be commented on by the Secretary of State of the Department of the Environment, the Deputy Prime Minister (see below).

The University appealed against refusal of planning permission, and that Appeal was heard at a Public Inquiry (PI), by Planning Inspector J. S. Nixon, from 16 November 2002, with 11 days of evidence, and international interest. Green belt issues were discussed but those were not the focus of the Inquiry. The main issue was that the police said that they would not be able to keep order in view of the promised demonstrations. The Planning Authority's main line was that another site would cause much less disruption. The University contended that wherever the site, the security and policing issues would be made just as bad by the protestors, as they had been at Huntingdon Life Sciences; that view was later upheld by the Secretary of State.

At the Public Inquiry, the most effective objectors came from the Animal Rights Coalition and from the British Union for the Abolition of Vivisection (BUAV) and, especially forcibly, from Stop Huntingdon Animal Cruelty (SHAC). The latter, to the amazement of participants and commentators, were allowed by the Inspector to threaten the University representative in person and to show a video of the violence perpetrated against staff of the firm, Huntingdon Life Sciences (including close-up shots of near-fatal head injuries from a battering by a base-ball bat by SHAC supporters). The US magazine *Science Week* under the headline 'Inquiry turns into OK Corral for UK primate research', described how the Public Inquiry as 'a hearing meant to be limited to zoning [planning] issues devolved into a circus-like referendum on primate research'.[9] The other member of the University to give evidence, a senior academic not personally involved in such research, gave evidence for a half-day near the start of the Inquiry. Other academic members of the University did not take part, even though some would have been prepared so to do, because of well-founded concerns over their safety.

Three months after the hearing, the Inspector decided against the Appeal. His refusal report at para 14.21 criticises the way that the University's case was presented in the calling of just two witnesses.[10] He wrote: 'Some clearly felt this to be an arrogant and high-handed approach to the public inquiry and to the openness in which the proceedings should take place.' Part of his stated complaint, despite the considerable level of violence inflicted on those targeted by anti-vivisectionists, was that those involved in such research, and the MRC supporting, it should have made themselves available rather than be so worried about anonymity. The government report on the application was later to dismiss this argument.

The Inspector's report[11] ran to 88 pages plus extensive annexes. In summary, he agreed that the building and landscape design were the best that could be achieved, but he decided against the Appeal. He believed that the argument for national importance of the research was not strong enough to weigh against security issues and some adverse effects on the green belt. The University and its backers, MRC, the relevant government departments, and Hefce, continued to believe and maintain that it did, despite the well-orchestrated attacks on that position.

The planning application being of national significance and international media interest, the next step under UK Town and Country Planning Act procedure was for the relevant Secretary of State to review the Inspector's decision; the procedure known as 'being called in by the Secretary of State (SoS) on grounds of national interest'. Much detailed consideration goes on before the politician involved (in this case the Deputy Prime Minister, John Prescott) gives his decision. That decision appeared on 20 November 2003, a year after the public inquiry, and the decision was to grant planning permission. The report ran to 13 pages plus annexes.[12] The most significant of many grounds given by the SoS for over-riding the decision of the Inspector were that:

- The Inspector had reported that there had been objections to the application from 'organisations including Government Departments' whereas that was not true; just two representations were made from Government Departments (DTI, and Prof David King, CSA/Head of OST) and they both demonstrated that the research to be done was of national importance (SOS paras 12–27).
- In regard to the key plank of refusal of the planning application, the stated view of the police that they could not properly keep order, SoS said that, 'In considering the threat of demonstrations the Secretary of State takes the view that the risk of unlawful activity should not dictate policy or decisions in this important area' (para 28 et seq). (It is worth noting a High Court ruling in July 2012 that policing of Leeds United football ground, which

had been suffering a lot of trouble, was part of standard police duty and must be paid for by the police not the football club.)

- Re the effect on local residents, SoS agreed with the Inspector 'that the health and public fear arguments are not compelling in this case and that the quiet enjoyment of life by the development's neighbours would be prejudiced to a moderate degree if the development were to proceed' (para 36).
- That the Secretary of State did not agree with the Inspector that the University could persuade owners of other land to reverse their refusal to sell their land for this purpose, and that the University had clearly demonstrated that other sites it owned were unsuitable.

So, the University had a planning permission to proceed, and it had the choice: to continue or to abort. Pressure by activists on the University continued. Several University officials, involved in one way or another with processing the planning and other aspects of the animal facility, were the target of personal harassment, including persistent e-mail and telephone messages and intimidatory visits, some late at night; the Director of Estates was followed.

The University had until then invested heavily to keep its animal facilities in good state,[13] and so was not wholly dependent on the new facility for continuance of cognitive research on primates; however, the new project was clearly needed for the promising research that was the basis of the bid to JIF, and expected to prevent or alleviate levels of suffering from Alzheimer's and other degenerative diseases.

The BBC News website of 23 May 2002 wrote about a speech to be given by the Prime Minister that day, 'His speech is likely to include attacks on protests against animal experiments …' and quoted Mr Blair as saying, 'It is time to speak up for science, stand up against those who seek to drown out rational argument and proper research through protests that have nothing to do with argument, everything to do with prejudice'. Specifically, Mr Blair said,

> We're faced with a current example, where Cambridge University intends a new centre for neurological research. Part of this would involve using primates to test potential cures for diseases like Alzheimer's and Parkinson's. But there is a chance the centre will not be built because of concerns about public safety dangers and unlawful protests. We cannot have vital work stifled simply because it is controversial.[14]

That was in line with the view held by many in the University that such an outcome would be a victory of prejudice over reason and academic freedom, not the first in history.

Late in 2003, the University reached the conclusion internally that the University might well have to cancel the project. The Estates Department was asked to re-work the capital cost of the project with 'highest and lowest' costing of risk; as was anticipated, the upper figures were those quoted. The main cost issue was the cost of running and supporting the operation of the building. It was felt that the legal injunctions against violence won by HLS for itself and its supporting companies would not give enough protection down the line for the planned project, and more particularly, despite the warm words of speeches by the Prime Minister and others in government, that in fact there would not be enough financial support to pay those costs. It would put an unacceptable strain on University funds, and indeed on the freedom of staff to work in an acceptable environment. On Monday 26 January 2004 University Council made the formal decision to cancel the project on the stated grounds of cost and affordability. The VC was asked by the Tuesday of the following week to finalise for promulgation a Notice cancelling the project, but such was the need to present the decision appropriately, and such were national and international pressures, the Notice was posted, and promulgated at a press conference later that same day, 26 January. During that week, word of the cancellation spread among those who had worked so hard and at such personal risk on the project over the previous six years.

There was widespread discussion, in the UK and USA especially, but little discussion within the University: among some, a sense of relief; among many, mixed emotions; among some, a sense of capitulation to violence, and a sense of failure, a sense rarely felt in the University.

Some further commentary is set out in the Annex.

Projects reported in Chapter 9

		£M
University Library		
Aoi Pavilion/West stacks basement		7.40
Entrance and exhibition refurbishment/extension		0.90
NW corner		5.44
SW corner		6.00
Infill of west side		5.70
		7.00
Shelving/storage refurb		2.70
(At prices the current)	Total UL	35.14
Maths		58.58
University Farm		
Dairy Unit		1.40
'Spudulike'		1.17
Replacement of three buildings		1.53
Avian Unit Zoo Dept		0.80
	Total	99.00

Designed, planned, legally pursued but not built

	£M
307 Huntingdon Road	
fees paid	0.32
project budget	24.00

Annex

The chronology of the project to put up a building for the purposes of neurological research on Alzheimer's and other ailments is set out above. The underlying story, of course, is about the use of animals in medical research, about the 'academic freedom' to follow such research paths, about the balance in a democratic country between the rights of those who hold strong views against certain activities and the rights of those who in pursuit of the common good seek to carry out those activities and, further, the rights of those untold numbers whose lives might perhaps have been saved or made much better by such medical research. It also is about the extent of the duty of police to maintain the law and protect those whose lives and property are seriously threatened and/or attacked, and about where costs of that should be borne.

In this case, University estate managers, and the lives of some of those designing the University project (as well as, of course, those of the brave scientists involved in the research), were materially affected by how those issues played out, so there is an estates issue within the much bigger issues. Therefore, some further commentary is given about what happened leading up to the cancellation of the project.

The most significant specific factors noted in the above decision by the Deputy Prime Minister (DPM) about whether it could happen were the definition of 'the public interest' in the research to be carried out, and the likely disruption caused if the project went ahead. Whether a university should be allowed to put up buildings for research is almost invariably a matter for that university and the Local Planning Authority, usually after a great deal of discussion between senior academics and other senior members of the University. In some cases, matters of ethics need to be discussed and decided; for example, research which may be funded from sources that may not have sufficient ethical standing, or research relating to certain aspects of the use of armaments, possibly for export. Such discussions are almost invariably well aired within the university and, except in very rare circumstances, a generally acceptable way forward is developed. This project was in many ways different. There were, or certainly there appeared to be, few within the University who believed that this research should not go ahead. It appears that there was general, although not universal, agreement that the benefits of the research associated with this project, particularly the expected advances in treatment for Alzheimer's and Parkinson's, were well justified, and that they justified the use of the lower primates in testing procedures and drugs. The research team at Cambridge was undoubtedly at the highest international level and if their research work was disallowed, then this raised issues not only of the balance of ethics, but also the sense of 'academic freedom' which is central to university research.

Comment in the media of other countries focused particularly on the matter of evidence given by experts as to the medical aspects of public interest: for example, in the US journal *Science Magazine* of January 2003, it was noted that

> the University had left itself in an awkward position to rebut such charges [that the research was scientifically flawed and unreliable and was neither of national importance nor necessary] in that it had banned its own scientists engaged in primary research from testifying, out of fear for their safety. That left the University's chief academic witness … virtually alone on the firing line.

In an article on this project, the *New York Times* of 8 August 2004 quoted Dr Aziz of Oxford University, 'Most people don't realise how their health and well-being would be affected without animal research. It's an intrinsic part of modern research'.[15] Anti-vivisectionists, however, argued forcibly that such research could be done equally effectively using computer modeling.

Regarding public interest, besides the hopes of those suffering from the diseases being researched, there was for the government the matter of the economic implications of pharmaceutical research being carried out in UK. Drugs companies were spending over £8M per day in the UK on research for new medicines.[16] Pharmaceutical companies were making it clear privately to the British government that unless they were able to conduct their business in a reasonable way without intimidation by those acting outside the law, then they would take their research to other countries, notably to the United States. The *New York Times* of 8 August 2004 quoted Jean-Pierre Garnier, CEO of GlaxoSmithKline, saying that 'his British employees were being terrorised' by the militant welfare groups.[17] 'I take it very personally,' he told the *Daily Telegraph*, adding that several unnamed companies looking to invest had decided against Britain because of the intensity of the animal welfare campaigns. At the time of writing, it would appear that in fact definition of public interest in regard to development of universities and university laboratories has matured since this project.

How did those protest groups operate so effectively? The forces marshalled against this project were considerable. One of the main organisations, SHAC (Stop Huntington Animal Cruelty), claimed to have nearly 5,000 committed supporters in the United Kingdom. It carried out a protracted campaign of violence over several years, not only against directly employed staff of Huntingdon Life Sciences (HLS), but also against those who did business with HLS. For example SHAC infiltrated into HLS a 'mole' who obtained email and telephone numbers of 135 managers and secretaries at DeLoitte and Touche, and those were circulated to some 8,000 anti-vivisectionist activists who jammed emails and telephone numbers using phone-jamming software. At that point DeLoitte and Touche gave in and stopped working with HLS, as did many companies. There were many other instances of such intimidatory tactics being used against companies across the world from Japan to the United States. It came to a point where the UK government had to give financial assistance to HLS to allow it to continue to function. While that campaign was going on, SHAC also turned its focus on the University project at 307 Huntingdon Road, and although such a high level of violence was not immediately shown, it was certainly threatened. For example, one of the two university witnesses at the public enquiry, the Director of Estates, who had previously had a bomb put into and detonated in his office building at another university by anti-vivisectionists, was overtly followed and threatened.

The umbrella body for such protest against vivisection in general was the British Union for the Abolition of Vivisection (BUAV). In fact, nearly all of the campaigning was against the use of lower primates in experiments. Use of other primates in research had long been banned by the government, and of all experiments involving the use of animals, primates were used in just 1 per cent. Of all animals used in research laboratories 85 per cent were rodents, mostly rats and mice, and their use raised no discernable protest.

Opposing these groups was a small but very articulate and brave organisation, the Research Defence Society; its executive director Mark Matfield deserves great credit for the expert witness role he played and his courage in doing so.

Despite the major speech by the Prime Minister, Tony Blair, on 23 May 2002, little was in fact done in support for some years. In particular, it was expected that following his particular reference to the University of Cambridge when he said that 'we cannot have vital work stifled simply because it is controversial' the Home Office would take primary legislation through Parliament, or would at least via secondary legislation issue directions to police forces to maintain law and order more rigorously around sites where there were protests against primate research. Finally, on 17 April 2003, HLS won a ground-breaking injunction preventing animal rights protesters from approaching within 50 yards of employees' homes. They also succeeded in banning protesters from the immediate area of the company and work sites, and reducing the intimidation caused by demonstrations insofar as only one demonstration was to be permitted every 30 days in such exclusion zones; numbers protesting were not permitted to exceed 25, parking at least half a mile away from the site and the demonstration not allowed to last more than six hours. The injunction also prohibited publication of names, addresses, vehicle registration numbers, email addresses or any information identifying 'protected persons'. Harassment was also deemed to include any 'artificial music such as klaxons or hooters'. Breach of the order was to be an arrestable offence, and that allowed the police to remove anyone immediately; those found guilty of offending against this injunction were liable for prison sentences of up to two years. The interim injunction was granted against nine animal rights activist groups, including SHAC, the Animal Liberation Front and London Animal Action. When a few months after Cambridge called off the project, work started on similar facilities in Oxford, injunctions were put in place, and the Oxford project did proceed to completion, albeit at high cost. The Oxford project had the benefit of being on a site used for many other activities so staff were less exposed than they would have been at Cambridge. The research at Oxford is said to be successful.

Public opinion in the UK has generally become less tolerant of those who resort to extreme intimidation or terrorism in pursuit of their beliefs. Police, the Home Office and other intelligence agencies are much better informed about radical anti-vivisection groups, and injunctions against them are being maintained with public support. Some public sympathy for certain extreme groups such as SHAC was lost when stories about the high life-style of their leaders leaked out (*The Observer*, 11 April 2004, spoke of 'comfortable lifestyle among the gin-and-tonic set'), and some involved in violent protest were jailed. Also, government and public awareness of the potential for research-based cures for feared diseases has increased through greater awareness as science generally is being somewhat more clearly set out to the public.

Regarding the key issue of policing and its cost, one of the core arguments of the Local Planning Authority, South Cambridgeshire, was the cost of keeping order for this research laboratory had it been built, including police time and use of helicopters. Although the Secretary of State in his ruling made it clear that an expectation of problems of maintaining law and order should not be a reason for withholding a planning permission, the government did not, in reality, do anything like enough to assure county police forces, and universities, that sufficient resources would be found to fund efforts to maintain law and order around research laboratories where there were protests. On the other hand, there were questions as to why when a small group of peace-loving Buddhists walked quietly along the pavement from Cambridge to the proposed site, the police mobilised a number of police cars, police officers and indeed the police helicopter at considerable expense.

The University spent a huge amount of staff time and dedication to the point of personal danger over six years on this project, and considerable design and legal fees, before calling it off in the face of intimidation. The University of Oxford carried through a similar project, at great expense, starting soon after Cambridge aborted its project.

In January 2007, seven animal rights activists were jailed for 4–7 years for what the judge called 'urban terrorism' and 'a relentless, sustained and merciless persecution' in their attacks on Huntingdon Life Sciences and related companies.

Notes

1 British Library, Cambridge, Oxford, National Library of Scotland, National Library of Wales and Trinity College Dublin.
2 The Legal Deposit Libraries Act 2003 extended the right to get digital books.
3 There is a Cambridge myth that Mr Aoi was visiting Cambridge, and cycling round on a bike. On seeing the cramped nature of the Library, he reached into his pocket, pulled out his cheque book, and wrote and handed over the cheque on the spot.

4 University Science and Technology Building Sub-Committee, 23 January 1997.

5 Quoted, for example, on the BBC website on 9 September 1999.

6 Prior to Ray O'Rourke's decision to buy, he and his brother, a stocky and powerful man with a 'clear' view of what he wanted, spent half a day in Cambridge discussing the construction market, much of it conducted sitting with the Director of Estates on a sunny bank overlooking the Maths site. Thus began a relationship that led, first via a suggestion by the Estates Director over lunch in Loch Fyne, that the huge and growing company of Laing O'Rourke should set up graduate training in Cambridge to ten years after those discussions at Maths and in Loch Fyne, new course being put in place, in the Engineering Department with an endowed chair.

7 Richard Saxon, *Be Valuable: A Guide to Creating Value in the Built Environment* (Constructing Excellence, 2005).

8 *Jump Up The New Primate Laboratory, SPEAK.* Nature Neuroscience **7**, *413* (2004), doi:10.1038/nn0504-413.

9 *Science Week*, Issue 23 (January 2003).

10 Report to the First Secretary of State J. S. Nixon, 7 March 2003.

11 The Planning Inspectorate, 7 March 2003.

12 Office of the Deputy Prime Minister, 20 November 2003, ref S/1464/01/F.

13 The University's five-year Planning Statement of 1989/1990 noted that,
> The University has made no provision for expenditure on the upgrading of accommodation required under new legislation. The cost of upgrading animal accommodation alone is estimated at around £7M, and recent visits by officers of the Health and Safety Executive in connection with genetic manipulation work and the COSHH Regulations will undoubtedly lead to substantial modifications being needed to laboratory accommodation, at a cost of at least £1M. The University cannot meet expenditure on this scale from its present resources, and the current level of the UFC capital-in–recurrent grant to the University is totally inadequate to meet the cost of upgrading accommodation to meet safety standards. However, failure to carry out such work is likely to lead to the closure of facilities and the cessation of academic work of the highest quality.

By 1999, eight projects for upgrade of animal houses at a total cost of £14.1M had been implemented or were planned.

14 The speech is set out in full in the 10 Downing Street Newsroom, 23 May 2002.

15 Lizette Alvarez, 'Animal Welfare Advocates Win Victories in Britain With Violence and Intimidation', *The New York Times*, 8 August 2004.

16 *Business Economics*, 22 November 2013.

17 Lizette Alvarez, 'Animal Welfare Advocates Win Victories in Britain With Violence and Intimidation', *The New York Times*, 8 August 2004.

Table 9.1 Overview of capital projects at the edges of the Central University

Chapter 9--Road to West Cambridge	Start	Finish	m2 gross	m2 ass	Architect	Svcs Engr	Contractor	£M
Athletic Ground Lighting	08/09/00							0.32
Univ Library Mr Aoi	1996	01/03/98			Harry Falkener Brown	RPA Samuelly		7.39
Entrance West stacks	1998	1998	Stacks	Stacks	Harry Falkener Brown	RPA Samuelly	R G Carter	0.91
Univ Library Extension--3NW	01/09/99	01/09/01	2,780	1,850	Harry Falkener Brown	RPA Samuelly	R G Carter	5.44
-- Phase 3SW	03/07/00	03/01/03	2,800	1,800	Harry Falkener Brown	RPA Samuelly	R G Carter	6.01
-- Phase 5W	01/12/03	31/05/05	2,520	2,000	Harry Falkener Brown	RPA Samuelly	R G Carter	5.69
-- Phase 6W	14/11/05	01/10/08				RPA Samuelly	R G Carter	7.01
Shelving and storage	14/11/05	01/10/08	Shelving	& storage	Hopper Howe Sadler	RPA Samuelly	R G Carter	2.69
Maths --Phase 1	01/08/98	12/09/00	8,896	5,650	ECA	Roger Preston	Laing	25.52
--Phase 2a	01/10/00	15/02/02	1,550	930	ECA	Roger Preston	Sir R McAlpine	5.28
--Phase 2b	01/02/00	01/06/01	3,370	2,000	ECA	Roger Preston	Sir R McAlpine	9.84
--Phase 3	01/04/01	23/01/03	4,755	2,900	ECA	Roger Preston	Laing/O'Rourke	17.94
Dairy Unit, Huntingdon Rd	01/06/97	01/06/98	3,000	2,550	Smith Woolley		R G Carter	1.39
Agronomy (potatoes)	14/07/03	15/03/04	486	370	JGP		Cocksedge	1.17
Farm refurb	30/08/05	14/11/05	1,315		Alun Design	RPA FirebraceR G Carter	R G Carter	1.53
Avian Unit	01/03/05			220	EMBS	EMBS	R G Carter	0.8
307	Aborted. Details conf							24
Cost of completed work, 1996 to 2006								98.93

10 The west side story

By the mid 20th century there was a general feeling in the University that in some golden age in the future new buildings for new and largely unknown purposes might find their home in West Cambridge. The astronomy building had been built on the site to the north of the Madingley Road, in Madingley Rise, there being a hill there, by the Cambridge standard of hills. And the astronomers were happy to go so far away from the University as it then was, if only to get away from the higher night ambient-light levels around the city. Then, when the University Library was set up at what was thought to be far from the rest of the University (see Chapter 1), there grew among the ranks of those tasked with looking to future land holdings more confidence in possible future expansion; land on the farms to the west of Cambridge, beyond the University rugby ground and the Officers' Training Corps (OTC) 'Rifle Range' (before it moved to Coldhams Lane), was gradually bought up before, during and after the Second World War. The Clinical Vet School was set up on those farmlands in 1949.

Following that, Professor Alec Deer and three other brilliant professors wrote the seminal 'Deer Report' of 1965/1966 (see Chapter 1) recommending that 'A development plan for the area of West Cambridge should be prepared in conjunction with St John's College [which owned land adjacent to the land bought earlier by the University] and Jesus College' [which owned the 'Rifle Range' land]. The report raised the question as to which science and related departments should move to West Cambridge, and clearly showed, in a business-like manner, that science departments and technology departments would do well to plan to leave their much-loved, overcrowded and rather dowdy sites in the University heartlands, and go west. In 1971, as noted in Chapter 1, there was a Report of the Council of the Senate on the development of the West Cambridge area.[1] In 1972 Council agreed that the General Board should be responsible for long-term planning. They appointed a committee to advise them under the chairmanship of Professor Peter Swinnerton-Dyer, resulting in 'The Report of the General Board on the long term development of the University',[2] a plan for possible future expansion of the University leading up to what was known as a 'steady-state University' based on an upper limit of 14,000 students. A lot of this was about the future growth of the University, with the conclusion that 'Future developments in physical sciences and engineering should normally take place on sites in West Cambridge'. First out of the blocks was the Cavendish physics laboratory which moved in

1973/1974 from its famous New Museums Site where the atom had first been split by Rutherford and his team, and the electron identified by J. J. Thomson; the move was to the near, the eastern, end of the West Cambridge site. The area of the site being developed was some 164 acres (66 hectares), about the size of the MIT campus.

More of the history of the early developments at West Cambridge is given in Chapter 1, and in Chapter 3 there is discussion of how the Town and Country Planning challenges were sensed and duly met so that development could take place at West Cambridge when required. This chapter will pick up the west side story during the decade of this book, project by project.

Astronomy

To start from the northern side of the area, the University observatory was set up in 1823, much of it later to be converted into the departmental library and offices. When the Institute of Theoretical Astronomy was set up in 1967 (by Fred Hoyle), it was designed for just 25 people, with a seminar room catering for 50. From the start of the decade being considered, the Institute of Astronomy grew like Topsy from its pre-Greenwich observatory beginnings. People are often surprised to learn that among University departments, Astronomy expanded about the most regularly, funded by private or corporate benefaction. In 1997, the University approved the expansion of the Hoyle building as a priority for fundraising. At that time, the project brief comprised both the construction of a new lecture theatre and also office accommodation for those staff transferring to Cambridge University from the Royal Greenwich Observatory when that was closed. The final brief was to accommodate also an additional 30 staff and to modify the observatory building to house additional library space. In the event, this proceeded in two phases and in the interim, space on the site was utilised in the Royal Greenwich Observatory Building, now called Greenwich House, when that transferred to University ownership.

In November 1999, in its first phase, a new and very well-designed lecture theatre with a capacity for 180 people enjoyed an opening ceremony. Sited at the back of the entrance hall of the Hoyle building, and named the 'Raymond and Beverly Sackler Lecture Theatre' in tribute to the initial donation from the Sackler Foundation, it was designed by Duncan Annand, Roger Parker and Hannah Reed and built by Rattee and Kett (Mowlem) at a project cost of £880k. (The architect was selected to lead this design largely because of the popularity of his design of the Isaac Newton Institute on Clarkson Road.)

Phase 2 was altogether more ambitious, and more difficult technically, and was made possible by a generous donation from Nick Corfield. Some parts of the Hoyle building were to be developed and expanded so as to provide open discussion areas, and to enlarge the common area at the heart of the building. The success of Phase I and the excellent relationships developed between the design team and the Department resulted in the appointment of the same team, plus David Brown Landscape Design. In October 2000 it was agreed to go out to tender on the basis of early contractor involvement (ECI), two-stage 'develop and construct' procurement so the contractor was involved in the design. The contractor selected was Sindall. The design was for a new two-storey extension to the rear (east) of the Hoyle building with the principal axis set at right angles to the existing single storey building. The new extension over-sailed the existing building at the intersection. As more funding came in, the scope of the project was developed, and the budget was re-set at £3.03M. The gross and assignable areas were 1,227 and 750m², quite a low balance ratio because of its nature, and comfortable in its 'feel'. It was not easy to design so much development within such an area of valuable trees: careful thought was given to those, and in discussion with the local authority's arboricultural officer, it was agreed that eight could be felled to allow for the new extension. Work on site started on 19 March 2001, and was completed £244k under budget but five weeks late because of site problems in the renovation of some parts of the existing building. Work was handed over on 27 March 2002 and there was an official opening by the then Chancellor, HRH Prince Philip, on 3 May 2002.

The design was hugely successful, not just in creating an excellent environment for the extension, but also in linking the previously disparate buildings of the Institute of Astronomy which had been segregated by a dense belt of conifers. The Corfield Wing extension of the Hoyle Building changed the architectural focus of the department and was the first deliberate element on that site to create an informal courtyard design: that was later completed by the construction of first the Kavli Institute for Cosmology for which the design team of Duncan Annand, Faber Maunsell and Cameron Taylor was being appointed in 2006 for a start on site of 16 June 2008 with a budget of £4.085M. That was completed in 2009. The Astrophysics Building, again with the same architect, was designed and built after the decade of this book.

Athletics ground lighting

Chapter 3 on 'Planning' described the planning aspects of the project to provide lighting for the athletics ground at Adams Road adjacent to the West Cambridge Site. The project did not go ahead during the decade of this book; some £300k was spent on fees over several years.

British Petroleum Institute

Adjacent to the Institute of Astronomy there sprang up a rather different sort of building, funded by a £25M benefaction from the oil company British Petroleum, which later became 'Beyond Petroleum'. The purpose was to research multiphase flow, and that involved the Departments of Earth Sciences, Applied Mathematics and Theoretical Physics, Chemistry, Chemical Engineering and Engineering, a true example of the modern inter-disciplinarity being encouraged by the University and by the Research Councils. Of the £25M, £2.196M went to build a modern set of offices and meeting rooms in a basement and two storeys above, with 1,156m² gross area, 910m² assignable, a balance ratio of nearly 80 per cent which is quite good. The building was to have high standards of sustainability for buildings of that time, with photovoltaic panels on the roof and really good insulation. The designers appointed in June 1998 were Cowper Griffiths as architect, Max Fordham as building services engineers and Hannah Reed as structural engineers. The constructor was Rattee and Kett, soon to be taken over by the national company Mowlem (which later was swallowed up by Carillion). There were several significant problems in the construction. Designing the waterproofing of the basement was difficult given the high water-table. The roofing sub-contractor had supply problems and got far behind programme; the main contractor was unable to stay on schedule. The work started on site on 25 October 1999; it was due to finish in November 2000, but it took until 7 March 2001 with a total out-turn cost of £2.48M: when the building was finished, the incoming BP users wanted to extend the basement, which was something of a nightmare because the main University 'backbone' data cabling ran within a couple of metres and so the excavation work had to be done slowly and by hand; a shovel out of line would cut off the University from the outside digital world, not popular. The official opening, on 29 May 2001, was conducted by the former Government Chief Science Advisor and President of the Royal Society, Sir Robert (Bob) May. Everyone seemed happy with the new facilities, and indeed they worked well. Sir (later Lord) John Browne, then Group Chief Executive of BP Amoco, spoke of BP's reason for choosing Cambridge,

> It is one of the few places in the world where it is possible to gather people who are at the leading edge of different disciplines, present them with a challenge and, by solving it, break through the current limits of knowledge.[3]

Relocation of the Department of Earth Sciences Geology

A project which was prepared but never delivered mostly because of lack of funding was the relocation of Earth Sciences (and Geography) department from its dispersed and unsatisfactory accommodation on the New Museums Site to a site adjacent to the Department of Astronomy to the north of the Madingley Road. The Planning and Resources Committee approved a Concept Paper in July 2003, and authority was given to spend £60k on design such as would be necessary for a fundraising and consultative document. Cowper Griffiths was appointed as project architect for this work. The estimated project cost was £84M, with a gross area of 17,700 m². The project catalysed a great deal of discussion within the Department about how it should configure itself for the future, and it proved difficult to get a consensus, particularly one that could justify the proposed size and complexity of the building. Providing a convincing business plan with discounted income and costs out for ten years after occupation of the new building was a considerable challenge. A great deal of management time was spent on this embryo project, and in the end the decision was taken, probably correctly, to mothball it. It was a healthy reflection on the committee and project management systems that a good and informed debate could take place with a negative decision accepted as a fair outcome.

Infrastructural development of main West Cambridge Site

The main West Cambridge Site is 66.45 hectares (164.2 acres), some 1.15km from the east end to the west end at the M11; it is 650m from the southern edge along the footpath to Coton and the farmland owned by St John's College, to its northern edge along the Madingley Road. There is a slight slope down to the south. Not far away, near the Travellers' Rest pub, Paleolithic (70,000 to 12,000BC) remains of Acheulian axes had been found earlier; the earliest archeological evidence, flint tools, found on the West Cambridge Site, however, dates back to the stone' and iron ages. Later, there was a series of regular enclosures and semi-open Roman fields. During pre-construction excavation, remains from two timber buildings, some quarry pits and two cemeteries, featuring forty burials, plus five isolated burials were found, with images of Minerva or Pallas Athena; also, a cache of third-century coins was found, an enclosure containing butchered animal carcasses apparently for some ritualistic purpose, spearheads, brooches, coins, pottery and a copper bust. This settlement seems to have been part of a much larger system that probably formed a ring around the Roman town. The site was abandoned in the fifth century AD, with no further activity there until the late medieval period when ploughing of the area began. A dyke was constructed known as Willowes Ditch, but it seems that there was no more building development until the construction of Vicars Farm in the nineteenth century.

At the west end of the site there is the High Cross Research centre. Within that, long land-leases were sold before the period of this book to the British Antarctic Survey (1976), to CADCENTRE (1978) and to the oil company Schlumberger (1984); the World Heritage Schlumberger building was designed in 1983/1984 for a 15-year life (still going strong but expensive to keep serviced). Also, as described in Chapter 8, the geotechnical research centre of the Department of Engineering was developed on a small area at the north-west of the site, remote because of the need to have a base relatively free of vibration.

Other than these developments, and those mentioned above at the east end of the site, the whole West Cambridge Site was ripe for development by the mid-1990s.

Development of the Vet School

The Cambridge Veterinary School was founded in 1949 with eight under-graduate students, but its origins go back to 1909 when the University Department of Pathology set up an outstation to study diseases of large animals. In 1935 the University entered into an arrangement with the Royal College of Veterinary Surgeons whereby it ran a pre-clinical course and a postgraduate diploma. In 1948 it was decided to start a Vet School. The course being a year longer than rival University vet schools, it struggled for students and cost-effectiveness, but the facilities of the School grew markedly in the 1950s and 1960s with many new facilities being built, in part to accommodate increasing student numbers. The 1980s saw further developments in the school, which were abruptly halted in 1988 by a UGC/UFC review conducted by Ralph Riley recommending closing of the Cambridge veterinary school (a critic at that time called it 'a club for horse-lovers') and closing the Glasgow School in Scotland.[4] This was one of the more politically unfortunate reports of the time as the Conservative front bench had some Cambridge alumni and the Party was desperately trying to staunch loss of support in Scotland. Mrs Thatcher was rumoured to have knocked it on the head for those reasons, but it was easy to do so in the face of a widespread pro-vet school campaign. The follow-up report took better account of the implications of the rapidly increasing legislation by EU regarding animal checking, and a better assessment of future needs for vets in UK (and better knowledge of what Mrs Thatcher would not throw in the bin), so the closure decisions were reversed and, indeed, student intakes increased with the Cambridge Vet School being successfully reformed and going on from strength to strength.

Through the 1990s, there was an increased emphasis on research and scholarship in re-furbished and newly built state-of-the-art vet facilities. Upgrades costing in total £830k built by R. G. Carter refurbished the Clinical Veterinary Medicine facility from 1998. Subsequently, further work designed by IBS and Hannah Reed provided 360m² asssignable at a cost of £2.21M. A modern post-mortem facility was built in 2000, at a cost of £2.11M. This was one of the projects designed by the talented architect Adrian Pettit of the Estates Department of the University. The area of the facility was 805m² gross, also built by R G Carter. An Equine Diagnostic Unit, construction cost £830k, followed; then a new Small Animal Surgical Unit, including five new operating theatres, and a Farm Animal Unit were completed in August 2003. Designed by Saunders Boston, they were built in 2002/2003 by R G Carter at a cost of £1.725M. With 1,720m² gross the unit cost of £1k/m² was surprisingly inexpensive. The official opening by the Chancellor was greatly enjoyed by all, especially, it appeared, by him. In 2006 the refurbishment of the X-Ray facilities in the hospital and the installation of new small and large animal X-Ray equipment were completed and equine outdoor clinical facilities were provided.

Master planning of the development of West Cambridge

In 1996 a Public Inquiry finally confirmed that the West Cambridge site could be developed as an extension to Cambridge (see Chapter 3). Consequently, selection of a master planner went ahead; the decision was made to appoint MJP, MacCormac Jamieson Prichard. ECA, Edward Cullinan Architects, which was developing a very good track record of planning university development in UK and USA, and was to design the widely-acclaimed Centre for Mathematical Sciences nearby, was not selected. An Outline Planning Application for the West Cambridge Site was submitted on 12 September 1997.

The master plan has been seen as reflecting in style a science park form rather than the more natural form of an extension to the University itself (see Chapter 12). The plan gave considerable capacity for large new buildings, but there was some criticism from among the architectural commentariat: the journal *Cambridge Architecture* of Autumn 1997 under the headline 'Is that it?' suggested that 'the MacCormac schematic has no more life in it than John Cleese's parrot – it is an ex-master plan'. (The Cambridge alumnus John Cleese is not known to have responded.) However, the report 'Cambridge Futures' (see Chapter 3), coming on top of the favourable outcome of the Regional Planning Guidance (RPG 6), made it easier for the local planning authority, Cambridge City Council, to grant planning permission following the master plan's circulation. Speaking for the development at West Cambridge, the Vice-Chancellor, Professor Sir Alec Broers, said

> you can't carry out world-competitive research without proper investment into the best equipment and facilities. The British science and industrial base depends on the important research work universities conduct, and we ignore its development at the whole country's peril.[5]

A big issue was the negotiation of the Section 106 Agreement, which set out how much infrastructure the University would have to deliver as a condition of planning permission, and how much would have to be paid towards local support facilities. Those negotiations were getting bogged down in ever more detailed discussions between lawyers and consultants until those meetings were stopped late in 1998 after a particularly over-detailed meeting with wall-to-wall consultants; agreement then came quickly in one-on-one meetings between the head of the City's planning department and the Director of Estates. There was remarkably little requirement imposed on the University beyond development of local infrastructure and ecology, which the University was happy to do. Full planning permission was granted in March 1999: so the development of the main site at West Cambridge could start.

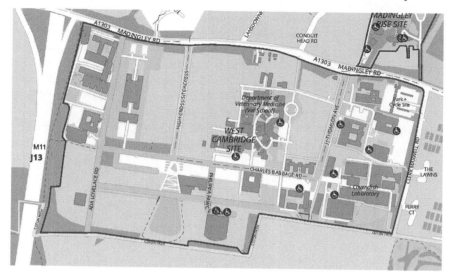

Figure 10.1   West Cambridge Site plan

Computer Lab

And it started fast. Three projects swung into development as soon as that permission was granted. The first and highest profile in the University was a new building for the Computer Lab, more properly the Faculty of Computer Science and Technology. Founded formally as such in October 1970, one might trace the history of the Computer Lab back under different names to 1937, or to 1949 when the Cambridge Electronic Delay Storage Automatic Calculator (EDSAC), the world's first fully operational computer, did its first run. Much of the credit can be attributed to the Lucasian Professor Charles Babbage (1792–1871) who was even then thinking about automatic computation. While still a student at Trinity College he first thought about what he called his 'difference machine' which would change the way people looked at arithmetical tables by calculating values and printing results directly. The mathematical laboratory was set up in 1937 in a spare area which previously had been part of the old Anatomy School in the New Museums Site: that can be thought to mark the beginning of computer science per se in the University. The head of the mathematical laboratory for a crucial period of its history was the great computer pioneer Maurice Wilkes, the designer of EDSAC.

The Computer Lab had been in cramped accommodation in Mill Lane along with the mathematicians, and then, thriving but in poor working accommodation in the small floor-area Arup Tower, in urgent need of more space. For some years it was thought best that it should be included in the new development for the mathematicians and allied workers at Wilberforce Road (see Chapter 9), but by 1997, with pressure to reduce the extent of that development, it was decided that instead it should get a new and large building in West Cambridge.

In October 1997, Bill Gates, the chairman of Microsoft, helicoptered into Cambridge and told the media why the William H. Gates foundation was donating £12M to the University, and why Microsoft had chosen Cambridge to be its first research base outside the United States. In particular, he forecast how exciting developments on which the University of Cambridge and Microsoft were already collaborating would be 'feeding into the fabric of everyday life'.[6]

Soon after the decision to proceed, it was further decided, not uncontroversially, that designers for the new building should be selected from an interviewed short-list, rather than directly appoint the master planning architects MJP; the multidisciplinary practice RMJM was appointed. The Lab building site was to be at the east end of the site, quite close to the houses along Clarke Maxwell Road (which was the only bit ever built of the long-planned western bypass for Cambridge: the bypass scheme was killed off the evening that a local election saw a political change in the Council).[7] Of the building, which would have 10,196m^2 gross, 6,702m^2 assignable, some 2,000m^2 gross were going to be released to Microsoft. This was a particularly exciting time for Microsoft: it was expanding very fast, breaking into a lot of new markets, and leading and following very rapid technological advances. Overheard at an evening drinks party was a comment by a senior Microsoft executive that the company was growing so fast that the space in the computer laboratory would not be enough, and that they would need around about 5,000m^2. A response from another Microsoft gentlemen was that the UK Town and Country planning system was so cumbersome that it could never keep up with the expanding requirements of Microsoft, and perhaps the research facility should go either to Munich or revert to Seattle. A University representative present asked for a hold on such a decision for a few weeks. Within just a couple of months of quite intense negotiation with the planning department of the City Council, starting with a degree of alarm that the University should seek to make substantive change to a major planning permission, and so soon, it was agreed that there should be a major amendment to the recently won planning permission whereby instead of Microsoft having most of the top floor of the new University building, it would have an entirely new building which had not been in the master plan or in the planning permission, adjacent to the University's Computer Lab building. Those negotiations and that crucial decision reflect the business-like relationships between the city's Planning Department and the University: that a major change should be agreed, and agreed so swiftly, reflected the degree of mutual trust that had been restored since the debacle over the nearby athletics ground flood-lighting planning application. Microsoft relinquished its option on the Lab's top floor (with compensation agreed) and negotiations to achieve a ground-lease on the site adjacent began.

Following the University's decision in October 1998 to make fundamental changes in its capital procurement systems (see Chapter 5), the Computer Lab project was the first full project to be procured under the new system of two-stage 'develop and construct' procurement with the constructors joining the design team and consultants part-way through the design process. Regarding design, RMJM as a multidiscipline practice providing both architectural and building services design was able to produce an integrated concept for the new building, with a very functional and efficient floor plan and elevations, and specified to a standard of sustainability which was very high at the time: in fact, in that regard, it was particularly notable that there was to be minimal heating in the building, other than around the entrance area to make arriving people feel that they were going to be quite warm. Heating was largely to come from humans and from computers, with heat exchange by circulating water in pipes embedded in the concrete ceilings: a system known as chilled beams.

Regarding the construction programme, the changes caused by the decision for Microsoft to have its own building delayed the project by a few weeks; a letter of intent was issued to the selected contractor, Shepherd Construction, on 20 August 1999. This appointment was one of the first contractor appointments to be decided by a normalised balance of assessed quality and price. The adjusted and reconciled tenders had Shepherd Construction as the second cheapest and they came top of the quality score, which was worth 30 per cent of the total scoring. (The cheapest tender, by about £50k, came lowest in terms of quality.) With the constructor on board, detailed design and works tendering of the 18 works packages proceeded apace until there was 80 per cent price certainty at which point the contractor was awarded the second stage contract, i.e. construction. The contract sum was within the budget of £21.80M including the element of VAT required to be paid, and plus enabling site works.

Work on site started on 7 February 2000; shortly before that there was a meeting on site as early-morning mist swirled around and a wan sun grudgingly edged over the flat Fen horizon, between the VC, the recent Secretary of State for Trade and Industry Peter Mandelson, who said, mostly jokingly, that such was the pace of local industry that there should be a 'Minster for Cambridge', the PVC Professor Roger Needham and the Director of University Estates.

The official-start, 'ground breaking', ceremony on 22 March had the VC operating a JCB; clearly enjoyed by him more than by those standing adjacent. (The University Newsletter was perhaps over-stating a tad with the description that the VC 'marked the occasion by helping the contractors dig the foundations for the new Computer Laboratory'.) The Vice-Chancellor declared that

> the start on site at West Cambridge is a significant moment for the University and the City. West Cambridge will allow the University space and modern facilities it desperately needs in order to compete with other leading international universities. It will act as a pioneering centre for research collaboration with local and international industry, not only boosting the British economy, but positioning Cambridge as one of the top technical players on the world stage.[8]

In turn, the director of the constructor Shepherd, said that

> this project has been set up by the University as a total team and partnering approach to construction. Cambridge University must be congratulated on giving leads to the construction industry in adopting the principles of Egan [John Egan had fronted a review of the construction industry following that of Sir Michael Latham] for construction projects. The total team has been working together for over six months prior to the start on site and the benefit of this can be seen by the progress after only six weeks on site.

The University Newsletter of April 2000 noted that 'as well as being the largest building scheme ever undertaken by the University, the 60 hectare site has the potential to become one of the largest building sites in Europe'. The opening ceremony, and the project generally, got extensive coverage in the national trade press. Partway through the project, the Minister of State for Local Government and Construction, Nick Raynsford (formerly of Sydney Sussex College), came to visit the site to form a view about the new collaborative ways of working and their possible wider application. He asked the contractor's site manager for his view. The site manager was widely known to be not only highly effective but extremely frank and, as one might describe it, 'direct' in conversation. When the Minister asked him what he thought of 'this new-fangled way of procurement', all present braced themselves for what was expected to be a very blunt and probably painful reply. What he actually said, after, as usual, wiping the back of his hand over his nose, was, 'well Minister, it's like this: when I started this job, I knew a lot more about it than I've known about most jobs when I finished them'. Breaths were exhaled.

The work was completed on 4 October 2001 with partial handover on 6 Aug. The total project out-turn cost of £21.8M was close to its budget of £19.3M plus the infrastructural element for the site, with unit prices of £1,845 per square metre gross and £2,807 per square metre assignable: these costs are approximately half of the unit capital costs of the Centre for Mathematical Sciences (CMS) project going on at the same time. Both were to very high standards of sustainability and low running costs, the more complex CMS buildings being constructed to higher architectural and landscaping specifications and of materials of particularly long life.

This was the first project ever to use the Soft Landings procedure, which was developed in and by the University (see Chapter 5). (The Soft Landings procedure is now nationally required for all public buildings of any size, and has spread through several other countries.) So, when this building was completed and handed over to the users, one office near the entrance was used by one representative of the designers and one representative of the contractors; they worked together, meeting and talking with users of the building, sorting out teething problems and explaining how best to operate the building so as to make it most productive, efficient and sustainable. Most public sector building in the UK operate far below their possible performance (typically by about 30 per cent). They remained in that role for about a month, and the benefits of personal contact, rather than electronic communications which soon become less collaborative and more contractual, worked extremely well. Also, the two individuals learned a great deal about how well the designs were working in practice, and how in a few aspects they were not satisfactory: good training at that stage of their professional lives.

The users liked their new building. One issue though was that the external timber weathered unevenly, giving an appearance of staining; that problem was overcome on the rear elevation by the application of marine oil (done just prior to the official opening of the adjacent Microsoft building). A more temporary problem was that the representative user when pushing for priority for the project was adamant that all the users would be recruited by the time of the opening of the new building: the heating calculations assumed just that, but when in fact far fewer staff were around to move in, and there was a cold winter, the building was consequently colder than designed as the heat from humans and their computers made up a significant part of the heating system. The building got an RIBA award commending its sustainability.

During the summer of 2002, a follow-up project was carried out to fit out the top floor where Microsoft had originally planned to have its new research centre. The area fitted out was 1,217 m² and assignable area 924 m². The budget, including quite a lot of work for new activities, was £1.1M; the out-turn cost for the works was £790k, with some expenditures on equipments.

Figure 10.2 Computer Lab (external)

Case study – Computer Lab: William Gates Building

The Computer Lab was one of the first buildings to be erected in Richard MacCormac's West Cambridge master plan and is close to the entrance to the whole West Cambridge site. Design started in 1998 following a limited competition.

It presents as a rectilinear building with an over-sailing flat roof on glu-lam beams and slender steel columns. The entrance is clearly signalled by festive fabric roofs over bike parking and a generous overhang to the three storey porch with terracotta and iroko boarded cladding, the terracotta being peeled away to create sideways windows into teaching rooms on the first and second floors.

The building is conceived in two parts: the front 'Public' wing with the 3+ storey 'street', an impressive space off which open reception, the cafe, lecture theatres and seminar rooms. The galleries and bridges linking to the offices at upper floors and the glazed walls and doors leading to the internal courtyards enliven the space. This public wing is used for teaching but also provides a venue for sponsored conferences, training events and trade fairs, often international ones. Off the north end of the street are clustered seminar and meeting rooms (including one set up as a boardroom for video-conferencing) and surprising, internal cupboards for supervisions of three to four people. Not only do they have no view out, they do not appear to have a vent for extract.

Behind this front wing, three fingers of double banked offices lead to the rear run of offices, forming two internal courtyards, the larger one with decking, gravel and planted with trees. Construction is of cavity masonry, with cavity fill, the outer brick leaf exposed on the outward facing facades and rendered white on the two internal courtyard facades, to increase the feeling of light. Floor plates are 12m wide with clear spans which allow for flexibility. The corridors are decently wide and in the original layout had some open spaces between offices to form break-out spaces. These are now being enclosed with a glass partition incorporating venetian blinds to create meeting rooms with some acoustic and visual separation. There is a slight range of office sizes, working well for individual research and for supervisions. The second floor has larger rooms used as labs and workshops for projects such as Physics for Medicine. Where the corridor changes direction there are recreation or seating areas, one of which has been colonised by plant-lovers.

The library is tucked away on the ground floor of the north private wing, rather small but according to the librarian not yet out of shelf-space. (Interestingly computer e-books have not grown in importance as they have apparently in Medicine and Economics.)

The demarcation between public and private works well: the noisy, social tea room is located on the outside at the end of the street, doors are sound-proofed and secure and the lofty street gives a coherence to the building. The computing teaching rooms with work-stations arranged in doughnut formation have views of full-height glazing to the south and west: the wide roof overhang provides shading from the sun on computer screens and anti-glare blinds are deployed.

Individual offices on south facades have very narrow timber shelves fixed externally at the head of windows to act as shading devices against low winter sun. It is unclear how effective they really are and probably the internal blinds would have been sufficient.

With such heavy computer usage the main problem was considered to be one of cooling and the strategy was for a highly insulated envelope. The air handling units (AHU) around the building are able to call for heating or cooling and this air is then blown under the suspended floor along each corridor. The set-point for the building is 21 degrees and each AHU works to adjust the flow of air in order to maintain that temperature around the building, by use of heating or cooling, whichever is required. In the street there is no cooling, only under-floor heating (gas fired) with the granite paving providing thermal mass. This wet system works well to keep such a large expanse of space as close to 21 degrees as possible.

This strategy seems to work successfully, except for the very early days when the building was not yet fully occupied and there were insufficient people and computers to generate enough heat. In the offices, the precast concrete floor planks provide exposed ceilings below with the benefits of thermal mass, with carpeted floors reducing impact noise transmission. Cooling is provided by 'chilled beams' which are actually heat exchangers above suspended ceilings: coolant is pumped round the building to each chilled beam, air is drawn over the chilled beam to be either extracted and re-mixed and returned through floor diffusers or simply diffuse from the gaps along the 'ceiling' and fall back into the room. The airflow is very slow and therefore no feeling of draughts; however although the control valves in each office for the coolant into the chilled beams works immediately there is a time lag of approximately two hours before the delivered air is perceived to be sufficiently cool. Energy recovery uses thermal wheel technology, cutting edge when built, now pretty mainstream. The BMS is controlled centrally and the facilities manager has a workstation in front of five screens with the ability to monitor and adjust temperature and humidity in the different zones of the building as well as open and shut the 16 roof vents above the street. The lecture theatres and large teaching rooms use the same chilled beam system as the offices but the chilled beams are larger. In the lecture theatres the system is linked to CO_2 monitors which control whether the heating and cooling is activated.

The facilities manager reports only two major problems: with the weathering of the timber cladding which has stained unevenly and with the roof membrane which was reduced in specification in Value Engineering.

Projected energy consumption figures were under 100kwh/m²/pa. Air-tightness targets were 3m³/hr/m² at 50Pa. Actual figures for energy use between October 2013 and 2014 were 151 kWh/m².

There are regular energy and environmental audits which indicate energy use over the last 10–15 years has been dropping. It is interesting that energy use was rising from 2001 to 2005 at which point it plateaued and then started to decline. What was the reason? 2005 was when the current facilities manager arrived, someone clearly very interested in energy use and comfort. The approach to 24 hour lighting and heating and cooling was changed with the realisation that very few people really wanted to work at odd hours so 24/7 provision is not necessary.

The approach to the design of this building was pleasingly holistic with environmental comfort and energy use a central issue. Because of the very early requirement for expansion to provide space for Microsoft, the adjacent building (the Roger Needham Building) followed on and was built almost at the same time, with the same architects and contractor, making it possible to carry over many of the lessons learnt in developing the design of the William Gates Building.

Architect, Services, Structural Engineering, Landscape, CDM: RMJM
Acoustics: Sharps Redmore
Fire engineering: Arup Fire
QS: Gardiner & Theobald
Contractor: Shepherd Construct

Microsoft Research Cambridge

Progress on the new Microsoft Research Cambridge Centre had been going very quickly; a deal was struck after fast but comprehensive negotiations whereby Microsoft (through its project manager AYH) would commission the design and construction of the building, on which it would take a 15-year lease with a 10-year break point, the building to be bought by the University's Amalgamated Fund on practical completion of construction. The cost of the building was provisionally set at approximately £12M. The rental rate was based on a satisfactory yield of 7.5 per cent applied to the total of design and construction costs, to be reviewed on an upwards-only basis at intervals of five years to a procedure that was mutually agreed. The same team that had designed and constructed the Computer Lab was taken on for this project, and that worked well for both projects. Work began on 3 July 2000 and at the start of works the managing director of Microsoft Research Ltd, Prof Roger Needham, (formerly a pro Vice-Chancellor of the University, and hugely respected by all) said

> the new building represents a significant investment by Microsoft and re-affirms Cambridge's status as a global centre of excellence for research. It will ensure we continue to recruit some of the best researchers in the world, providing them with an environment that will allow them to push forward the boundaries of computer science.[9]

The work on the 6,098 m² gross (5,000m2 assignable area) was completed on 10 September 2001, four weeks behind programme, but on its budget of £13.696M. The design and construction reflected the Computer Lab satisfactorily. There is a dignity and a feeling of efficiency, especially around the spacious entrance area. The staff of the Microsoft European Research Centre moved there from a small temporary home above the Lion Yard shopping centre, and revelled in the extra space and new facilities. Professor Ian Leslie, head of the University Computer Lab next door to the Microsoft building, said

> now we will benefit from Microsoft having their own building next door and from the greater opportunities for research this will provide. There has been much talk about our brain drain from Britain to the USA. Well, this is definitely a plug.[10]

In 2013 Microsoft moved to a larger building near Cambridge rail station. And nearer to pubs and social facilities.

Site infrastructure

Meanwhile, work had started on the infrastructure necessary to support and facilitate such extensive development of West Cambridge. Interestingly, stimulated by a government initiative for 'developing cities' in 1996, the first infrastructural project was to design and build a park-and-cycle facility: this was to provide 292 car-parking spaces, each with a secure locker, a box (a rather expensive box) where a bicycle could be kept overnight. The idea, which proved quite popular after a somewhat slow start while people hesitated about such a new concept, was that University staff would drive to the facility, park up, get their bikes out of their lockers and then cycle to their departments in the University; at the end of the working day, the procedure was reversed. This was thought to be the first park and cycle scheme in UK (apart from a small scheme in the old Cattle Market in Cambridge) and had a capacity for 450 bikes.

The design of the project was by landscape architect McQuitty Landscape Design, with Hannah Reed as engineers. The first thing to be done was an archaeological survey. National legislation had hugely increased the amount of archaeological work that had to be done prior to the start of work on site for any major development in UK. As noted in Chapter 7, only a small number of companies got onto the list recommended by local planning authorities. Fees charged by such companies were high; companies from within relevant departments at Oxford and Cambridge had secured significant places in that market.

Tenders for the survey were received, within budget, for completion by April 2000: although a huge amount of archaeological remains were discovered on the adjacent parts of the West Cambridge Site as noted above, little of interest was found on the park and cycle site.

One design issue was about lighting. Although it had been a very long time since the general means of getting departmental research observations had been from their telescopes, some longer-established members of the Astronomy Department were concerned about light pollution: a little research did still use the telescopes and they were popular with amateurs. After a lot of discussion, a compromise was found, and a more efficient and less polluting downward-facing high-tech form of lighting was agreed.

Work on site was started by contractors May Gurney on 7 August 2000, and completed at £1.21M, a bit below budget; the official opening, with the local MP, was carried out on 26 April 2001. There was much good-natured banter as a large gaggle of mostly rather unsteady cyclists rode off towards the cameras, led by a proctor in his robes and mortar board, and a happy-faced Vice-Chancellor. The police were also happy: the project was given a Secured Car-parks Award by the Association of Chief Police Officers, Chief Superintendent John Fuller saying, shortly before moving out of the way of the advancing cyclists,

> I'm extremely pleased to present this Secured Car-parks award to this commendable parking facility on behalf of the Cambridgeshire Constabulary and the Automobile Association. This is in fact the third such awarded to the University of Cambridge in less than two years, and is a further confirmation that a primary consideration for the University is the well-being and security of their staff and property.[11]

The sun shone, everyone was given a green Park & Cycle T-shirt, and the event ended with no one being run over.

With the first two buildings and the park-and-cycle park in place, the general infrastructure was being developed to cope with other new buildings on the site. In addition to the infrastructure that had been constructed in association with the computer laboratory project, and the park and-cycle-project, design and costing of the first phase of the main elements of infrastructure started in 1997. With a budget of £5.29M, those works included:

- Roadworks connecting the site onto the adjacent main road
- Off-site drainage, mainly on the West Cambridge sewer

- Internal site services including surface water drainage
- Modelling, site leveling and access roads within the site to the building areas.

Starting from September 1999, those works were put out to competitive tender, and work on site began later that autumn. The biggest single element of the first phase infrastructure works was the off-site foul water sewer. This ran from the West Cambridge Site, along the main Madingley Road, past Magdelene College and onwards along Chesterton Road to a main collection point. At the bottom of Castle Hill, alongside the College, excavation came across a substantial Saxon burial site, which had a lot of bones and a hoard of coins. Work stopped while archaeologists studied what was found. After much discussion lasting some months about the hugely complex issues of ownership, work continued, an attractive monument being placed on top of the burial site at the main road junction by Magdelene College.

In the negotiations with the water company, Anglian Water Services Ltd, it had been agreed that the University would pay two-thirds of the total cost of the works: that ended up at £2.773M, £427k under budget. Work on the sewer started in August 2000, and was completed in May 2001. This was the first University project to have a 'pain/gain share' incentive built into the contract, with the contractor sharing cost savings brought by good design and/ or construction practice, and, conversely, a sharing of cost-over-runs. Generally, the incentive was found successful for this sort of project but there are pitfalls with more complex projects, not least about who the money actually goes to on the supply-side: company or individuals?

The main road into the site, named after J. J. Thomson who first isolated the electron, was constructed by the company May Gurney, just £17k under budget, between 3 July 2000 and 30 April 2001, at a budget of £1.328M. The total cost of the Phase 1 infrastructure works, other than the park-and-cycle site, and the works associated with the computer lab, was £7.881M; including those, £10.881M in all. Around this time, a temporary catering outlet, known as 'Wests' was built (at a cost of £350k) near the main entrance to the site; it was popular, and offered some, appreciated, parking.

Phase 2 infrastructure works, costed at £18.6M, were designed to provide infrastructure for buildings for the Materials Science and Metallurgy Department, various engineering buildings, a sports centre and a West Forum. In addition, there was infrastructure to be developed for the East Forum, which was research and residential accommodation as described below: this infrastructure included the East Forum pond and canal (£554k), a footbridge and cycle ramp access (£507k), a Plaza (£423k), the Northern approach road (£423k): total, £3.82M costed into the Residences and Nursery project budget.

The PV colonnade

The most technically advanced, and in many ways the most exciting project for the West Cambridge development never happened. It was to build a colonnade parallel with the canal along the south side of the site, marking the boundary between the site and the green belt. The colonnade was to be some 400m long and 9m high, giving cover for the major pedestrian and cycle routes. There were quite a few people in the University who never much liked the concept of the colonnade, partly because of its considerable size (it had to be physically significant in impact as it represented the boundary with the green belt, and could be seen from afar), and because the height of the colonnade meant that when there was wind and rain, people walking underneath the colonnade would get wet. The master planner remained reluctant to address this scale issue, and that increased the resistance to the concept.

However, a cunning plan was developed to fit photo-voltaic cells all along the roof of this colonnade, the electricity generated being used to split water to produce hydrogen for fuel-cell powered buses. The installation would have been the largest PV installation in the UK at that time, the first PV/hydrogen bus service, and the project would have provided the University not only with the potential for research in an area which was rapidly becoming extremely important internationally, but also it would also have been a high-profile statement on renewable energy use and the hydrogen economy.

An application was submitted for EU funding on 15 March 2001. The bid was favourably reviewed by the European Commission, achieving a technical score of 88 per cent, the highest score awarded at that time to any project submitted to the Commission under the energy programme (FP5 Energie Programme). The EU saw this project as supplementary to another which had three buses in each of ten European cities running on hydrogen: that programme wasn't based on PVs, and had some shortcomings which the West Cambridge project was seen as addressing. In September 2001 the University authorised the selection of a colonnade designer by way of a limited design competition, a means of procurement normally avoided by the University, but justified in this case when time was becoming tight. Three invited entrants were each paid £6k towards costs.

It was very difficult at that stage to estimate the cost of the project. Of the total project cost of £3.28M, €2.2M (then £1.38M) would be funded by an EU grant; the preliminary estimate of what the University might have to pay was £1.9M. The overall project consultant from the start was the exciting practice Whitby Bird, an innovative and far-sighted practice which had long been very supportive of this project, and the architect Marks Barfield was appointed to do the design for the colonnade. Further promises of funding were achieved, in part from the hypothecated landfill-tax fund (the only hypothecated tax then in UK) whose proceeds were required to go towards innovative energy supply or waste reduction projects. The EU funding was time-limited to 42 months from the date of the grant being approved. Very difficult technical and commercial discussions to make provision for three buses to be run on the hydrogen fuel cells continued, and much technical development of the photovoltaic system and hydrogen production. Although a great deal of progress was made on all of these, with much support for the project coming from all directions nationally, and a very good design presented by Marks Barfield, the project nevertheless ran into difficulties. Firstly, and perhaps in hindsight the most significant, was the timing. The decision for the University to make up the relatively modest funding deficit came at a time when the University was spending £0.5M per working day on buildings (including their maintenance and utilities). The rate of capital expenditure was unprecedented in University history and there was an increasing concern, pointed out not least by the Estates Department, that if the current rate of building construction continued, then at some point in the foreseeable future, estate running costs would become unacceptably high in the overall university budget. Also, procurement staff were at full stretch. For the first time in this expansion programme there was, therefore, a degree of nervousness among key people in the University hierarchy about the pace of capital development. Secondly, and more prosaically, the lead time for ordering the buses turned out to be two years; that seemed extremely long, but it was the fact of the matter. The timing hence proved to be critical with all the technical, managerial and financial decisions to be made in order to ensure funding drawdown in sufficient time within the EU 42-month deadline. Discussions with the EU department administering the promised grant were extraordinarily frustrating and unproductive. Although the relevant EU committee had noted that this was the best proposal they had had that year, the system nevertheless did not allow for any extension in time for the project to proceed before the offer of the grant was withdrawn. There was a considerable bureaucracy within the department for advertising, assessing and judging the grants, and its director clearly wanted to maintain that turnover rather than lengthen deadlines of good projects: jobs might be at stake if the flow was slowed down by good projects rolling over. The Director of Estates, who had had previous experience of dealing at that level in Brussels, went to meet with a senior official, but he achieved little, just a very minor extension of a few weeks.

Alternative plans were drawn up whereby to save development and fund-raising time, the PVs could be mounted on the ground or on top of existing building roofs, but important elements of funding were by the DTI which needed an integrated scheme, and the Carbon Trust, which administered the funding from the landfill tax, regarded the colonnade as an essential part of the scheme. In the context of the unprecedented rate of capital procurement for buildings, the intransigence of the EU in refusing to significantly extend the deadline, and the concern about the University finding its share of the cost, it was decided in 2003, with huge reluctance and to the acute disappointment of the many who had been technically involved in the project, to abort the project.

Disappointing from a technological point of view, and certainly disappointing to those many who, nationally as well as internally, had been involved, it was also disappointing to technical commentators (there had been a great deal of supportive media coverage, both in technical journals and indeed in broadsheets).

Cancellation of the PV colonnade project did not actually much affect the progress of development of the West Cambridge Site. Other, much bigger, and to most people, far more significant projects were going ahead at considerable speed.

Tech-transfer and incubator buildings, and residential accommodation

The development of residential accommodation for post-doctoral and other members of staff of the University, with a crèche, some shops and a large amount of built space for spin-out companies and other entrepreneurial research activities was a central part of the development, in some ways the 'heart and stomach' of the master plan. Provision for these functions originally were all lumped together in a project with a budget of some £30M. Negotiations with a potential benefactor, the giant figure of the Cambridge Phenomenon, Hermann Hauser, for facilities for entrepreneurial activities including spin-out companies continued for some years; these often went well, but for a few academics there was an issue relating to the effect that there might be on those who benefited from profitable research done in their University-provided labs by University or externally funded staff, the issue of sharing the value of intellectual property rights (IPR). This cannot be discussed here in any detail. There were matters about declaration of interest. There was never any suggestion at all of any graft or corruption, just the defence of favourable intellectual property arrangements. The situation regarding long-overdue reform of such IPR matters was a factor in the way that the University was managed, and it was steadily and successfully reformed, bringing Cambridge more into line with IPR rules in other universities

East Square

Largely as a consequence, a decision was made in 2001 to separate out the different functions into two projects: one project known as the East Square would have the buildings for the entrepreneurial and spin-out activities, and a second, separate project would provide residential accommodation, nursery and shops. For the first project, known as East Forum and Plaza, a budget for a building of some 7,000 m² gross area was set in 2002 at £25.5M. In March 2002, the matter was discussed by the University's Finance Committee within the context set out in the previous paragraph; there was stated a concern about the requirement for a proposed level of rent needed to support the likely financial position. Hermann Hauser, one of Nature's enthusiasts, made a formal and generous offer of support and it was at that time thought that proposed benefactions might total £8M with a further £4M being underwritten. A formal meeting was arranged for the signing of a cheque for £12M on 23 April 2002. However, as was noted at the Buildings Committee (BC 02/43),

> difficulties concerning use of the building had just arisen and this meeting has had to be postponed. No future date has yet been arranged for the cheque signing and it is not known at this stage how long the postponement will be.

The project was effectively delayed for several years for the reasons set out above. A design was done by Richard McCormack for a single, large building which would have been too expensive.

Tactics changed. A report went to the University from University Council in May 2004 noting a significant change to the basis of the project from one managed by the University to an outsourced form of development whereby the University would grant a lease, probably around 125 years, to a commercial developer who would at their own risk fund, design and arrange construction for a building that would provide a total of 2,418m² of gross space for Cambridge Enterprise, 3,170m² gross of lettable space for incubator units, plus catering and other space with a total gross internal area of 6,850m² gross in a 'Broers building' on one side of the open square joining the main site at the bottom of J. J. Thompson Avenue to the southern edge of the site at the foot/cycle path which bounds the green belt to the south. On the other

Figure 10.3　Aerial view of (left to right) Physics for Medicine Building, Hauser Forum and Broers Building and Residences

By courtesy of University of Cambridge

side of the square is the 8,000m² gross, £12.2M budget, Hauser Forum for Cambridge Enterprise funded mainly by a gift of £8M from the Hauser-Raspe Foundation, and £2M from the East of England Development Agency. The design of this split-building was done by Wilkinson Eyre (concept) and Dunbar SMC Covell Matthews, White Young Green and Mott McDonald. This was for two connected buildings and had a reasonable degree of sustainability, with ground source heat pumps which are effective when well maintained. The project finally started on site in October 2008, with the work carried out by the constructor Willmott Dixon, one of the last big family businesses in the construction industry, and with a high reputation for sustainable and sound building work. The building took 61 weeks to build, and was opened on 20 April 2010 by the Chancellor. It has, perhaps, a 'science-parky' appearance with brightly coloured panels, but it has excellent and popular space and facilities inside, and a fine view from the main entrance to the site. The building works well for its occupants; by mid-2013 there was a six-month waiting list for space there.

Residential and nursery accommodation

The other part of the East Square project was the provision of residential and nursery accommodation primarily for research staff. The proposal was to provide a mixture of one, two and three bedroom flats with a child-care nursery and shared facilities, including some convenience shops, within 13,830m² gross area. This was refined into having 206 flats, 62 of them with either two or three bedrooms, and 144 single bed flats. The nursery was to provide 88 places for pre-school children of University staff and grad students, with the shared facilities to provide an area of some 700 m² for the management office, caretaker's workshop, launderette, shops (such as cycle repairs, stationer, printing, a cafe and perhaps a hairdresser). The project was funded on the basis of a business plan showing that rental streams would underpin the capital cost which was set at £22.74M: the project kept within that budget. The business plan supporting the project was on the basis of an internal loan from the University, at a rate of 6 per cent, which supported both the capital cost and whole-life costing.

After much soul-searching, it was decided that the master planners, MacCormack Jamison and Prichard (MJP), should be the lead designer; MJP had expected to be designing a lot of the buildings at West Cambridge, but because of their reluctance about collaborative design and procurement with early contractor involvement, it was decided that this was to be the only project thus far in which they should have a design role. Of course, as master planners, they were invited to make comments on the designs of all projects coming forward; their comment was useful.

The other designers on this residences and nursery project were Oscar Faber for building services and Price and Myers as structural and civil engineers. Procurement was by the methods by then standard in the University, develop and construct, whereby the designers would take the design through concept stage, and then once a planning permission was obtained a contractor was selected on the basis of quality, overheads and profit margin to join the team; then before design was finalised, the designers were novated to work for the contractor via a pre-agreed contract. MJP continued to campaign widely against this form of procurement, even although it was by then becoming increasingly popular nationally. The head of practice wrote privately to the Vice-Chancellor asking to meet him, the purpose being to persuade him to have the University revert to the single-stage form of procurement whereby the constructors were kept out of the process until quite shortly before they went on site. The Vice-Chancellor invited the Director of Estates to be present at the meeting. The outcome, after some tense exchanges, was that the University would continue to procure on a team basis.

The contractor selected on a quality-price basis was AMEC, a powerful national, now international, construction company. Work started on site on 24 February 2003 and was completed on 16 August 2004, the nursery having been delayed by 12 weeks and the residences by 14 weeks; no extensions of time had been granted to the contractor and so liquidated damages were imposed (the University stopped any use of 'ascertained' damages from 1998, and from then had just liquidated damages clauses because ascertained damages in an HE context are difficult to quantify). The final account was agreed within the original budget: the out-turn cost was £21.744, approximately £1M under budget.

The design of the single-bed flats was somewhat bleak: one member of the Buildings Committee unkindly compared them to a Stalinist housing complex. There was a hope to get away from the unattractiveness and maintenance problems of flat roofs, and start establishing pitched roofs for the West Cambridge development with sloping roofs between 3° and 6° for the residential buildings; however this suggestion came forward a bit late, and for reasons of cost and compliance with the master plan design guide, the decision was made in October 2002 to keep to the design of a pitched roof just to between 1° and 2° slope.

One building has Oxford blue/Cambridge blue coloured paneling which, with the tree landscaping filling out, has become increasingly acceptable, and certainly the internal design of the flats is extremely popular: very quickly a substantial waiting list for renting built up. The design of the two and three bedroom flats on the other side of the parking square is much softer in appearance, timbered, and generally well appreciated from the start. Again, these flats were hugely attractive to tenants. The nursery had some teething problems with the increased requirements for health and safety, and some issues which had to be overcome. However, overall, like the rest of this project, it was after a while seen as a considerable success. The only significant failure in this project was that the area provided for the shops, for internal managerial reasons, was never taken up, although in other universities such space is snapped up for convenience shopping to the benefit of all. That failure to provide shops and facilities was a significant error.

Nanoscience Centre

In April 2001, a JIF bid was submitted for new buildings for an interdisciplinary research centre for nano–fabrication. The bid was for £7.1M, and the project would provide 1,750 m² of assignable space: that included 675 m² of clean rooms, 695 m² of laboratory space and 780 m² of office space. A leading reason for the success of this JIF bid was the brilliance of the scientists working in that department in the University, particularly Professors Sir Richard Friend and Sir Mark Welland.

The multidisciplinary practice Building Design Partnership (BDP) was appointed to develop the designs. They came up with a particularly efficient and elegant set of offices behind an embankment at the eastern end of the site, tucked away behind the Physics Department, and a particularly efficient and ugly clean-room and laboratory building alongside. Work by Shepherd Construction (who had done a good job in constructing the new Computer Lab and Microsoft European Research Headquarters building) started on 2 January 2002. Completion of the works was contracted to be 2 November 2002, but due to delays in design information, progress slipped and construction was completed, exactly on budget, on 2 December 2002. Both buildings were completed to a good standard, and have worked very well in operation.

After this project there was a refurbishment of 350m² of the microelectronics area of Physics at a cost of £1.124M.

Marconi Centre/CAPE

The next project to be discussed was another project that never happened, or rather, it never happened in anything like the way it was envisaged. During 1999, there were extensive and deep discussions between senior members of the University and the company Marconi plc: the outcome was that Marconi committed £40M to fund the construction of a 'Marconi Communication Centre' with an extensive programme of research into communication technologies. The site was to be strategically placed between the new Computer Lab and the Nanofabrication Centre. Marconi made it clear that they had strong views regarding the building procurement process and wanted to lead that, or at least to work closely with the estate management department. After extensive and constructive discussions with Marconi, they agreed that a two-stage procurement process with early involvement of the contractor would be best, and that procurement should be led by the University's estate management department: the latter decision was in part dictated by the tax situation in that the University as a charitable organisation was in a better position than was Marconi.

The design team, selected in the autumn of 2000, on a quality/price basis, was T. P. Bennett, Hoare Lee and Oscar Faber, with AYH as external project manager and quantity surveyor. That team set about designing a building of some 4,100m² gross floor area, mostly generic space similar to that which was being provided in the nearby Computer Lab (the William Gates building). By March 2001 the designers had prepared an outline design. As per normal procedure in the public sector, the invitation for contractors, as for designers, was advertised in the European Union Journal, *OJEU*; there was a very large response as usual (sometimes there were over 100 applicants through *OJEU*) and it took ages to sift through them. (Most of the sifting was done by project quantity surveyors; some were easier to sift out than others: in this case, one applicant for this West Cambridge appointment was what seemed to be the owner of a fish and chips shop in Mumbai who appeared to have software that made applications to all *OJEU* adverts.) Five contractors (not including him) were interviewed and assessed, and Shepherd Construction was selected on the usual quality/cost balance basis.

Negotiations continued concerning the leasing arrangements that would be made with Marconi, including the rent that would be paid: it was hoped for a similar deal to that which had been achieved with Microsoft.

On 19 September 2001 the project was given planning permission by Cambridge City Council. As the design, planning and lease negotiations all continued apace through the earlier part of 2000, it gradually became apparent that Marconi as a company was under great stress. The danger sign was that Marconi was agreeing to just about everything being asked, so it began to be thought that something was 'Rotten in the State of Denmark'. Then, quite suddenly, there was major restructuring in the company, with news of considerable overspending. It became clear that the re-organised company was not going to be able to proceed with the Cambridge project. After a lot of pressure, the restructured company agreed to reimburse the University for most of the expenditure that it had incurred on designing the building and planning the project.

Despite this set-back, a revised project went ahead. By April 2002 all works packages had been negotiated with a tender report defining an expected contract sum of £14M. As that sum clearly was not going to be forthcoming from Marconi, it was then proposed to re-plan the project to provide a building for use as a Centre for Advanced Photonics and Electronics (CAPE), a division of the Engineering Department. Happily, the building as it had been designed suited the purposes of CAPE remarkably well if the available space could be increased, maybe by addition of a third floor. At that point the expected cost of the building was estimated at around £16.5M if a start on site could be achieved by the third quarter of 2002: inflation was beginning to run high. (As set out in the previous chapter, for many years, around this time and southern England especially, inflation in the construction industry was running consistently well above RPI and CPI nationally; HM Treasury carried out an internal study in 2006 as to why that was.)

Through 2003, design progressed well; the building would provide an area of 4,675m² gross and net (assignable) area of 3,850m² assignable. For this revised project, the engineers Hoare Lea were succeeded by services engineers

Zisman Bowyer and Partners, and the contractor Marriot was awarded the first and then the second stage contracts. A new planning application was submitted and approved. Work on site started on 15 November 2004 and completion was on 5 December 2005. Over the earlier troubled years of the project, the budget had been increased from £14.3M to £17.2M, brought back to £14.0M, and then up to £14.625M, £3.8k/m² assignable which was quite good. The funding was mainly £12M from SRIF 2, and £2.25M arranged by the Engineering Department with the research arms of industrial companies which would occupy parts of the building. £1.45 million was required to fund large and very sophisticated research lab clean rooms. Procurement of those was extraordinary. The Department had become aware of a company in the south of France which had gone under and had a set of clean rooms which could be for sale. These were inspected, and bought. Reference to this was made at the ground-breaking ceremony at the start of construction: acquisition of the clean rooms was described as rather like buying a set of second-hand clean rooms from eBay on a wet Sunday afternoon. There was considerable risk in this procurement, notably how well or badly it would fit with the space and service provision in the building once it pitched up on its trucks in Cambridge. It could have gone horribly wrong, but due to the good design, particularly by the services engineers, with early, professional involvement of the services sub-contractor CRC, and close management of the project, all was well.

Figure 10.4   Centre for Advanced Photonics and Electronics (CAPE)

The project was handed over on 5 December 2005 to its final budget of £14.40M in the programmed 55 weeks, with the clean rooms being handed over two weeks later.

The building is externally very attractive, with clean lines and a well-defined, indeed exciting, entrance area, complete with a water feature which simulates photons of energy with spurts of water arcing through space, coloured at night. Some of the inspiration for such water features came from US universities, most of which have excellent water features. Inside the building, the large atrium is effective, and the commercial security of offices and other research areas is efficiently achieved. The large clean room suite on the second floor works well. Control of vibration through the structure of the building was a considerable design challenge throughout the project, and resolving it would have been easier if it had been identified as such a problem earlier on; similarly an earlier resolution of radio frequency shielding would have been helpful as the main RF lab caused design and detailing problems, in part because it was located in an external corner of the building. In the end, however, both of these issues were resolved to the satisfaction of those moving into the building as noted in the post-occupancy evaluation.

Re-housing/expansion of the Physics Department

Earlier it was noted that in the 1960s the world-famous (29 Nobel prizes) Cavendish Physics Lab, then located in the New Museums Site, was expanding so fast that it had to move if it was to continue as a world-famous lab; so, under the leadership first of Professor Mott and then of Professor Pippard, the Department of Physics 'upped-sticks' to a rather isolated corner at the south-east of the West Cambridge site, completing the move in 1971. The accommodation built for them formed a huddle of buildings designed by the CLASP system, pre-fabricated building components assembled Lego-like on site. The CLASP system was popular nationally at the time, and indeed throughout its life with the Physics Department because of the considerable flexibility in being able to move internal walls quite easily. However, buildings using a CLASP system were designed before the world oil crisis of 1972 with its consequent escalation in the cost of energy. As a consequence of that timing, the buildings were, and are, extremely inefficient in terms of insulation, and therefore in terms of the costs of energy and of pollution by CO_2 emissions. The Physics buildings had also begun to leak extensively by the turn of the century, so there was the unhappy combination of a lot of heat leaving the building and water entering it. The Estates Department therefore began a study of the relative costs of continuing to run the buildings as they were compared with the cost of replacing them sequentially with new and much more efficient buildings with lower running costs and much lower carbon dioxide emissions. The first conclusion was that the unit cost of running the Physics Department buildings was almost double the average unit cost for running University buildings generally. Secondly and significantly, the whole-life assessment showed that even without entering any specific costs for CO_2 emissions (those were to follow later) in a period of about 15 years, maybe 17 years, new buildings would pay for themselves through reduced heating and maintenance costs. The existing buildings had a design life of 30 years, and had been built about 30 years earlier. Therefore a proposal was put through the Buildings Committee to the Planning and Resources Committee of the University that a programme of building replacement for Physics buildings should be worked up in more detail, and that was agreed.

The first matter was to decide how much space the Department would need in the new buildings over succeeding years. The normal space analysis, based on the University estate plan expansion figures of undergraduates at a maximum of 0.5 per cent, graduates at 2 per cent to 2.5 per cent, and research staff planned at 5 per cent but up to a maximum of 11 per cent annually was carried out. The answer was that the departmental needs would be reasonably accommodated over the years of continued expansion by about the same amount of space as it presently had; that was because of currently low space occupancy, even by the standards of universities at the time.[12] Although there was overcrowding among graduate student areas, many of the academic staff had an amount of space considerably greater than the University space norms would suggest, some of it for storing equipment used for their PhD work some years or decades earlier.[13] The Department denied that there was under-use of space. An offer of an independent review of space-utilisation was accepted by the Department and carried out, and the result confirmed current space total was a satisfactory base-line for setting the capacity for the new buildings. Once this was agreed by all, design work could start. And it did.

The architect Building Design Partnership (BDP) was appointed to do a feasibility study for the sequential re-housing of the Department into new buildings and the results of that went to the Planning and Resources Committee on 19 March 2003, and were approved in principle. The indicative cost was £137M. The Department of Physics was encouraged to develop a full business case, as was by then the standard procedure (see Chapter 5) and to start identifying sources of funding. It was proposed around this time by the Estates Department that because of the very high rate of inflation in the construction industry, it would in fact be cheaper and certainly much more efficient for the University to raise a bond to cover such costs and then to proceed with design and construction, rather than wait some years for fundraising and have to pay a much higher construction price. Although many supported that case, it was not implemented until the later development of NW Cambridge.

After much design and even more discussion, most of it very constructive, it was agreed that the first phase of the rebuilding of the Department of Physics should be to provide a new building for the function of 'physics for medicine'. This would be essentially multidisciplinary with physicists, chemists and medics. The site for this building being at the southern end of the main access route, JJ Thompson Avenue, it would be a building that people arriving would see prominently and therefore it had to be visually attractive, as well as being efficient for its users and built to high sustainability standard.

Using the normal processes of appointment on the basis of quality and cost balance, a design team of architect BDP, K. J. Tait as mechanical and electrical services and Whitby Bird as civil and structural engineer was appointed. Design for a building of nearly 6,000m² assignable area with a budget of £25.7M continued into 2005, but by the summer of that year it was becoming apparent that there would be insufficient funding, and therefore the scale of the building was reduced to 4,434m² assignable area with a budget of £20.56M. By the summer of 2006 a £10M funding bid to Science Research Infrastructure Fund (SRIF 3) was secured, plus £2.5M from the Wolfson Foundation. The project budget was then set at £12.5M for a reduced area of 1,348m² assignable, 2,358m² gross area. The unit cost had gone from £3,543 to £5,301/m² gross as building size reduced and design developed. The contractor Willmott Dixon was selected for stage one, pre-construction services. The stage D, now Stage 4, report was issued on 11 August 2006; the planning application, however, was delayed by a new local plan requirement that 10 per cent minimum of energy for new buildings should be generated by renewable sources. This caused some delay in design, and added some £200k to the project capital cost, while reducing whole-life costs with an acceptable pay-back period. The planning application was submitted on 7 September 2006. Work started on site, on programme, in May 2007 and despite a number of construction problems, finished on time after 59 weeks of work on 23 June 2008, at a construction contract cost of £9.5M.

Construction had started with provisional sums amounting to less than 2 per cent of the total contract sum, remarkably low for a complex building, and the team's work-ethos continued throughout construction; the final account was agreed with the contractor within two weeks of practical completion, most unusual in the generally contentious UK construction industry, beating even the third phase of the CMS (Maths) project.

There were few post-completion problems, and those were mostly fixed quickly. One issue on a project for multidisciplinary use is that it is more difficult to identify and appoint a single representative user, which time and again had been found to be an extremely important ingredient of success in a project given the advanced form of democracy common in universities, especially in ancient universities. In this project, however, this issue did not really become critical, although as the post-occupancy evaluation notes at paragraph 10.3,

> there had been some concern during the design development stage that no appointment had been made in respect of the academic leader of the facility, and so it was not possible to take a specific brief for the scientific requirements. Thus, the design was for generic laboratories, but in subsequent occupation there have been no requests from the users for any modifications in the building, demonstrating the users' satisfaction with the layout and facilities provided. Users have been successful in attracting research grants in respect of the scientific work which is undertaken in the building.[14]

After completion, underspend from the construction contract was used to meet the renewable energy obligations intrinsic in the planning permission. Generally, once the building was fully occupied, both gas and electricity consumption were higher than predicted during design and ten-year costing although the building secured a BREEAM rating of 'Very Good'. There was a gas leak in 2010 which may have skewed the numbers, and the overage in electricity use is almost certainly due to greater 'plug-load' for research equipments rather than the servicing of the building; for such a research building one might expect energy use through plug-load to be 2–3 times as much as the energy needed to service the building.

Looking back on the project, there was a considerable management challenge with the hoped-for funding, and hence the consequent budget and building size rising and then having to fall again. Despite all that, a very satisfactory building was produced, although at a high unit capital cost.

The project *per se* was clearly successful. A more Olympian view of the project, however, might recall that the original idea was to build new premises for the Department to allow it to decant from one of the existing buildings into this new building, thereby allowing, sequentially, each building in turn to be demolished with a new building replacing it, so that finally the entire department would be re-housed in new and much more efficient accommodation. What happened in practice was that that aim was lost/changed along the way, so that the new building was used for expansion of the world-class research activity, while the rest of the Department remained in its existing outdated accommodation, beyond its design life and in time needing decanting space to allow for the replacement of the old buildings to take place as some day they must be; at the time of writing, that process has re-started. Balancing short/medium-term needs, all of which are entirely commendable and pressing in themselves, with longer term estates needs was always difficult.

Further infrastructure

Another phase of infrastructure was constructed, completing late in 2008. That included the demolition of a cooling tower[15] and removal of the adjacent car-parks so as to construct the haul road for the Hauser Forum project and the extension to Charles Babbage Road and associated car-parks. There was also a review of signage throughout the site and the removal of the temporary catering facility, and replacement of trees long dead. All that work was carried out by Jackson Civil Engineering, within the budget of £6.42M.

Institute for Manufacturing

Several projects at West Cambridge were conceived and to various degrees planned and designed before 2006 but built after the decade being studied. A project that took a bit over ten years to achieve was the expansion and relocation of the Institute for Manufacturing, part of the Department of

Engineering. For many years some of the funding, £5M, was promised by the Gatsby Charitable Foundation. The cost-estimate was £12M, and further support came including a benefaction from Dr Alan Reece. As the activities continued to expand within the Institute, with more teaching, more research and wider connections with industry, space in its Mill Lane home became ever tighter and unsuitable. A plot to the west of the residences and nursery site was available. A good design was worked up by the design team of Arup Associates, a multidiscipline consultancy providing architectural, mechanical and electrical building services, and structural and civil engineering. On 20 October 2004, PRC accepted the business case under the Capital Procurement Procedure (see Chapter 4). The develop-and-construct contractor was Kier/ Marriot Construction. The design celebrated the concept of the courtyard, which by then had been firmly restored in the University's architecture whenever possible, partly as a good way to provide outside area for meeting and, more particularly here, to facilitate natural ventilation; the building was designed to BREEAM Excellent standard, especially important for a building with labs. The wide and open spaces within the building, with a very large central area for all staff and students for meeting/discussion and catering, was redolent of that which had been shown to be so successful in the Centre for Mathematical Sciences project, and worked every bit as well. The bright colour of the exterior walls was certainly stimulating, even if the particular shade was not to the entire liking of all passers-by. The size of the building is 4,400m² gross area; the total project budget was £15.1M: 3,080/m² assignable space so £3,432/m² gross area, £4,902/m² assignable space. Work started on site on 21 January 2008 and 14 months later was completed, within its budget.

Materials and Metallurgy Department Building

The West Cambridge project with an even longer gestation period was that for Material Science and Metallurgy. A bid to the first round of the Joint Infrastructure Fund was worked up during 1998, and was on a list prioritised by the Resource Management Committee which comprised the six chairs of the Councils of the Schools, chaired by the Vice-Chancellor. The bid for Materials Science and Metallurgy, then priced at £40M, was prioritised as the top project immediately below the line that was drawn. It was agreed that should there be further rounds of JIF, and there were, this project should be placed above the line: however, by the time of the next round of JIF, priorities had shifted and funding from that source was not secured.

The design team of NBBJ architects and Ramboll as engineers was appointed, with contractors Willmott Dixon appointed on a 'develop and construct' procurement basis. The large building, 10,600m² gross, officially opened in October 2013, for a construction budget of £48M, was designed to extremely high standards, particularly in relation to prevention of any vibration in certain parts of the building, necessary because of the sensitivity of the experimental work by the department; some labs are very sensitive to any vibration, others house experiments which generate very considerable vibration. The foundations were therefore constructed with two metres of concrete. The building works well for its users, and the external appearance is impressive in a solid sort of way, with strict lines enhanced by attractive recessed indented brick design representing a crystalline structure, and in the evening light the building glows, like the brickwork of Robinson College.

Sports centre

From the very first master planning for the West Cambridge Site in 1996, it was envisaged that there would be a major sports facilities building. This would have a gymnasium, various indoor sports facilities such as tennis, and a full-size international-standard swimming pool. The colleges all provide sports facilities, to varying extents but because the number of Colleges hasn't increased as student and staff numbers have, an increasing proportion of University post-docs and research staff and teaching staff were without any college connection, particularly in the sciences with its immediate pressures to 'publish or perish', and to avoid time-consuming teaching commitments with a college.

In 1999, the consulting designer Arup Associates was appointed to do a feasibility study for the sports centre, and £60k was budgeted for that. They came up with a very impressive design with a striking appearance and high energy efficiency. It was thought that a combined heat and power plant (CHP), which is highly efficient in large-scale use and has good sustainability credentials, could be based at the sports centre and provide energy not only for that but also for much of the rest of the West Cambridge Site.

Given the clear need for such facilities at University level, and the excellence of the design, it was surprising that there was little response to appeals for funding. Support from the colleges was disappointing, and some in the University felt that a high priority was not given to the fundraising campaign for this sports centre. Whether that feeling was or was not justified, insufficient promise of funding came forward, and so the project was put on hold indefinitely.

One possibility that emerged soon after the turn of the century was a proposal to include an ice hockey facility, and there was hope that this might bring some funding; this hope however did not sufficiently transpire.

Outline Planning Permission for the sports centre was obtained in April 2002, and there was an application for renewal in March 2007; that was granted in June 2007, and some £350k was allocated to take the scheme to detailed planning application stage. The project budget set in 2006 was £44.8M. The proposal to include a swimming pool was dropped because of cost, and a review was carried out on the sports hall and the extensive tennis hall. The area of the sports centre was reduced to 14,500m². Enough funding came in to allow work to start by contractor SDC Builders Ltd in 2012; it was finished in July 2013. The facilities include a vast sports hall, housing two full-size courts for basketball, badminton, netball and other court games. The outturn cost was £16M. The appearance of the building was as striking as the original concept, and as encouragingly open in showing to passers-by activities going on inside.

Overview of the West Cambridge development 1996–2006

Taking an overview, by historic standards of the University of Cambridge, and certainly by the standards of other established universities, the development of the West Cambridge site was considerable and certainly greater than was envisaged when the site was first mooted: £145M of buildings and infrastructure were managed, designed, built and commissioned at West Cambridge during the decade of this book, and a further £91M were planned and largely or partly designed by 2006 and built by the end of 2013.

The master plan for the site has drawn criticism for having the appearance more of a science park than what would be expected as part of the University of Cambridge itself, and the persistent absence of support facilities such as convenience shops and bank facilities has been a disappointment and an irritation. (A disgruntled academic said that the support facilities were so poor that he was surprised that the Greater Crested Newts were happy to stay on the site.) However, more importantly to the academic staff, students and those from industry working in the buildings, the operational capabilities of the individual buildings has been deemed excellent. Of course, some things could have been done better, and some lessons learned, are set out in the final chapter of this book.

Projects covered in Chapter 10

	£M
Astronomy Hoyle	0.9
	2.8
Athletics flood-lighting – fees and design	0.3
BPI	2.5
Master plan for West Cambridge	1.7
Vet School	0.8
Clin Vet diagnostic	2.2
Post Mortem	2.1
Farm Animal Unit	1.7
Computer Lab	21.8
Fit-out of top floor, incl equipments	1.1
Microsoft European Research HQ	13.7
Park and Cycle	1.2
Residences and nursery	21.7
Nanoscience	7.1
Physics Microelectronics refurb	1.1
CAPE	14.4

East Forum: Broers/Hauser	12.2
Phase I, Infrastructure	10.9
incl Catering –'Wests'/Dairy unit demolition £531k	
Phase 2 Infrastructure	18.6
Phase 3 Infrastructure	6.4
Manuf Eng	15.1
Physics for Medicine	12.5
Materials and Metallurgy construction cost	48
Sports Centre	16.0
Total	237

Aborted

Colonnade: £3.3M
Dept of Earth Sciences and Geography: £84.0M
Total: £87.3M

Notes

1 Reporter 26.4.71 No. 4749, Vol.101, No.25, 28.4.71, p.668.
2 Reporter No. 4884. Vol.105, No.13, 17.12.74, pp.54–580.
3 One of John Browne's houses was in adjacent Madingley Road, Cambridge; when he took the title of Lord Browne of Madingley, a wag augmented it to 'Lord Browne of Madingley Road'.
4 See Hansard 7 March 1987.
5 University Newsletter, June/July 1999, p.4.
6 University Newsletter, Summer 1998. Later the Gates Foundation was to donate $210M for scholarships, said then to be the biggest donation ever made to a European university.
7 Of more concern was that in one of those houses lived the City councillor who chaired an environment committee and had views on getting design changes to the planning permission master plan, for example, pitched roofs rather than uniformly flat roofs.
8 University Newsletter, April/May 2000, p.6.
9 *Cambridge Evening News*, 13 April 2000.
10 *Cambridge Evening News*, 13 April 2000. Twelve years after taking on the lease, Microsoft's European research headquarters moved to a still larger building of over 8,000m², built by Wates as contstructors, on the CB1 site adjacent to Cambridge rail station, a mixed development with a total area of some 1,500,000 m² of floor space.
11 University of Cambridge News Release, 23 April 2001, p.1.
12 A concerted HE-wide campaign from the early 1990s had led to quite a lot of improvement in university space management, especially in the 'new' universities, but that was from a low base, and the pattern of undergraduate teaching over three relatively short terms. In the USA many urban universities use their teaching buildings till 10pm, and with much longer terms.
13 The matter came to a head at a meeting of the West Cambridge Development Committee when the Head of Department was strongly maintaining that space was being used very efficiently; the estates department representative noted that on a recent walk around he had observed that one professor had three different rooms, in addition to the room that he had in his College. When the Head of Department explained that that was necessary because of the different functions of that Professor, the reply was that in fact this professor had been dead for some months.
14 Poe Physics for Medicine EM.
15 During these works, a chemical got back-flushed killing the fish in the adjacent lake: duly to the authorities, some wrists got slapped.

Table 10.1   Overview of capital projects at West Cambridge

Chapter 10-- West Cambridge	Start	Finish	m2 gross	m2 ass	Architect	Svcs engr	Struct engr	Contractor	£M
West Cambridge Masterplan	1997	1998			MJP				1.67
Athletics Ground Lighting	1997	1999			EMBS				0.29
Astronomy--Hoyle	1997	Nov-99		180 seats	Duncan Annand	Roger Parker Asscts	Hannah Reed	Rattee &Kett	0.91
Astronomy--Hoyle	Mar-01	Mar-02	1,227	750	Duncan Annand	Roger Parker Asscts	Hannah Reed	Sindall	2.78
BP Institute	01-Nov-99	07-Mar-01	1,156	910	Cowper GriffithsMax Fordham		Hannah Reed	Rattee & Kett	2.48
Comp Sci Dept--William Gates	07/02/00	04-Oct-01	10,195	6,700	RMJM	RMJM	RMJM	Shepherd Constr	21.74
William Gates top floor fit-out	01/07/02	01-Nov-02			EMBS				0.79
Microsoft Research HQ	01/07/00	Sep-01	6,098	5,000	RMJM	RMJM		Shepherd Constr	13.70
West Cambridge Park and Cycle	Aug-00	Mar-01	300 spaces		MJP/ McQuitty		Hannah Reed		1.21
Site Infrastructure					MJP/ McQuitty		Hannah Reed		32.15
Infra--West Cambridge sewer	01/08/00	May-01					Hannah Reed	Anglia Water Svcs	2.77
Coton Footpath							Hannah Reed	May Gurney	0.47
West's Catering Dairy demolition	Sep-01	Feb-02							0.51
NanoScience	01/02/02	01/12/02		1,750	BDP	BDP	BDP	Shepherd Constr	7.09
Marconi	aborted				T P Bennett	Hoare Lee	Oscar Faber		
Centre for Photonics & Electronics	03/03/04	31/05/05	4,675	3,850	T P Bennett	Zeisman Bowyer	Faber Maunsell	Kier Marriott	14.39
Residences/ Nursery	24/02/03	01/10/04	62 2/3 bed	114 single	MJP	Oscar Faber	Price Myers	Amec	21.74
Vet School-Clin Vet Diagnostic	Jan-02	Jul-02	830		EMBS			R G Carter Thetford	0.83
--Post Post Mortem	May-99	Apr-00	1,720		EMBS			R G Carter Thetford	2.11
--Farm Animal Unit	Aug-02	Aug-03	805		SaundersBoston			R G Carter Thetford	1.72
Clin Vet Med			805	360	IBS		Hannah Reed		2.21
Physics Microelectronics refurb	01/09/04		350						1.12
Institute for Manu Eng	21/01/08	15/03/09	4,400	3,080	Arup Assoc	Arup Assoc	Arup Assoc	Kier Marriott	15.09
Broers/Hauser	2009	2010			Wilkinson Eyre	WYG	Mott Macdonald	Willmott Dixon	12.15
Materials and Metallurgy	From 1998	Sep-13	10,600		NBBJ	Ramboll	Ramboll	Willmott Dixon	constr 48.00
Physics for Medicine			5,512	4,434	BDP			Willmott Dixon	12.49
Sports Centre	2012	July 2013	14,500		Arup Assoc	Arup Assoc	Arup Assoc	SDC	16.01
Relocation of Earth Sciences/Geol	Aborted								84.00
Colonnade, aborted.	Aborted				EMBS	Marks Barfield	Whitby Bird		3.28
Cost of completed work 1996 to 2006									236.42

11 North West Cambridge – the planning stages

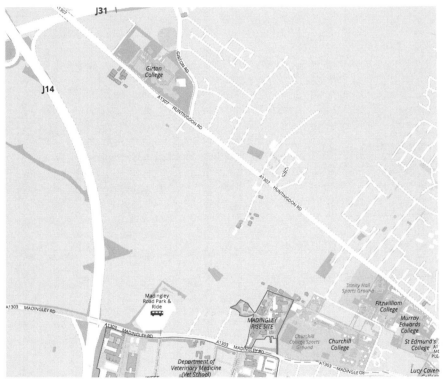

Figure 11.1 North-west Cambridge Site plan
By courtesy of University of Cambridge

As soon as the University estate began to be managed in a coherent way there was 'thought for the morrow'. The reports by Professors Swinnerton-Dyer and Mott (see Chapter 1) showed that the University, particularly the science/technology departments, should look to the west for future expansion; thereafter, steps were taken, albeit spasmodically, to consolidate and then develop the West Cambridge site. In the late 1990s the West Cambridge site was thought to be able to take about £800M worth of buildings and their infrastructure; at the time of writing nearly half of that has been achieved. The planning processes of the North-west Cambridge site were set out in Chapter 3; what follows in this chapter is a description of the actions taken to support those processes so as to get to a position whereby the North West site could be developed in the way that the University wanted.

As noted in Chapter 3, the Regional Planning Guidance (RPG) Public Inquiry was held in Ely early in 1999 and published in November 2000; the University argued strongly for the ability to expand to the north-west and the outcome, RPG6, included Policy 26 which stated that:

> Development Plans should continue to include policies for the selective management of development within the area close to Cambridge, discriminating in favour of uses that have an essential need for a Cambridge location.[1]

That opened a line of logic which would allow development at North West Cambridge provided that the land could be taken out of the green belt and designated for development, specifically for the University, and included in the subsequent County and Local Plans in accord with the 'plan-led' system.

For the next stage, the Plan for Cambridgeshire County, a Deposit Draft Plan was issued by the County early in 2002, and an Examination in Public took evidence before three government-appointed Inspectors from October to December 2002. Again, and in more detail, the University argued its case for expansion to accommodate its rapidly expanding research programme, and house its increasing number of researchers. The Estates Director argued and was cross-examined over a couple of days as to why such expansion should take place in the green belt at the 150-acre North West Cambridge Site, and whether in truth the policies of the University and the adjacency of the site would lead to sustainable development and transport links.

One key argument put by the University was that the definition of 'affordable housing' should include accommodation for 'key workers' as well as 'social housing': that was not hitherto the case in UK, and it would be hugely to the University's advantage if it could provide housing for its staff, especially its less well-paid staff, and to have 'key-worker' accommodation accepted as its full contribution to the local need for 'affordable' accommodation in planning applications. The acceptance of that by government was welcomed by the University, and more widely, and it made a crucial difference to the viability and nature of the North West Cambridge development.

The resultant 'Cambridgeshire Plan' set out policies which were entirely satisfactory for the University.[2] In particular, there was an acceptance that the University should be allowed to develop the land at North West Cambridge and that there should be a release from the green belt to allow for that, subject to certain conditions, all reasonable, one of which was that prior to such release, the University would need to show current need for the area having accounted for the capacity of existing development land. Policy 26 of RPG6 (above) is reflected in the Plan in Policy P9/8 which requires new employment proposals in and close to Cambridge to demonstrate that they 'fall within one or more of the following categories'. The first on the list was:

> high technology and related research and development industries and services which can show a need to be closely related to the University or other research facilities or associated services in the Cambridge area …

The recommendation safeguarding green spaces and separation was sensible:

Recommendation 8A
The vision for Cambridge is of a compact, dynamic city with a thriving historic centre. Apart from its unique historic character, of particular importance to the quality of the city are the green spaces within it, the green corridors which run from open countryside into the urban area, as indicated on the Key Diagram, and the green separation which exists to protect the integrity of the necklace of villages. All of these features, together with views of the historic core, are key qualities which are important to be safeguarded in any review of Green Belt boundaries.

This Plan provides for three expanded communities within the context of the overall vision. These are focussed on the University in West/North West Cambridge, on Addenbrooke's in the south, and on the airport site to the east. With that near-assurance of green belt release for University development, the major hurdle for development was cleared. What remained, in Town and Country Planning requirement terms, prior to starting with master planning, transport planning, business planning and applications for planning permissions, was to get set out in the Local Plans for Cambridge City and South Cambridgeshire District Council the extent, nature and layout of the development in terms of housing, academic and other research buildings, infrastructure, site services and public open space: substantially, that matter was decided following a Public Inquiry (PI) October–December 2005.

In the years leading up to that PI, the University had encouraged the setting-up of one local residents' association which could develop a consolidated view representing local opinion, rather than work with the existing, rather fragmented local community associations. Briefings and discussions also were opened with a wide range of local and national representative and pressure groups, in particular those representing the conservational aspirations around Cambridge, and nationally. At that time a leading body representing local opinion was the Cambridge Preservation Society: nationally, there was inter alia the Council for the Preservation of Rural England. As well as routine briefings, an interactive website was established so that all and sundry could express their views and argue their cases; generally, this was a very positive and helpful process, not least because those with diverse views could argue their views against each other while the University took careful note, forming a view as to where the centre of gravity of opinion lay. At that time, an inter-active website was innovative. A series of three all-day discussion events over a period of months was set up in Murray Edwards College (then known as New Hall), and those proved to be crucial. Attendance included representatives of local and national community pressure groups and local authority councillors and officers. The first of those three events started with a description of what the University was seeking, and why, and a note that the aim of the day was to move towards a solution that would best meet University aspirations taking account of the views and positions of those present and others. It was made clear that the University would, of course, expect all concerned to fight their own corners, but it was generally agreed that it would be worth at least developing a solution which had the best chance of winning the most acceptance across the board: for opponents, the 'least bad solution'. The majority of each of the three days, which were spread over a few months, was spent with a number of cross-sectional groups, each of about eight attendees, sitting around a plan of the area with wooden pieces representing housing areas, schools, transport links, shops and so on with the challenge of putting what the University needed in the best positions practicable. Almost immediately people were saying things like, 'No, it would be better to put that number of houses here'; the conversation was what to do rather than what not to do. At various points each group would present its optimal solution to the other groups for mutual discussion. Mid-way through, one delegate said to his group that development should start from the centre of the site, not from the edges, starting around 'a park about the size of Parker's Piece' in central Cambridge: in fact that concept was mirrored in the detailed planning approval in 2013 with those words used by the planning officer. Debate continued on the interactive website between sessions, and after the last, right up to the Public Inquiry for the Cambridge Local Plan.

Soon after the end of the third day of discussion, a consensual view emerged, indeed it was emerging earlier in embryo, discretely encouraged by the University. In essence, that concept was to start the development around the centre of the site, first developing there an identity, with housing to about three or four storeys around a large public open space, Parker's Piece being held as the model, with housing and facilities then radiating outwards towards the three sides that bound the site, keeping a good distance from the adjacent motorway network and from the housing along Huntingdon Road, houses with large back gardens. The senior officers of Cambridge City Council had been arguing for the converse: devel-oping from the existing city boundary outwards across the site; but over time with much discussion, that position weakened, until it was no longer proposed.

This way of conducting discussions and negotiations with local, council and national groups was seen as being successful, and was held by the relevant government department as a model for development consultation nationally.

Within the University, the planning for North West Cambridge was handled by a new committee chaired by the VC with College and academic representation, serviced by the estates department which led the agenda: the North-west Cambridge Development Group. This committee was effective and decisive, and gave the officers the authority and input they needed. (Only one procedural issue arose and that was 'declaration of interest': no members of the committee should have undeclared personal outside related interest. The issue got sorted within a few months.) It was important that development of planning for North West Cambridge was kept close to the University committee and decision-making structures so as to ensure that the development was and was seen as being intrinsic to the University rather than just a development scheme as its edge. That stimulated clarity, transparency and support.

Colleges, especially those which bordered North-west Cambridge, had potential for taking up opportunities for expanding their graduate accommodation as part of the plan. That dialogue with Colleges worked well, though it lost momentum after 2005 as overall output progress for North-west Cambridge planning slowed and the expected date of completion of the first buildings fell back by a couple of years.

At the Public Inquiry for the Local Plan for Cambridge City starting in October 2005, the University had again decided not to be represented legally, or by consultants other than one planning consultant, EDAW; this decision was made mainly for tactical and presentational reasons insofar as the University felt that during preceding months and years, the argument was sufficiently strong, and that the University could present its own case directly to greater effect, and less confrontationally, than through lawyers. And a good deal less expensively.

When the opening day of the Inquiry arrived, although North West Cambridge was said to be one of the largest releases of green belt regionally, there was not a single third party objector: just the City Council officers and the University present. Nearly all matters of substance had been agreed prior to the Inquiry. The appointed Inspector commented on the lack of third party objectors and joked that 'we could be finished by tea-time': in fact it took only a few days for the University's aspects of the PI.

The case for more space for research was put by the University's Pro Vice-Chancellor for Research, and the rest of the evidence and cross-examining given and taken by the Estates Department officers, supported by the one planning consultant. The case was made for development that would meet the University's research and housing needs for at least 20 years. The Cambridge Local Plan was agreed on 20 July 2006, and all that the University had asked for was included in the report (See Annex A). The key conclusions include:

> Ownership of the land in NW Cambridge is seen by the University as critical to delivering the provision of key worker housing and college accommodation as well as enabling funding of the required infrastructure.
>
> 8.101. We see merit in the future needs of the University being met in a comprehensively planned urban extension which delivers high quality buildings in a high quality landscaped environment. We accept that in terms of delivery there is considerable benefit to be derived from using land in the single ownership of the University. On the basis of there being a need for the University proposals and that this could not be met

elsewhere, we are satisfied that there would be justification for the release of Green Belt land in NW Cambridge to meet that need.

The Inspector had indicated that it might be he that would preside over the PI for South Cambridge District Council SCDC Plan (and so he was) so there was confidence that a similar result would be achieved for that part of the site. Once the Cambridge Local Plan was adopted by the City Council, the University focused on developing its plan for North-west Cambridge in greater detail, with wider support, and with greater input into the emerging thinking of the community about the forthcoming 'Area Action Plan' for the developments to the north-west of the city. This area planning was led by the two local councils, Cambridge and South Cambridge District, under the guidance of the 'Joint Planning for Growth Areas Organisation' which co-ordinated and advised the local councils; they were more specifically working together than they generally had before, even though it did not appear as though they were particularly enjoying that, having rather different approaches and political representation. By this time the Planning and Compulsory Purchase Act of 2004 was triggered insofar as the local plans were to be replaced by a local development plan (LDP), which comprised a series of key documents relating to development policies and requirements. There was at the time a view that neither of the two councils were showing any sense of urgency in progressing the plans from North-west Cambridge; rather, they appeared to be responding, without proactive enthusiasm, to government requirements for considerable additional housing provision of 47,500 more houses over 1999–2016. And there was a view that from 2006 University planning for North West Cambridge lost momentum.

In September of that year, 2006, the two councils issued a paper 'Issues and Options'. This was a modified plan for North-west Cambridge. It was generally a good document, but showing less development and more green, open space. The University considered this change to be unacceptable; indeed, it was against the run of policies in Regional, County and Local Plans. And so it made representations.

In May 2007, the two councils produced a paper entitled 'Preferred Options'; this in fact contained a range of options with the University's master plan at the top end of the range for development. There followed a further consultatation, and in May 2008 the two councils produced a 'submission draft' which, when once consulted, would go to the Inspector in charge of the long-continuing Examination in Public. The recommended development plan proposed that the western edge of the development be drawn back towards the centre of the site, leaving a very thin strip for buildings. Also, the plan offered development on an area near the 'Park and Ride' site by the Madingley Road: this would not have been useful for development as the dreaded greater crested newts had been sighted there (whether they were living there by choice or had been put there is not known). There was yet another flurry of consultations and meetings and in the end a compromise was cobbled up, published in May 2008 for yet another round of consultations during September and October 2008, with the plan going back to the Inspector prior to another Examination in Public (EiP) starting on 25 November 2008. The Inspector, who had earned widespread respect from the start of the Cambridge City Local Plan in 2005, set out three aspects on which he would concentrate: the needs of the University, the mix and viability of development, and matters concerning the green belt, environmental footprint and open space. The Pro Vice-Chancellor, Professor Ian Leslie, gave a convincing account regarding the increasing research and other needs of the University, as he had done in November 2005 at the first in the series of Examinations in Public. Consultants addressed the other two aspects: the use of external consultants by the University, rather then hiring-in professionals as University staff as per the previous policy, had much increased over the previous three years.

After due consideration, the Inspector reported with a plan which was very similar to that proposed by the University in the first EiP late in 2005; that was opened to yet another round of public consultation in early 2009. The EiP resumed for one day, 9 June 2009, after which the Inspector finalised that plan. What emerged after three and a half years, and huge fee bills, was not much different to what had first been proposed in 2005, but it had had a great deal of public scrutiny and consultation. This process might be seen as a prime example of what the incoming government in 2010 saw as over-consultation causing unacceptable delays to development and associated economic stimulus; the Localism Bill quickly following that election was designed to bring the UK system more into line with the less protracted systems used in other developed countries.

By this time, many years had elapsed since the consultants were first appointed. EDAW, which had been taken over by AECOM, which now has 100,000 employees worldwide, (as so many other UK and worldwide practices were swallowed up by that US-based company, with varying effects on those companies' morale and standards) was re-appointed. Thereafter, with changed management in the University, there was a long period while the masterplan and development ideas were reviewed by external consultants and confirmed. A transport study showed an increase of just 4% above the already predicted 32% increase in Cambridge traffic. The University's Regent House confirmed approval for the development in January 2013: Regent House had been regularly consulted since the early 1990s and, with the exception of a small number of academics, it was widely supported from the start. Planning Permission (see Annex B) was granted by the Councils in February; the planning officer's report noted:

Officer report by the City Council head of planning services, Patsy Dell, on University planning application for NW Cambridge August 2012.
This proposal represents an urban expansion that will connect to Cambridge, but also respect its relationship as a gateway to the adjoining villages and surrounding green belt.

It will provide for a development that will ensure future residents can enjoy a good quality and sustainable lifestyle, our meeting the long-term development needs of Cambridge University.

The development would also include a new green space, the size of Parker's Piece, a primary school, a community centre, plus a health centre, sports facilities, and an energy-centre generating power and heat. The new road junctions on Huntington Road and Madingley Road would cause additional delays during peak times, but of less than one minute. New bus route would be introduced to discourage car use, and cycling and walking would be encouraged, but nevertheless continuing congestion fears triggered an objection from Girton Parish Council which had for several years taken a stand against development of NW Cambridge despite the considerable benefits it would bring to the village.

(To be fair, the Girton inhabitants themselves did not generally object to the development and mostly were either unconcerned or else were in favour of the added facilities it would afford.) The first phase included 100 affordable homes for University staff, 325 rooms for students, 400 houses for sale, a school, community centre, shops, a GP surgery and public open space.

Enabling works on site were started by the contractor Skanska at the 'ground-breaking' ceremony at 12.30pm on the very wet Thursday 20 June 2013, with speeches by the Chancellor, Lord Sainsbury, the VC and others. Then the Director of Estates Strategy led the contracted-out design and procurement teams off on a team-building run through the rain.

It had been a long haul, longer than it need or should have been, from the crucial Regional Planning Guidance Examination in Public early in 1999, then the favourable review of the Cambridge Green Belt (the most crucial step in the development of North-west Cambridge) following acceptance of the County's Structure Plan, the local plans of the two councils, and then the protracted review and acceptance of a plan similar to the original through the Area Action Plan, and consequent final report by the Inspector. The process would be considered by many as a very thorough and necessary review of development of an area of the Cambridge Green Belt. Others could describe the process as a 14-year slow-rolled and cumbersome consultation while the acute need for housing near Cambridge, with desperately needed accommodation for young research staff, got worse and worse and those deprived of such accommodation suffered accordingly for so long. Falling behind the programme held closely until 2006 caused a significant increase in the cost of development as construction prices went up steeply around 2013–2014, probably by about 15 per cent in 18 months, as contractors and suppliers sought to restore their balance-sheets as the volume of work in Cambridge and South-east England increased dramatically.

However, whatever one thinks of some aspects of the process, the outcome was and is an excellent plan for development which will meet the research and accommodation needs of the University and the aspirations of its city and its district starting with completion of the first phase in 2017, and do so in a very sustainable way with a high level (Code Level 5) of sustainability for housing and the second highest rating, BREEAM Excellent, for public buildings.

The North-west Cambridge project was the stuff of the dream-weavers of the 1990s onwards, and their dreams for a well-integrated University-based community looked like they were coming real. It is the biggest development of any UK university. Its comprehensive nature and the way it meshes with its parent city has widespread support. Its 15-year pre-construction planning, and its £bn size make it a fitting way to end the project chapters of this book.

Notes

1 DETR, *Regional Planning Guidance for East Anglia to 2016* (Department of the Environment, Transport and the Regions, 2000).
2 Cambridgeshire and Peterborough Structure Plan 2003.

12 Overview and lessons from the estate expansion 1996–2006

After any large construction project it is well worth the trouble to stand back after a few years and assess what the project achieved, and what it cost; what went well and what didn't. For a programme of projects an overview in hindsight is even more worthwhile, but very rarely is that done. The hundred major projects described in this book are of course part of the continuum of development of the Cambridge University estate, but the intensity of projects was greater than ever before in its history and was accompanied by management changes to cope with that and other challenges.

In such an overview of this capital programme, it is best to look first at the output, the physical legacy of the decade, and then consider the processes behind that. And then to ask: was what was achieved worth the cost?

University sites

Until the government brought in 'plan-led' development regulations (see Chapter 3), there was little or no pressure on universities to produce coherent development plans and to stick to them, updating them as appropriate. Had the government brought in such a 'plan-led' system earlier, then the New Museums, Downing and Sidgwick sites would all have been more attractive, open and efficient. In the absence of such legislation the University should have done at least as well as was done by the civic universities in the first half of the twentieth century. The University should have had its own plan-led site master-plans: clear, widely discussed and stuck to.

A senior member of the City Council has noted that some sites, notably Downing and New Museums, remain 'inaccessible', and that little was done during the decade under review to improve the

> canyons of tall buildings; subsequent to their construction the City Council as Planning Authority strongly discouraged University buildings being taken to more than three storeys. The quality of the buildings is good but they look inwards and the sites do not contribute to the historic area of Cambridge; one quarter of central Cambridge is in the ownership of the University. In particular, the New Museums site is disaggregated: the University should have pursued the opportunity it had in 2005 to

apply for the demolition and replacement of the Arup building; beyond that, there was less scope for addressing the issue.

Local councillor John Hipkin has said that:

> Comparison of site-planning and architectural style of the various University sites is interesting and important. The Sidgwick site initially had coherent planning; it now has an esoteric coherence as a decidedly 'faculty area', and in that sense it works well, despite the lack of site coherence in the architectural styles of the individual buildings.

It is fair to conclude that of the three main pre-existing sites, only Sidgwick was much improved during the decade being discussed.

During the decade, two new sites were created and one planned. The Maths site (see Chapter 8) is an interesting case of isolation, and lack of integration with the rest of Cambridge: however, it is difficult to see how the University could have done much better in this regard when the local community opposed any degree of openness or civic inclusion of the site, and the local planning authority supported that campaign in 1997–1998. The West Cambridge site (see Chapter 10) is more isolated than it need be, and there is very little sense of 'street scene' with community and social life. It is argued that at West Cambridge the plot density should have been at least twice as much; that however might not have been acceptable to the Local Planning Authority under pressure from neighbours who at that time had neither been extensively consulted nor taken along in the preparation for site development. The Addenbrooke's site is more dense and works better than West Cambridge. Alex Reid, architect, consultant, former Director General of RIBA and former County Councillor, has noted that:

> In fact, the amount of building at West Cambridge could have been accommodated on about a quarter of the site, with the balance of the site being kept for future development, but there would have been risk that the balance of the site might subsequently have been denied planning permission.

The option of trying to get approval for densification at West Cambridge was not considered openly although by late 1998 some estates officers had it in the backs of their minds; any open consideration of that option would have been nugatory and counter-productive at that stage. Also, it was thought by those officers that the wide spaces between buildings (mostly surface car-parks) could be built on before long provided there was support from the City Council regarding environmental, particularly preservation of the few green areas, and transport issues; support that could be expected, subject to planning conditions and 'planning gain' (through S106 agreement; see Chapter 3). Generally speaking, one can densify a site; one can't easily 'de-densify', especially not in Cambridge.

City Councillor John Hipkin has commented that,

> the West Cambridge site certainly had a unified master plan, but that was for a mixture of different types of building, research/teaching, institutional, and residential, and the wide spacing of the buildings and the lack of any 'street presence' makes the site unsatisfactory and soulless. It would have been better to have had more residential accommodation, and facilities for those working and/or living on the site, as well as denser development.

The site has been unpopular with many of those working there, partly because of a sense of isolation, partly because of the lack of facilities. There was a mistaken view in 2001–2002 that no firms would take on leaseholds of shops or such facilities at West Cambridge because of lack of footfall; this is a circular matter in that better facilities would have led to greater footfall, as successfully happened at other universities. Provision of convenience shops along the Madingley Road could have been viable. There is at least anecdotal evidence that a major factor in Microsoft's decision to leave West Cambridge was the lack of support facilities and nearby social venues (such as pubs). Alec Broers has stated that interaction between Microsoft and the University's Computer Lab, while helpful and productive, was less than had been hoped or expected and that social infrastructure would have helped.

The lessons from West Cambridge planning were quickly learned, and the planning of the third new site of this decade, North West Cambridge, was designed right from the start, from the Examination in Public of the Regional Planning Guidance in 1999, to be well integrated into the fabric of the city. John Hipkin continues:

> NW Cambridge site was planned from the very beginning to be a mixed site with a balance of residential, academic, business, and infrastructural facilities. The public consultation for that site from the start has been excellent and the people of Cambridge feel that they have had a good and meaningful input into the form of development.

The absence of any local or national objectors to the University's plan for North West Cambridge as set out in the proposed master plan at the Public Inquiry in 2005 was perhaps the most notable aspect of estate planning during the decade.

Regarding integration of academic communities with non-academic aspects of the city, there never has been very much sharing of facilities through the ages. That criticism applies to a degree to this decade also, but there were exceptions such as the new University bus service which was widely welcomed, and the refurbished West Road music facilities which have been well managed to the benefit of the citizens of Cambridge. At West Cambridge, sports facilities have become available, although later than hoped, and somewhat remote for the majority of Cambridge's population; the athletics ground is well-shared.

University buildings

Building design suitability

After some 100 construction and refurbishment projects, a total capital expenditure of some £765M and an increase in the built area of the University of 33 per cent, the first question to pose is whether or not, or to what extent, the new construction and refurbishments were suitable for the users of those buildings, and for those who visit or pass by them. How will they be for our grand-kids?

Generally, the building and refurbishment projects are thought to have been both necessary and appropriate for the development of research and teaching and were planned through a rational process. By and large, the sizes of the new buildings were about right, and by the end of the decade there was general (if neither universal nor enthusiastic) acceptance that the developed Cambridge University space norms based on extrapolation of approved staff/student numbers and activities for a period of ten years after occupation were a suitable basis for deciding the size of new buildings: some buildings such as the Computer Lab and CMS Maths development were indeed getting full at around the ten-year point. Some were under-used in early years.

Regarding design, until this decade Cambridge had traditionally appointed, often with little competition, high-flying and internationally recognised 'iconic' architects, keen to achieve particularly impressive individual 'landmark' buildings which would be identified with them (and look good in photographs often taken in late evening, and without people). The decision to draft, agree and promulgate criteria for the appointment of designers whereby the highest priorities were to be given to the suitability of the design firstly for the users and secondly for the immediate physical environment resulted in selection of designers from less stellar but widely respected design practices. As Alex Reid, put it,

> about the start of the decade there was a new generation of growing architectural practices such as Allies and Morrison and Edward Cullinan and Associates; they had a very high standard as designers, and the University plugged into that, rejecting the 'starchitects' such as Foster. That led to well-designed and reputable buildings.

Or, as Gordon Hannah, joint founder of the national engineering consultancy practice Hannah Reed, and long-time, eminent member of the University Buildings Committee put it, 'you don't need monumental architects especially on sites where there are already monumental buildings, for example the Sidgwick Site. Over the decade, the University got it about right'. (It should, however, be noted that there was some pressure from City Council officers to appoint high-status architects.) Designer appointments were implemented by a broadly based group from Buildings Committee and building users and that gave wider credibility than the previously much less open procedure.

For most people the quality of architecture of individual buildings in Cambridge University, as everywhere, is to a large extent a matter of personal taste; however, despite the speed and intensity of the capital programme during the decade, and the economies in design and construction costs, there are at the time of writing no signs of the social revulsion that there was subsequent to the British building boom of the 1960s: time will tell whether the legacy of the new University buildings should have been better. For those who see architecture more objectively and more professionally, opinions though mixed are generally favourable. A recent comment by the then-Chief Executive of Cambridge City Council, Rob Hammond, was: 'The view from the Chief Executive's office was that the quality of buildings commissioned by the University during this decade represented good modern architecture, and was not guilty of cheapening'. John Hipkin said that: 'Regarding some individual building developments, the conversions of Hotel du Vin, Old Addenbrooke's/Judge, and Fitzwilliam Museum have rightly been widely praised: they might be praised as "brilliant"'. Sunand Prasad of PenoyrePrasad, and President of the RIBA 2007–2009, recently commented:

> The University brought a rigorous and entirely fresh approach to procurement combining the Latham and Egan agendas with a belief in design quality. Cambridge University buildings of the last couple of decades generally are exemplary of their type both in design and construction – the sure sign of the better-informed client.

So, overall, the size and design of buildings of this decade are generally thought to be more 'well-mannered' and more 'user-friendly' than those of other post-war decades of University history. A social issue was noted in the article quoted in Chapter 6, from the journal *Architecture Today*:

> Against so much gravity [Law and History buildings] Divinity is a jaunty building. The Divinity Faculty after two months' occupation is more at home here in a way that neither the History Faculty nor the Law Faculty have ever managed in their respective buildings. If the Divinity building is a building that makes its occupants feel 'at home', one might speculate as to why, at the end of the twentieth century, Cambridge should have decided to set aside a 40-year old tradition of 'special' university building? Perhaps we should see this change as connected to the University's need to make itself seem a less exclusive institution and to show that 'elite' need not mean inaccessible'.[1]

There is, however, an understandable view that it would have been worth investing in greater flexibility in the new buildings, even although that would have made them more expensive: it is always difficult to quantify how much flexibility should be bought during design and contract stages. The accelerating rate of change, particularly in scientific and medical and technological research, means that laboratories have to be upgraded at increasingly shorter intervals: in a review of Oxford procurement during the 1990s it was said that on average science labs had to be refitted about every seven years. Therefore, there is a fairly strong case that a large, generic research facility could have been built during this decade, possibly at West Cambridge. There would have been alternative options for procuring such flexible research space: possibly a large and rather sophisticated building, perhaps a research hotel (such as was discussed for Arts, Social Sciences and Humanities on the Sidgwick Site and was later, in mutated form, constructed there); research hotels are capable of being used for a number of years for one purpose and then relatively easily converted for another purpose. Alternatively, there could have been cheap buildings made from components that could be recycled with the buildings being demolished after their useful lives, or replaced more routinely every 10 to 15 years or so. Neither possibility was pursued. In part, frankly, this was because no one other than members of the Arts, Social Sciences and Humanities faculties asked for such facilities and the huge pressure on procurement staff reduced lateral thinking to some degree. More particularly, there was no identified generic capital funding available other than by borrowing capital or borrowing against the promise of research grant funding to cover interest and the appropriate element of repayment. There was the argument that borrowing was not the 'way things were done at Cambridge', but there was the counter-argument that the construction inflation rates were higher than capital repayment rates, and, anyway, residential units at West Cambridge albeit with their more certain income stream, were satisfactorily funded through internal borrowing.

Generally speaking, apart from a limited number of roof-leaks (for example Maths and Criminology), the physical performance of buildings procured during the decade appears to be generally good, as it should be for new/ refurbished buildings. The standard of University buildings, as defined on the Royal Institute of Chartered Surveyors assessment scale, is high in the range for research universities in the UK, and maintenance costs are well within the best (i.e lowest expenditure per unit area) quartile of Russell Group research universities. As for building services, operational layout and acoustics, criticisms have emerged relating to some lack of local control over air and light quality, notably initially in the Maths development, and harsh acoustics in the Faculty of Education building; otherwise, building users generally say that they are relatively content with new and refurbished buildings.

Benefits of University development to the city

Some of those Cambridge citizens, and councillors, who express views that areas of the University estate are not particularly accessible or welcoming for the people of Cambridge, question what benefits are brought to local people by the development of the University estate, apart from employment. Clearly, there were benefits from the community and infrastructural work required to be carried out under planning permissions, so-called 'planning-gain', and it is worth noting that in the majority of cases the University was content with city requirements for 'planning-gain' work as being work they would wish to do anyway.

The first big opportunity to address feelings that University development brought little benefit beyond Section 106 local contribution and infrastructural requirements from planning permissions, was in the development of the North West Cambridge site. Right from the start at the Regional Planning Guidance Examination in Public in 1999, the University was at pains to note that the new development, if approved, would have shops, doctors' surgery, one or more schools and considerable shopping and public transport provision, as well as well-managed green spaces at an early stage.

Research into the effects of University expansion of local and regional employment was very helpful in Cambridge as it was in other university cities in giving statistics on employment and other benefits that resulted from approval of university development planning permissions.[2] The total output generated by UK universities by 2013 was estimated at £65.55 billion (2.8 per cent of GDP, up from 2.3 per cent in 2007).[3] The employment multiplier indicated that for every 100 jobs created directly within an institution, another 117 jobs are generated elsewhere in the economy. So, as a rough estimate, an expansion of the University by some 2,000 staff could represent a direct increase in employment of some 2,340 other jobs, plus the huge amount of indirect employment and inward investment resulting further afield from the added research carried out in the new buildings.[4]

Environmental sustainability

In 2003 a projection of future maintenance and utilities costs if the estate went on expanding at the current rate led to an estates case being made for tapering off the rate of expansion, but high-priority projects continued to come forward and to attract capital funding. The estate therefore continued to expand. So also did its buildings become ever more highly serviced. Hence its environmental footprint grew. For example, there was an increase in carbon dioxide produced as a result of increased energy use of some 10 per cent between 2003 and 2005, to 55,728 tons. However, the increases were significantly less than they would have been had not the University moved into more environmentally sensitive procurement by the middle of the decade: it was a significant step to require cost and energy-use plans over at least ten years as a precursor to project authorisation and procurement. It would have been good to have gone for a longer period than ten years, and that same discussion went on in HM Treasury in 2006, but after ten years the whole-life cost numbers get too vague to be of much use: what will be the cost of energy in ten years' time? What will be the value and the cost of carbon emissions?

It could well be argued that the University should have moved earlier and more quickly towards whole-life procurement, and have overcome earlier the opposition to increasing capital costs in order to reduce whole-life costs and CO_2 production; early opposition was on the usual grounds that when budgets were so tight there were higher priorities than what was sometimes referred to as 'saving the planet'. Case studies earlier in this book comment on the varying levels of success in actual achievement of better standards of sustainability. More generally, it is expected that in the next year or two, more data analysis will become available on trends in energy consumption across the estate (developing the sort of analysis on the Cambridge University estate done by the Carbon Trust in 2005/2006 (see Chapter 5)).

In hindsight, there should perhaps have been more determination to carry through the PV colonnade project at West Cambridge (see Chapter 10). To have developed the technology and a bus service running on hydrogen produced using solar energy would have done a lot for sustainability and more generally for the University: the decision came at a hectic time, and the bureaucratic rigidity in the part of EU that managed environmental grants didn't help, but giving up was probably a bad decision.

Richard Saxon CBE, member of the Buildings Committee since 2004 and former Chairman of the then largest architect-engineer practice in Europe, BDP, said,

> At present (Autumn 2014) the University, like most other owner-occupier clients, is most concerned with controlling capital cost (Capex) and with minimising operating carbon emissions (Opcarb). Both these focuses are incomplete. Whole-life-cost (Capex plus Opex) and whole-life carbon emission (Capcarb plus Opcarb) can both be optimised to advantage with the tools available and emerging. I hope that the University will soon set budgets based on benchmarks for Capex and Opex, and for Capcarb and Opcarb. It may well emerge from current studies that the University's largest single carbon emission source is that from its building programme (Capcarb) rather than the Opcarb from its existing stock.

There remain very exciting managerial possibilities in energy reduction.

Process

Management and organisation

The Registary during the decade, Dr Timothy Mead, recalls,

> by the mid-1990s, it was clear to the VC that the University should assert itself as a corporate entity: there are things that needed to be

done which the University could do but that individual departments/faculties/colleges/museums could not do so well on their own without University co-ordination. That underlined the long-running need to balance encouragement of individual achievement, with the need to co-ordinate and focus centrally so as to help both individual initiatives and also the University more widely: that applied also to capital matters.

And, in the words of Geoffrey Skelsey, chef du cabinet in the VC's office, the conversations in Colleges, High Tables, and in The Eagle are invaluable, but autonomy has to be modified by experience and information and, in the end, central authority. Cambridge can attribute its success to creative anarchy, but we can never be sure whether and to what extent. The University can be seen as an academic co-operative.

Over the centuries, people understandably had thought in terms of college estates and departmental buildings funded and largely managed by the people who worked in them. Management of what became thought of as 'the University estate' did not start significantly until the mid twentieth century. Once established as a self-standing department, the estates function until the late 1990s was and was seen as part of the Treasury division of the University administration. However, early in the decade being reviewed, it generally became directly responsible to the committee structure for operational activity, and, in practice in line-management, to the Vice-Chancellor. Later, when the University was re-organised under a Unified Administrative Service (UAS), the estates department came under UAS for administration, but not for capital projects or such operational matters. The numbers of PVCs increased, from two to five, but their roles were essentially to oversee rather than manage, unless a senior administrator in charge of function was not adequately fulfilling their duties. The switch to a deeper role of PVCs could have adversely affected the standard of capital project management but that was avoided. By conscious effort, internal tensions over the nature, extent and speed of change diverted attention and focus on programme and project management to a remarkably small extent in the context of the history of University organisation. The management arrangements for the capital works at Addenbrooke's also worked well, in particular the composition and *modus operandi* of the joint committee. The main constituent reason for success was the clarity of the complementary, different but dove-tailing, aims of Trust and University.

These organisational changes, and others set out in Chapter 4, were fundamental to the capacity of the University to manage the wave of capital work; they were the right changes, and they were achieved just about in time to cope with the rapidly increasing pressures of the capital programme. Had they not been just about in time, the wheels would have come off the University's expansion programme. That the reforms were the right ones and came just in time is one main conclusion of this study.

Planning

Planning can be thought of on two levels: holistic University planning for its estate, in particular for the large-scale development programmes, and the planning needed for individual construction projects. The decisions precipitated by the Vice-Chancellor in 1999 and agreed by Council regarding future expansion of the University at undergraduate, graduate and post-doc levels were absolutely crucial, and were timely in allowing the substantive start to a worthwhile plan for the development of the University estate. As Rob Hammond noted,

> prior to the master plan for West Cambridge in 1997, the University appeared to have no master plan for its estate. This made planning more difficult for the City Council insofar as there was no adequate information about how much the University was planning to expand, of where or how. The first University-wide estate plan to be made available, in 2000, was in that sense, very important, especially as that coincided with the improved collaboration after the resolution of the divisive planning application for the lighting of the athletics ground. [See Chapter 3.] Decisions about the numerical growth of the University allowed the substantive start of a worthwhile plan for the development of the University estate.

The decision to base the estate plan on a gap analysis (what the University had, compared with what it wanted to have five years, and then ten years on, hence showing the amount and quality of space to be achieved for each department) was sound, even although in some cases allowance for expansion proved to be insufficient for some departments around the ten-year point. The drafting of the estate plan brought about progressive, and sometimes unwelcome, discussion about the management, and real costs, of space within departments and was an important factor in the management of the expansion of the University. An important shortcoming of using this estate plan as a basis for capital development, however, was that there was then no central capital reserve of money that could be used to effectively optimise the prioritisation of proposed projects: the start of most projects relied on current availability of capital funding as and when it could be found. During the periods of government capital funding for research-based science/medical/technology projects, the University could bid for capital funding according to its priorities; beyond that, progress largely depended on funding availability rather than on true University prioritisation as came about later with the eventual raising of a bond as previously proposed.

One criticism, albeit minor in the scale of things, was that the estate plan and its interpretation did not make allowance for the unexpected preference of some departments to stay in disaggregated, even low-standard accommodation because of its convenience and familiarity, and/or the potential for grabbing adjacent space as that became available, rather than go for the well-regulated and less informal, and maybe less flexible space of a new, and perhaps geographically distant, building.

One lesson worth recording is the value of the agreement with the City Council to incorporate a clause in the Outline Planning Permission for West Cambridge to make provision for a biennial review so as to keep the plan current without the need to apply for changes to the Permission. Councillor Nimmo-Smith observed this process in action initially as a member of the Planning Committee and then as Leader of the Council. He has said,

> the possibly unique system brought into the Outline Planning Permission for West Cambridge to have a mutual two-yearly review to update the Permission by mutual agreement was known about at Councillor level, and appreciated as a means of keeping the Permission live and optimised without the need for application for amendment to the original Outline Permission.

Regarding the wider planning aspect of future site development, the most important challenge of the decade by far was getting North West Cambridge out of the green belt and declared for University development; that was achieved more quickly and more extensively than most had anticipated, and the master-plan produced for the 2005 Local Plan Inquiry was remarkably similar to what was in the outline planning application for the North West Cambridge site seven years later. That said, the subsequent delays to the 2005 planned programme associated with the re-validation of the master plan and Area Action Plan caused not only a longer wait for the facilities and residences, but also a considerable increase in cost, estimated at an extra £40M, to the University: construction prices went up by 12–15 per cent locally in the couple of years prior to signing the contract for Phase 1 which was some two and a half years later than had been planned.

Regarding the planning process for individual projects, Councillor John Hipkin has said that:

> University developments since 1996 represent the latest chapter of Town/Gown relationships: how each perceives the other. There is still a feeling among some citizens of Cambridge that the University thinks that it 'owns the City' and gets its way over development proposals.

Former Cambridge City Council Leader, Councillor Nimmo-Smith has said that,

> Since the decision to develop the hi-tech aspects of research work in the University, and by others, there have inevitably been tensions between the University and the City as the Cambridge Phenomenon got going and grew hugely. However, the energy which the University has put into the growth of Cambridge has been and is well directed and responsible, and the quality of that influence has steadily improved over the last two decades.

Getting planning permissions was never easy, and certainly not in Cambridge. Up to the end of 2005, the University was making over 50 Planning Applications most years, and only one was refused, with that refusal later to be over-turned by the Deputy PM on Appeal. As time went on and consultation became better, Councillors, especially new Councillors with more experience of universities, came to have more understanding of the University's aims, mechanisms and constraints. Ian Nimmo-Smith has commented about the political environment within which planning applications were considered,

> In May 2000 the Lib Dems took control of Cambridge City Council. There had been a number of Labour Councillors who had tended to regard the University with some reserve occasionally bordering on cynicism, there being some feeling that the students were particularly privileged, and that apart from the provision of employment, the University did not contribute a great deal to the citizens of Cambridge. By contrast most Lib Dem Councillors had personal experience of higher education, and an appreciation of the wider role of Cambridge University in particular. This change of local government happened to come soon after the University assessed and took action following the planning application for lighting of its sports-ground off Adams Road.

Transport planning

Some believe that a lesson to be learned was that the University did not have a coherent internal plan as to the extent or the way it should be involved in transport planning decisions. Relevant planning committees within the University should have transport as one of their many terms of reference, possibly with an internal professionally qualified transport officer. Rob Hammond, former Chief Executive of Cambridge City Council, has suggested that it would have been helpful if the University had had a more positive involvement in local infrastructure, albeit within the constraints of avoiding inter-political party and inter-factional debates, such as those surrounding the guided bus project. The University input into the joint University/ Local Authority 'Cambridge Futures' certainly was highly beneficial: that cut through party and local authority boundaries and frictions, balanced and helped to integrate public and private areas and was very influential, and could be a model for future infrastructure debates.

Funding

It is noted above that prioritisation of capital planning was less effective in the absence of a central capital fund, prior to the October 2012 £350M bond issue. A recommendation to borrow was made in 2005 through the New Museums Site/Mill Lane planning committee, chaired by Professor Malcolm Grant. That proposal was lost partly through insufficient continuity of business, and partly through a reluctance to borrow even although it was shown and accepted that construction inflation was for a long time higher than borrowing costs, so it would be cheaper to borrow and go ahead and minimise construction inflation. Had borrowing been started in 2005/2006, a good job could have been done to improve the New Museums and Mill Lane sites, with a good connection to the Galleries shopping precinct, and an attractive way through from the New Museums site to the river. And the Arup Tower could have been replaced by a much better new building and at far lower cost than its subsequent refurbishment. It is however at least possible that had such a proposal to launch a bond been pursued in 2005 there would have been serious opposition from some academics, and maybe some colleges.

Capital infrastructure funding programmes of £2.6bn by government and certain large-scale charitable trusts channelled through Hefce during the decade were exploited very successfully by the University; while there will always be arguments about which projects did and did not receive University backing in the bidding process, it would appear that these programmes were taken up by the University as well as was possible, certainly with huge success, and that gave impetus to the capital programme at an optimum time. Beyond such programmes, and funding directly from research organisations, capital funding was by benefaction and undoubtedly that was a huge success for the University during this period. (See Chapter 2.) And the success of early construction projects led to further capital benefactions as people saw what was emerging out of the ground and the research programmes being enabled; success bred success.

The one capital funding which went badly wrong was that promised by Marconi, which company went belly-up; the University however did manage to recover its costs, immediately learned relevant lessons and got replacement funding from other companies. The flexibility of the Engineering Department and of the Treasurer's office was remarkable.

Procurement

Procurement lessons were set out in Chapter 5; as an objective view, Colin Jones, construction lawyer with the legal practice Hewitsons, commented:

> The key issues in procurement over that decade's work were the quality and the time given by the consultants, and adoption of the two-stage procurement with an early contractor involvement, which for many projects, is ideal. Its introduction was not rushed: it was introduced early, and then took time for all levels of estate management staff and consultants to adopt this model of procurement. Although the button was pressed early and quickly on innovation, because of the huge quantity of work, some projects may have progressed under more traditional procurement models. Regarding forms of contract, adoption of the Engineering and Construction Contract (ECC) was very important. And regarding the few projects that used another form of contract such as GC Works, at English

and Criminology projects for example, if ECC had been imposed early enough then that may well have helped pre-contract work given the ECC emphasis on project management. The uptake on the adjudication process was very good, and it was wise to find an early practice run.

As noted in Chapter 5, designers and contractors sometimes commented that they had to work harder for the University than for other clients, but they were fairly and punctually paid. The chairman of the national contractor, Willmott Dixon, Rick Willmott commented,

> The work the University led in construction procurement was a fine example to other clients and we have seen many follow similar procurement routes over subsequent years. Working with the University and Sir Michael Latham has certainly influenced the thinking and development of Willmott Dixon.

Gordon Hannah notes:

> Regarding procurement and value for money, the University needs to be congratulated in achieving what it did with such little dispute. I know that there were some problems, but nothing like those made on so many public sector contracts (e.g. the Cambridge guided bus project). This came about by the introduction of two-stage tendering, and introduction of quality/cost assessment of tenders. This must have resulted in saving a great amount of capital (unfortunately difficult to quantify), and also a reduction in full-life costs.

Overall, the unit cost, discounting construction inflation, of non-laboratories was reduced by about 15 per cent in real unit costs for non-lab buildings (see Chapter 5), but the inexorable rise in the cost of health and safety related legislation exceeded the economies achieved for laboratories and caused their unit cost to rise.

The use of 'develop and construct' procurement, rather than design-led to tender, or 'design-build', clearly was crucial to achieving the quality, cost and absence of litigation of the £765M programme, and to the overall near-budget compliance of just +0.1% overall. However, although it sounds good to come in overall on budget over a large number of projects, as a public sector client it isn't as simple as that because for under-budget projects funding money doesn't get drawn down, whereas for over-spent projects extra money has to be found. Usually, private-sector clients such as developers can trade off under- and over-run projects.

Risk was substantially reduced in the 'develop and construct' form of procurement not just because the constructors were involved in design and there was a better team-work approach; a key matter was that sub-contract packages of work were being tendered as design proceeded, and a clear picture of the total contracted cost built up. The decision to proceed to second-stage, the construction contract, was made when the total of works-package tenders reached 85–90 per cent of the total works; that is, there was 85–90 per cent cost-certainty on the main contractor contract before it was agreed. Hence, the chosen early contractor involvement procurement by 'develop and construct' procurement much reduces the cost over-run problems generally associated with full design before contractor involvement, and reduces the incidence of quality problems, and the loss of professional input from able and visionary designers, that often result from simple design-build procurement.

Much improvement was derived from the compiling and dissemination of information through the Soft Landings process (now a national programme for whose invention and early development the University, along with Mark Way of the architecture practice RMJM, can take credit), and through post occupancy evaluations (poe's) which were done for all projects of over £1M. About 50 poes were produced up to 2005, and by making them public, as is the fundamental intention of poe's, the industry generally, and in particular those with whom the University worked, notably designers and constructors, also learned a great deal. Soft Landings, and publicly available poe's for all projects, are lessons to be taken from this decade.

Another key change made was the definition of a sole Representative User (RU) who had considerable responsibility, authority and a voice throughout procurement. Previously, and among many public sector clients, the future users of a building spoke with many voices to the designers and to the official client. With ever-more extensive social media now being increasingly used by building users to communicate their views to all and sundry, it is even more important not just to capture and harness those views but to channel them effectively into the procurement process. That is a challenge very few clients have faced up to, but they must.

Perhaps the most helpful way to assess how the University looked at procurement was that the National Audit Office in its generally critical report of March 2005 on public sector procurement (as noted in Chapter 5) noted that the way that Cambridge University was procuring was 'best practice'. That led by secondment to influence in government procurement, and the principles set out for procurement of the 2012 Olympics infrastructure followed similar lines.

Staffing and training

Recruiting during much of the period was difficult and sometimes not very successful as salary rates rose sharply in the industry while University salaries were constrained. Specialist professionals like those in estate management were tied down to trans-University pay grades: fortunately there was during the decade little staff turnover despite the siren voices of greater salaries elsewhere, and despite the pressures of increased workloads. In fact, despite concern at the time, recruiting restraint of that decade is now thought by senior academic and administrative members of the University to have had little effect on the quality of buildings achieved. It is impossible to gauge to what extent, if any, strain caused by over-work by some officers led to any adverse effects on the quality of decision-making and hence on any of the new buildings; it certainly had effects on their lives, some good, some bad. Great care was taken to avoid positive or negative discrimination of any kind, the ills of which were clear.

There was, as there always should be, discussion about the extent of contracting in or out of staff functions. The policy throughout the decade was to contract-in by the employment of qualified staff wherever that was more cost-effective than paying fees to external consultants. John Hipkin has said.

> The 1998 decision of the University to employ its own estates planning and property staff considerably reduced the costs of external consultancy during the decade, to a far greater extent than the cost of employing in-house staff, and also led to a more trusting and a closer relationship with the regional, county and district authorities. The employment in the Estates Department of a highly-effective external communications officer, and certain mechanisms such as setting up interactive websites for big developments, were good decisions, although as so often for good decisions, should ideally have been made earlier.

The main training issues were to spread understanding of new ways of working, and to show their advantages to internal procurement staff and to those working for the University. New ways of reducing and managing budgets and risk had to be learned, and lessons taken from mistakes that were made: mistakes are to be learned from, not to be hanged for. Despite increasingly high work and activity levels, many days/half days were taken out for training, especially for those in executive positions. Also, personal training incentives made a significant difference: individuals were subsidised to do extra training of value to them and to the University. Thirdly, there were biennial very-open Open Days, always with high-level external as well as senior University speakers, including the VC, on the programme, with wide attendance from senior members of the academic and external communities. These activities and events were important factors in the procurement of the capital programme.

Conclusion

What then did all this activity and use of resources achieve? The procurement management systems that were adopted reduced waste in capital procurement, as was beginning to happen nationally among some clients. The overall programme was achieved within a fraction of a per-cent of the total of set budget. There were under- and over-spends but none were huge. Budgets set became steadily tighter in real terms with closer bench-marking, greater team-work through 'develop and construct' procurement, much lower legal costs and often, but by no means always, better assignable/

gross area ratios. Taking out construction inflation rates, average unit costs of completed projects[5] in real terms fell by some 13 per cent for non-lab buildings. Unit costs stayed about the same for laboratories, for which building standards, especially regarding health and safety, had particularly increased.

The money-cost the University paid for the projects discussed was some £770M, nearly all from benefactions, the government-led JIF and SRIF capital funding programmes and grants from major Trusts. Although the efficiency of the estate was improved, the medium and longer term costs of maintaining it went up in real terms in accordance with its increased size and higher servicing levels; income associated with the higher levels of research and education should more than underpin those costs. The total costs of energy and the amount of CO_2 produced in the University have increased as both the size and the servicing levels of the estate have increased, but less than they would have if the standard of environmental design had not been raised by adoption of whole-life procurement. Considerable economies in consultancy fee costs came from bringing much more work 'in-house', and that bred a sense of teamwork and trust. The additional maintenance and refurbishment output by the minor works programme improved the quality and efficiency of University buildings, even if the old sites aren't much better in appearance or much more welcome to outsiders.

Over a hundred projects were completed in the decade 1996–2006, or were designed and managed during that decade with construction soon after. Fourteen major projects were planned and designed but then aborted, all for good reasons and all but one (Experimental Psychology) aborted at relatively low cost. There was a net increase of over 160,000m² of built space of the University, an expansion of 33 per cent. As a consequence, the University had the facility to pursue its chosen path of considerable growth in research and teaching as a world-leading university.

In an article entitled 'Taking the Long View', Prof Sir Leszek Borysiewicz, VC from 2010, wrote:

> So what now are the choices before us? What do we want our University to look like in 10 or 20 years time? Here, I offer three questions that we must address in the coming years. The first set of decisions before us concerns our physical growth …[6]

So, what's new?

Notes

1   *Architecture Today,* Issue 114 (January 2001).
2   See the CVCP report in 1997, *A New Partner: Universities, Students, Business and the Nation.*
3   *HESA and the Universities UK Economic Impact Modelling System* (2013); UUK report *The Impact of Universities* (April 2014).

4   See K. Kirk and C. Cotton, *The Cambridge Phenomenon* (Third Millennium Information (TMI) Ltd, 2012).
5   Project cost less VAT, furnishings and fittings, and off site infrastructure.
6   Prof Sir Leszek Borysiewicz, 'Taking the Long View', *CAM 70* (October 2013).

Index

Printed and bound by CPI Group (UK) Ltd, Croydon, CR0 4YY

24/10/2024

01778281-0008